Optimization Techniques and Recent Innovations in Mechanical Engineering (ICOTRIME 2020)

AA007402

IOP Conference Series: Materials Science and Engineering
Volume 1104

Ghaziabad, India
11 - 12 December 2020

Editors:

Sanjeev Kumar
Vinay Sharma
Pawan Kumar Arora

ISBN: 978-1-7138-3398-7
ISSN: 1757-8981

Printed from e-media with permission by:

Curran Associates, Inc.
57 Morehouse Lane
Red Hook, NY 12571

Some format issues inherent in the e-media version may also appear in this print version.

Institute of Physics
Dirac House, Temple Back
Bristol BS1 6BE UK

Phone: 44 1 17 929 7481
Fax: 44 1 17 920 0979

techtracking@iop.org

Additional copies of this publication are available from:

Curran Associates, Inc.
57 Morehouse Lane
Red Hook, NY 12571 USA
Phone: 845-758-0400
Fax: 845-758-2633
Email: curran@proceedings.com
Web: www.proceedings.com

TABLE OF CONTENTS

IOP Conf. Series: Materials Science and Engineering 1104 (2021) 011001 doi:10.1088/1757-899X/1104/1/011001

Main theme of Scientific activity

International Conference on Optimization Techniques and Recent Innovations in Mechanical Engineering (ICOTRIME 2020)

International Conference on Optimization Techniques and Recent Innovations in Mechanical Engineering (ICOTRIME 2020) is an inventive event organized with motive to make available an open international forum for the researchers, academicians, technocrats, scientist, engineers, industrialist and students around the globe to exchange their innovations, interact and share the research outcomes which may lead the young researchers, academicians and industrialist to contribute to the global society. ICOTRIME 2020 is being organized on 11th – 12th Dec 2020 at Krishna Engineering College, Ghaziabad, UP, India.

Optimization techniques incorporate both the recent inventions, innovations and exceptional outcome of research findings in various fields of Mechanical Engineering as well as software fields. Optimization technique is very fast emerging multidisciplinary research fields which recognize and provide optimum solutions by using advanced computing proficiencies. It is a recent area research which includes many disciplines of technological science, but mainly it bridges the gap between industry and academia to reduce the wastage of material and time to design a new product. This conference will provide an excellent platform for the young researchers, scientists, academicians, engineers, mathematicians around the globe to showcase their innovations; peer reviewed their work and collaborates with international community to achieve multidisciplinary research outcomes. ICOTRIME 2020 will spotlight the renowned world-class keynote speakers, special sessions, along with the regular research presentations.

The conference invites research paper submissions all around the world and will include myriad tracks in the area of Optimization techniques and recent Technologies in Mechanical Engineering.

About US

Krishna Engineering College

Krishna Engineering College (KEC), Ghaziabad was established in 2004 by Friends Charitable Society (FCS) Ghaziabad. FCS has eminent Businessmen, Industrialists and Professionals as its members. FCS members are also associated with other leading Educational Institutions.

KEC has been established as an institution for the millennium with a vision to turn out scholars who will be Global Players when they leave the portals of KEC. With this vision in the background, the college endeavors to impart core technical knowledge, help students to develop requisite soft skills, instill adaptability for change and builds cross cultural compatibility to its students. The college is approved by AICTE, Delhi and is affiliated to AKTU, Lucknow. Currently, the college offers five programs in B.Tech. (Computer Science & Engineering,

Information Technology, Electronics & Communication Engineering, Mechanical Engineering and Civil Engineering) with an approved in take of 600 students/year.

Within a span of 16 years, the college has established its credentials as an institution of repute by figuring among top 20 colleges of over 400 Engineering & Technology Institutions of AKTU in session 2016-17. The college infrastructure is a planned endeavor built over a period with care and concern keeping all the needs of a premier Engineering Institution. The campus is located in Mohan Nagar, Ghaziabad, next to IAF Station Hindon and is strategically located vis-a-vis Delhi (5 Km from border), Ghaziabad (6 Km), 12 km from Noida. The campus is spread over an area of 11 acres with plans for 3 academic blocks covering over 25000 Sq. meter of built up area, 3 Hostels (2 for boys and 1 for Girls), an Auditorium, workshops, play grounds.

The Department of Mechanical Engineering

The Department of Mechanical Engineering is devoted to advance the knowledge and educate students in the area of Mechanical Engineering that will meet the needs of Industry and academics in the ever changing business dynamics and technological challenges associated with the modern world. Department strives for excellence in Teaching and Learning, Innovative Research and Community Service. The Department works with an aim to develop specialized knowledge in Mechanical Engineering, laying much emphasis on developing the students' practical skills in thermal, design, production, industrial etc. As we take note of our learners' professional needs ardently, so various In-house events such as Technical events, Seminars, Personality Development classes, Workshops, Conferences, Guest lectures are being arranged at regular intervals.

List of Editors

Name + Title(s)	***Address & affiliations***	***Email + URL***
Dr. Sanjeev Kumar	Dean(A) & HOD (Mechanical Engineering Department), Krishna Engineering College, Ghaziabad, U.P., India	Sanjeevkg9@gmail.com
Dr. Vinay Sharma	Professor, Birla Institute of Technology, Deoghar, India	vinay@bitmesra.ac.in
Dr. Pawan Kumar Arora	Professor, Galgotias College of Engineering and Technology, Knowledge Park-II Greater Noida, U.P., India	pawan.kumar.arora@galgotiacollege.edu

IOP Conf. Series: Materials Science and Engineering 1104 (2021) 011001 doi:10.1088/1757-899X/1104/1/011001

List of reviewers, those will effectively involve in the review process, who already gave their consent to review the research papers. (Reviewers are not limited to)

Name	Email id	Affiliation
Dr. R. K. Das	randip@iitism.ac.in	Indian Institute of Technology (ISM), Dhanbad, India
Dr. A. R. Dixit	amitraidixit@iitism.ac.in	Indian Institute of Technology (ISM), Dhanbad, India
Dr. Naveen Kumar	naveenkumar@dce.ac.in	Delhi Technological University, Delhi, India
Dr. V. K. Bajpai	vkbajpaifme@nitkkr.ac.in	National Institute of Technology, Kurukshetra, India
Dr. P. C. Tewari	pctewari1@rediffmail.com	National Institute of Technology, Kurukshetra, India
Dr. Ranganathan M. Singri	ranganath@dce.ac.in	Delhi Technological University, Delhi, India
Dr. Vikas Rastogi	rastogivikas@yahoo.com, rastogivikas@gmail.com	Delhi Technological University, Delhi, India
Dr. Pavan Kumar Kankar	pkankar@iiti.ac.in	Indian Institute of Technology Indore, India
Dr. Prashant K Jain	pkjain@iiitdmj.ac.in	IIITDM, Jalabpur, India
Dr. M. K.Paswan	mkpaswan.me@nitjsr.ac.in	National institute of Technology, Jamshedpur, India
Dr. R. C.Singh	rcsingh@dce.ac.in	Delhi Technological University, Delhi, India
Dr. Rajeev Choudhary	rajivchaudhary@dce.ac.in	Delhi Technological University, Delhi, India
Dr. Alokesh Pramanik	alokesh.pramanik@curtin.edu.au	Curtin University, Australia.
Dr. Mohit Tyagi	tyagim@nitj.ac.in	National Institute of Technology, Jalandhar, India
Dr. Gulshan Sachdeva	gulshansachdeva@nitkkr.ac.in	National Institute of Technology, Kurukshetra, Haryana, India
Dr. Harish Kumar	hodme@nitdelhi.ac.in	National Institute of Technology, Delhi, India
Dr. Akhilesh Arora	akhilesharora@dce.ac.in	Delhi Technological University, Delhi, India
Dr. N. Yuvraj	yuvraj@dce.ac.in	Delhi Technological University, Delhi, India

Dr. Saurabh Aggarwal	saurabh.agrawal@dtu.ac.in, agrawals.iit@gmail.com	Delhi Technological University, Delhi, India
Dr. Paras Kumar	paraskumar@dtu.ac.in	Delhi Technological University, Delhi, India
Dr. Pushpendra Singh	dr.psinghs@gmail.com	Delhi Technological University, Delhi, India
Dr. Ajay Singoli	ajay.igit@gmail.com	G. B. Pant Engineering College, New Delhi, India
Dr. Varun Goel	varun@nith.ac.in, varun7go@gmail.com	National Institute of Technology, Hamirpur, India
Dr. Siddharth	sidmech@nith.ac.in	National Institute of Technology, Hamirpur, India
Dr. Avdesh Yadav	avadheshyadava@nitkkr.ac.in, avadheshyadava@gmail.com	National Institute of Technology, Kurukshetra, India
Dr. D. K.Dwivedi	todkdwivedi@gmail.com	MMM University of Technology, Gorkhpur, India
Dr. Sachin Maheswar	sachinmaheshwar@gmail.com	Shershah College of Engineering, Rhatas, Bihar, India
Dr. K. L. A.Khan	klakhan1@rediffmail.com	Krishna Institute of Engineering Technology, Ghaziabad India
Dr. H. K. Rai	hod.me@ipec.org.in	Inderprastha Engineering College, Ghaziabad, India
Dr. Chandan Kumar	me@niet.co.in	Noida Institute of Engineering Technology, Greater Noida, India
Dr. V. R.Mishra	hod.me@glbitm.org	G.L. Bajaj Institute of Technology & Management, Greater Noida India
Dr. Ashish Soti	ashishsoti@gmail.com	Dronacharya Group of Institutions, Greater Noida, India
Dr. Vaibhav Jain	vaibhavjain@mait.ac.in	Maharaja Agarsen Institute of Technology, New Delhi
Dr. Parul Gupta	parulgupta197@gmail.com	MIT Group of Institutions,, Moradabad, India
Dr. Madhup Mittal,	madhup.mittal@thapar.edu	Thaper University, Patiala, India

Dr. Pawan Arora	pawan.kumar.arora@galgotiacollege.edu	Galgotia University, Greater Noida, India
Dr. Ajay Kumar	akumar@ddn.upes.ac.in	University of Petroleum and energy studies, Dehradun, India
Dr. Abhishek Saxena	talk2abhi3@gmail.com	MIT Group of Institutions,, Moradabad, India

Committee

Steering Committee:

Chief Patron (s)

- Goel, Manoj, Krishna Engineering College, Ghaziabad, India.

Conference General Chair (s)

- Tiwari, Sandeep, Krishna Engineering College, Ghaziabad, India.

Conference Chair (s)

- Chattopadhyaya, Somnath, Indian Institute of Technology (ISM), Dhanbad, India
- Kumar, Pradeep, Indian Institute of Technology, Roorkee, India

Conference Co-Chair (s)

- Sharma, Vinay, Birla Institute of Technology, Deoghar, India
- Sachdeva, Anish, National Institute of Technology, Jalandhar

Organizing Chair (s)

- Kumar, Sanjeev, Krishna Engineering College, Ghaziabad, India.

Technical Program Chair (s)

- Murtaza, Qasim, Delhi Technological University, Delhi, India
- Walia, R. S., Punjab Engineering College, Chandigarh, India
- Singh, Bhanu Pratap, Maharishi University of Information Technology, Lucknow, India

Publication Chair (s)

- Tyagi, Mohit, National Institute of Technology, Jalandhar, India
- Sachdeva, Gulshan, National Institute of Technology, Kurukshetra, Haryana, India
- Kumar, Harish, National Institute of Technology, Delhi, India

National Advisory Board

- Das, R. K., Indian Institute of Technology (ISM), Dhanbad, India
- Dixit, A. R., Indian Institute of Technology (ISM), Dhanbad, India
- Kumar, Naveen, Delhi Technological University, Delhi, India
- Bajpai, V. K., National Institute of Technology, Kurukshetra, India
- Tewari, P. C., National Institute of Technology, Kurukshetra, India
- Singri, M. Ranganathan, Delhi Technological University, Delhi, India
- Rastogi, Vikas, Delhi Technological University, Delhi, India
- Kankar, Pavan Kumar, Indian Institute of Technology Indore, India
- Jain, Prashant K, IIITDM, Jalabpur, India
- Paswan, M. K., National institute of Technology, Jamshedpur, India
- Singh, R. C., Delhi Technological University, Delhi, India
- Choudhary, Rajeev, Delhi Technological University, Delhi, India

International Advisory Board

- Hloch, Sergej, Technical University of Kosice, Slovakia.
- Kozak, Drazan, University of Osijek, Croatia.
- Pramanik, Alokesh, Curtin University, Australia
- Basak, A.K., University of Adelaide, Australia.
- Legutko, Stainslaw, Poznan University of Technology, Poland

Organizing Secretary

- Tyagi, Mohit, National Institute of Technology, Jalandhar, India
- Sachdeva, Gulshan, National Institute of Technology, Kurukshetra, Haryana, India
- Kumar, Harish, National Institute of Technology, Delhi, India
- Pandey, Shailesh M, National Institute of Technology, Patna, India.
- Kumar, Vikas, Krishna Engineering College, Ghaziabad, India.

Technical Program Committee

- Aroa, Akhilesh, Delhi Technological University, Delhi, India
- Yuvraj, N, Delhi Technological University, Delhi, India
- Aggarwal, Saurabh, Delhi Technological University, Delhi, India
- Kumar, Paras, Delhi Technological University, Delhi, India
- Singh, Pushpendra, Delhi Technological University, Delhi, India
- Singoli, Ajay, G. B. Pant Engineering College, New Delhi, India

- Goel, Varun, National Institute of Technology, Hamirpur, India
- Siddharth, National Institute of Technology, Hamirpur, India
- Yadav, Avdesh, National Institute of Technology, Kurukshetra, India
- Dwivedi, D. K., MMM University of Technology, Gorkhpur, India
- Maheswar, Sachin, Shershah College of Engineering, Rhatas, Bihar, India
- Khan, K. L. A., Krishna Institute of Engineering Technology, Ghaziabad India
- Rai, H. K, Inderprastha Engineering College, Ghaziabad, India
- Kumar, Chandan, Noida Institute of Engineering Technology, Greater Noida, India
- Mishra, V. R., G.L. Bajaj Institute of Technology & Management, Greater Noida India
- Soti, Ashish, Dronacharya Group of Institutions, Greater Noida, India
- Jain, Vaibhav, Maharaja Agarsen Institute of Technology, New Delhi
- Gupta, Parul, MIT Group of Institutions,, Moradabad, India
- Mittal, Madhup, Thaper University, Patiala, India
- Arora, Pawan, Galgotias College of Engineering and Technology, Greater Noida, India
- Kumar, Ajay, University of Petroleum and energy studies, Dehradun, India
- Saxena, Abhishek, MIT Group of Institutions, Moradabad, India

Organizing Committee:

Advisory Committee

- Kumar, Sanjeev, Krishna Engineering College, Ghaziabad
- Singh, M.K., Krishna Engineering College, Ghaziabad
- Mishra, A. N., Krishna Engineering College, Ghaziabad
- Ali, Rashid, Krishna Engineering College, Ghaziabad
- Kumar, Pramod., Krishna Engineering College, Ghaziabad
- Prasad, Jay Shankar, Krishna Engineering College, Ghaziabad
- Pandey, Archana, Krishna Engineering College, Ghaziabad

Technical Session Committee

- Shukla, Rakesh, Krishna Engineering College, Ghaziabad
- kumar, Anil, Krishna Engineering College, Ghaziabad
- Kumar, Satyajeet, Krishna Engineering College, Ghaziabad
- Prakash, Om, Krishna Engineering College, Ghaziabad
- Sachendra, Krishna Engineering College, Ghaziabad

Publication Committee

- Kumar, Vikas, Krishna Engineering College, Ghaziabad
- Singh, Shailesh Kumar, Krishna Engineering College, Ghaziabad

Publicity Committee

- Mishra, Ashish, Krishna Engineering College, Ghaziabad
- Kumar, Vikas, Krishna Engineering College, Ghaziabad
- Singh, Shailesh Kumar, Krishna Engineering College, Ghaziabad
- Kumar, Ashutosh, Krishna Engineering College, Ghaziabad

Hospitality & Food Committee

- Mishra, Ashish, Krishna Engineering College, Ghaziabad
- Garg, Nitesh, Krishna Engineering College, Ghaziabad
- Srivastava, Praween, Krishna Engineering College, Ghaziabad
- Sachendra, Krishna Engineering College, Ghaziabad
- Prakash, Om, Krishna Engineering College, Ghaziabad

Financial Affairs Committee

- Kumar, Vikas, Krishna Engineering College, Ghaziabad
- Singh, Shailesh Kumar, Krishna Engineering College, Ghaziabad
- Kumar, Satyajeet, Krishna Engineering College, Ghaziabad

Event Management Committee

- Mishra, Ashish, Krishna Engineering College, Ghaziabad
- Garg, Nitesh, Krishna Engineering College, Ghaziabad
- Kumar, Vikas, Krishna Engineering College, Ghaziabad
- Singh, Shailesh Kumar, Krishna Engineering College, Ghaziabad
- Kumar, Satyajeet, Krishna Engineering College, Ghaziabad
- Malik, Prashant, Krishna Engineering College, Ghaziabad
- Jaiswal, Aarti, Krishna Engineering College, Ghaziabad

Sponsorship Committee

- Mishra, Ashish, Krishna Engineering College, Ghaziabad
- Gupta, Arpita, Krishna Engineering College, Ghaziabad
- Kumar, Vikas, Krishna Engineering College, Ghaziabad
- Aggarwal, Ashutosh, Krishna Engineering College, Ghaziabad
- Singh, Hukam, Krishna Engineering College, Ghaziabad

Delegate Committee

- Chauhan, Ujjwal, Krishna Engineering College, Ghaziabad
- Rai, Hari Mohan, Krishna Engineering College, Ghaziabad
- Aggrawal, Ashutosh, Krishna Engineering College, Ghaziabad

Media Management Committee

- Tiwari, Bhawna, Krishna Engineering College, Ghaziabad
- Prakash, Om, Krishna Engineering College, Ghaziabad
- Rai, Hari Mohan, Krishna Engineering College, Ghaziabad
- Thakral, Gagan, Krishna Engineering College, Ghaziabad
- Garg, Kanika, Krishna Engineering College, Ghaziabad
- Saraswat, Pavi, Krishna Engineering College, Ghaziabad
- Aggarwal, Aayush, Krishna Engineering College, Ghaziabad

Registration & Help-Desk Committee

- Raheja, Neha, Krishna Engineering College, Ghaziabad
- Sanwal, Tanushree, Krishna Engineering College, Ghaziabad

Call For papers

The conference ICOTRIME 2020 is being organized at Krishna Engineering College, Ghaziabad, Uttar Pradesh, during 11th – 12th Dec 2020. It will feature world-class keynote speakers, special sessions, workshops, along with the regular/oral paper presentations. The conference welcomes paper submissions from researchers, practitioners, academicians and students, and will cover myriad tracks in the field of Mechanical Engineering and associated research areas. Original unpublished manuscripts, and not currently under review in any other journal or conference, are solicited in relevant areas including, but not limited to, the following:

Optimization Tools and Techniques

- Response Surface Methodology (RSM)
- Search Tool
- Neural Network
- Genetic Algorithm
- Grey Relational Analysis
- Taguchi's Method
- Unconstrained & Constrained Optimization
- Learning Algorithms
- Combinatorial Optimization and Mathematical Programming
- Numerical Methods
- Multi-objective Programming
- Meta-heuristic Optimization Algorithms
- Hybrid Algorithms
- Optimization techniques

Production and Industrial Engineering

- Applications FEA

- Manufacturing
- Material Science
- Metrology and Surface Engineering
- Non-destructive Examination
- Machining
- Virtual Manufacturing and Concurrent Engineering
- Manufacturing
- Industrial Automation, CAD/CAM and Robotics
- Joining Processes (Welding)
- Logistics and Supply Chain Management
- Modeling and Simulation
- Stochastic Modeling
- Non-Conventional Energy
- Product Design and Development
- Reliability and Maintenance Engineering
- Surface Engineering and Coatings
- Mechatronics
- Design for Manufacturing and Assembly
- TQM and Six Sigma
- Theory of Constraints

Thermal Engineering

- Multi-Phase Flow
- Heat Transfer
- Bio-fuels and Internal Combustion Engines
- Transport Phenomena in Porous Media
- Environmental Engineering
- Enhanced Oil Recovery
- Polymer Processing Science and Technology
- Waste Management and Waste Disposal
- Energy Management and Energy systems
- Air Conditioning and Refrigeration
- Renewable Energy
- Micro/nano-fluid and life science application
- Novel Materials
- Thermal Management
- Micro-fabricated Thermal Management Devices and Systems
- Hybrid Energy Storage Systems
- Ionic Liquids, Surfactant and Interface Science

IOP Conf. Series: Materials Science and Engineering 1104 (2021) 011001 doi:10.1088/1757-899X/1104/1/011001

Design Engineering

- Theory and research methods
- Design Methods & Processes
- Organization and management
- Information and knowledge
- Design support tools
- Design innovation
- System Engineering
- Engineering Design practice
- Industrial Design
- Computer aided design and simulation

Important Dates

Electronic submission of full papers:	1st Oct, 2020
Notification of paper acceptance:	20th Oct, 2020
Early bird Registration and submission of camera-ready papers:	15th Oct, 2020
Late Registration and submission of camera-ready papers:	25th Oct, 2020
Conference date:	**11th – 12th Dec 2020**

Registration

- **Indian Delegates (INR)**

Category	Early registration	Late registration
UG & PG students/ Research scholar	6500	7000
Faculty/ Academicians	7500	8000
Industry Delegates	8500	9000
Attendees	2000	3000

- **International Delegates (US $)**

Category	Early registration	Late registration
UG & PG students/ Research scholar	200	250
Faculty/ Academicians	250	300

IOP Conf. Series: Materials Science and Engineering 1104 (2021) 011001 doi:10.1088/1757-899X/1104/1/011001

Industry Delegates	300	350

Registration Fee includes

- Registration Kit
- High Tea
- Lunch/ Refreshment

Registration Detail of ICOTRIME -2020

- Candidate are advised to visit our website www.krishnacollege.ac.in/icotrime regularly for updates
- All Paper presenter & conference attendees require pre-registration.
- Only registered participants are allowed to attend the conference.
- Fill up separate registration form for each participant.
- For online registration delegate has to do NEFT first and update the transaction ID on online gateway or send Scanned copy of payment DD or Bank transaction receipt along with signed copyright form must send to the Organizing chair by email on icotrime@krishnacollege.ac.in.
- Students must attach the copy of identification card of their institute with the registration form.
- At least one author of each accepted paper must register by Oct 15, 2020 and present the paper at the conference in order to include the paper in the ICOTRIME -2020 Conference Proceedings.
- Bring the physical copies of the registration form, payment DD or Bank transaction receipt along with signed copyright form at the time of conference

Bank Transfer Details
Bank Name:

Account Name:

Account Number:

Branch Address:

IFSC Code:

MICR Code:

Author Instruction

Paper Submission Guidelines:

Researchers are welcomed to submit original high quality manuscript which has not been submitted anywhere for publication, it should not be under consideration and not be copyright protected.

Manuscripts will be checked for plagiarism using software.

Easy-Chair Submission: will be updated soon

IOP Conf. Series: Materials Science and Engineering 1104 (2021) 011001 doi:10.1088/1757-899X/1104/1/011001

Publication

ICOTRIME 2020 proceedings will be published in scopus indexed.

Invited Keynote Speaker

Prof. (Dr.) Pradeep Kumar

Professor, Department of Mechanical & Industrial Engineering Indian Institute of Technology, Roorkee Roorkee 247667, India Phone: +91-9837423860 (Cell), +91-1332-285602 (O), +91-1332-285130 (H)

E mail: kumarfme@iitr.ac.in, kumarfme@gmail.com,

Scopus: https://www.scopus.com/authid/detail.uri?authorId=55952759400

Thomos Reuters: http://www.researcherid.com/rid/O-3879-2017

Google Scholar: https://scholar.google.co.in/citations?user=sJy047QAAAAJ&hl=en

Dr. Somnath Chattopadhyaya

Thomson Reuters Research ID - K-7101-2012, ORCID ID - 0002-6814-7936

IOP Conf. Series: Materials Science and Engineering 1104 (2021) 011001 doi:10.1088/1757-899X/1104/1/011001

Google scholar profile: https://scholar.google.com/citations?user=x22n_KoAAAAJ&hl=en

Scopus Author ID - 24070120200

Associate Dean(Innovation & Incubation) Associate Professor,Department of Mechanical Engineering

Indian Institute of Technology (Indian School of Mines)

Dhanbad - 826004,office : 03262235618,Mobile - +919431954821,Residence - +9103262235731

Website : www.iitism.ac.in, Facebook name; Somnath Chattopadhyaya

Official Email Address : somnathchattopadhyaya@iitism.ac.in

Dr. QASIM MURTAZA, PhD-Ireland, PDF-UK
(qasimmurtaza@gmail.com; qasimmurtaza@dce.ac.in, M: 9654364948)
Google scholar Profile:https://scholar.google.co.in/citations?user=LRuTyiIAAAAJ&hl=en
Experience

Total Experience 16 years plus.

IOP Conf. Series: Materials Science and Engineering 1104 (2021) 011001 doi:10.1088/1757-899X/1104/1/011001

Dr Alokesh Pramanik

Google scholar profile: https://scholar.google.com.au/citations?user=-1cFdgkAAAAJ&hl=en

Department of Mechanical Engineering, Curtin University, Bentley, Perth, WA, Australia

Area of research: Machining processes and composite material

IOP Conf. Series: Materials Science and Engineering 1104 (2021) 011002 doi:10.1088/1757-899X/1104/1/011002

Peer review declaration

All papers published in this volume of *IOP Conference Series: Materials Science and Engineering* have been peer reviewed through processes administered by the Editors. Reviews were conducted by expert referees to the professional and scientific standards expected of a proceedings journal published by IOP Publishing.

- **Type of peer review:** Double-Blind review
- **Conference submission management system:** All papers are submitted through easychair online portal.
- **Number of submissions received:** 112
- **Number of submissions sent for review:** 84
- **Number of submissions accepted:** 42
- **Acceptance Rate (Number of Submissions Accepted / Number of Submissions Received X 100):** 37.5%
- **Average number of reviews per paper:** 5-7
- **Total number of reviewers involved:** 32
- **Any additional info on review process:**

The research papers are invited for the possible publication from all around the world and we have received 112 research papers in total. The first step we have taken is analyze all received research papers for the plagiarism. Further, 84 research papers are selected for the review process those have low percentage of plagiarism.

The review procedure is totally double blinded review process. Initially, a list of reviewers is prepared with their specialization and field of interest. List is also attached for your reference. Based on the list, we have sent abstract of the research papers to seek the consent for reviewing. If any of first two reviewers doesn't give their consent then the abstract is forwarded to the next reviewer. This way we have started the review process. A review report is also provided to all reviewers to ease the review process. Out of the two reviewers, if one reviewer has rejected the research paper and another has resent the paper for the major revisions then the research paper is forwarded to the third reviewer for the reviewing. But if the third reviewer has also rejected the research paper then the paper is rejected. We have given at least three chances to authors for the possible revision i.e., initially as per the plagiarism report and then after based on the reviewer's comments.

After receiving the final draft of the research paper, the research paper is again analyzed for the plagiarism. Finally, authors are requested to provide their research papers in the template of the IOP.

- **Contact person for queries:**

IOP Conf. Series: Materials Science and Engineering 1104 (2021) 011002 doi:10.1088/1757-899X/1104/1/011002

Dr. Sanjeev Kumar,
Dean (A) & HOD (ME)
Krishna Engineering College, Ghaziabad, India
deana@krishnacollege.ac.in

Optimization of NPIC Controller using Genetic Algorithm

Aditi Saxena[1] Jitendra Kumar[2] Vinay Kumar Deolia[3]

[1] GLA University aditeesaxenaa29@gmail.com

Abstract. This paper has investigated the NPIC (Nonlinear proportional integral Controller) and Genetic algorithm has applied to obtain the minimum value of error. The objective function has been created using Integral absolute error (IAE) This controller provides a unique combination of proportional and integral controller along with a nonlinear factor present in the controller. which makes the controller much more effective in comparison to linear controller, as it has always remained a challenge to control a robotic manipulator which is highly non linear.so this controller provides an approach to control a nonlinear manipulator through nonlinear controller. He abstracts should summarize the contents of the paper in short terms, i.e., 150-250 words.

Keywords: NPIC (Nonlinear Proportional Integral Controller), IAE (integral Absolute error), Nonlinear

1 Introduction

As we all know that today"s world is moving towards robotization and automation there is high necessity of robot in today"s world with the help of robot we can produce high quality of products in lesser period of time, skilledlabour shortage can be fulfilled by robot we can easily handle the pressure to increase the production rates to compete the market by robot we can produce best quality of products in shorter period of time with increased productivity the word robot is being originated from a Czech word robota which means slaved or force labor but in order to manipulate the robot we highly need a controller as without a controller robot could itself became a disaster,In terms of robot robotic arm is known as robotic manipulator .there are various controllers and techniques which are used to control and manipulate the manipulator.Controlling of robot has always became an exigent work for engineers, parameters of controller are needed to be tuned very finely accurately and precisely in order to control a robotic manipulator and also to achieve high trajectory tracking hence in the upcoming year it has became a very tedious and challenging task to develop such a control system which is able to handle complexities and non linearity"s . To cope up with today"s demand we need to design an intelligent control system which could be adaptive and robust to design such control system various controllers and control techniques are used which also took us towards interdisciplinary fields, for example suppose we are talking about to control a robotic manipulator so the first thing which we need is the physical model of the robot which falls under mechanical engineering next step involve designing of controller for the plant(manipulator) so these controller designing would fall under electronics and communication field and in order to make the controller more intelligent adaptive and autonomous we may can used artificial intelligence, neural network which falls under computer science and engineering so in short we can say that this research topic is itself a combination of various interdisciplinary branch,this interdisciplinary research has very widely increase the area of research by adding fuzzy control, neural network .fuzzy controller is a remarkable combination of artificial intelligence and control engineering these intelligent controller has make an successfully mark in the commercial applications as fuzzy controller has many advantages that it can easily handle all the non-linearity"s It has less rise time and maximum overshoot ,fuzzy logic provides certain sort of flexibility to the research engineering"s as there parameters are not just confined between true and false fuzzy logic provides a solution to a problem who somewhere or the other lies between true and false here we can take an example of controlling an autonomous car so for fixing the controllers for break fuzzy logic provide flexibility such that we can handle te break if suddenly some pedestrian or some vehicle comes in our waythis technique is being considered as one of the best technique among the controller as its result and performance are superior in comparison to the traditional controller and as the year passes by various controllers are designed in combination with various controlling technique like fuzzy, neural network and many more and for more accurate and precise tunning of the controller different optimization techniques are also being used which provide adaptability to the controller for their tuning feature in this review paper a detailed description of various controllers, controlling techniques and optimization techniques are being given with their Advantage, disadvantages and drawbacks. This controller is been investigated through genetic algorithm and various results are being obtained.

2 Literature Review

Controlling of robot has always became an exigent work for engineers these controllers were came into existence since 1950 when, To control the nonlinear and complex manipulator systems PID came in light the structure of this controller is very easy to design and its implementation is very unexacting this controller is very economical in terms of cost [1-4] but it was not delivering perfect results and was not able to handle the non-linearities and uncertaintyto overcome these issues another technique came into existence in 1960 the name of this technique was classic adaptive control technique . all these techniques are basically rely on the dynamic model of the manipulator system they also provide us with the best way to deal with nonlinearities and complexity[5-7-8]but as it has been says that nothing is perfect so there were certain drawbacks and disadvantage in these control techniques as for delivering perfect result it requires accurate mathematical model also it is difficult for gain scheduling design to handle the large operating point[9,10]. so to handeled those system who are being poorly dynamically analyzed a new technique was investigated that is artificial intelligence which was combined with these classic control techniques[11]after this to enhance the techniques and to make the system more intelligent and adaptive fuzzy logic controller (FLC) and artificial neural network and also the combination of the both that is neuro-fuzzy technique these merging of controllers give very good result in increasing the capabilities of sliding mode controller (SMC)[12].Fuzzy logic controller provides almost accurate results for uncertain system zadeh has proposed this fuzzy logic which is not just restricted between true or false but inspite of that help us to deal with the probabilities that lies between atmost true and atmost false [13-20]after this fuzzy logic controller came which was proposed by mamdani[21,22,23] as the years passes the elaboration in the work of fuzzy logic controller was kept on increasing as the work related to fuzzy was further carried out by Graham[24], Grahamand Newell[25}when these classic control methods combine with artificial intelligence (AI) and neuro fuzzy logic this gives the control engineering a new way and a new trend and just from control engineering it is switched to robotics control and automationthese fusion technique are very much helpful in controlling almost all the fields like chemical,food,textile and these controllers are playing a vital role in controlling the robotics a simple human hand in terms of robots is known as robotic manipulator they are very much non-linear and multi input multi output links of the manipulator can be minimum when but the links are being kept on increasing depending on the specific task for which the controller has been designed with the help of controllers we are able to reach those locations which are not easily accessible welding and assembling Intelligent techniques were arrive in 1990"s which has overcame the drawbacks of classical and traditional PID controller and the combination of PID controller with fuzzy logic completely resolve all the drawbacks of PID and Fuzzy controller and also increase the advantages as a whole and as a result it has been noticed that by adding both the techniques robust system can be formed .control system which are based on fuzzy logic are in detail analysed by wang and kwok and then they themselves suggest the combination of fuzzy pd and fuzzy I controller .further by making a small improvement in the fuzzy PI controller Li and Gatland proposed a fuzzy three term controller

3 Dynamic Model of Manipulator

The equations in form of matrix have been taken from Lin and they are further simplified in forms of theta which shows the position of the links.

$$\begin{bmatrix} S_{11} & S_{12} \\ S_{21} & S_{22} \end{bmatrix} \begin{bmatrix} \ddot{\theta}_{11} \\ \ddot{\theta}_{22} \end{bmatrix} + \begin{bmatrix} P_{11} \\ P_{21} \end{bmatrix} + \begin{bmatrix} f_{r1} \\ f_{r2} \end{bmatrix} + \begin{bmatrix} fn_{1p} \\ fn_{2p} \end{bmatrix} = \begin{bmatrix} \tau f_{1p} \\ \tau f_{2p} \end{bmatrix}$$

Where

$$S_{11} = I_{1p} + I_{1p} + m_{11} l_{c1}^2 + m_{22} l_{c2}^2 + 2m_{22} l_{11} l_{c2} \cos\theta_{22} + m_{vp} l_{11}^2 + m_{vp} l_{22}^2 + 2m_{vp} l_{11} l_{22} \cos\theta_{22} \quad (1)$$

$$S_{12} = I_{2P} + m_{22} l_{c2}^2 + m_{22} l_{11} l_{c2} \cos\theta_{22} + m_{vp} l_{22}^2 + m_{vp} l_{11} l_{22} \cos\theta_{22} \quad (2)$$

$$S_{21} = S_{12} \quad (3)$$

$$S_{22} = I_{2P} + m_{22} l_{c2}^2 + m_{vp} l_{22}^2 \quad (4)$$

$$P_{11} = -m_{22} l_{11} l_{c2} (2\dot{\theta}_{11} + \dot{\theta}_{22}) \dot{\theta}_{22} \sin\theta_{22} - m_{vp} l_{11} l_{22} \left(2\theta_{11} \dot{+} \dot{\theta}_{22}\right) \theta_{22} \sin\theta_{22} \quad (5)$$

$$P_{21} = m_{22} l_{11} \dot{\theta}_{11}^2 l_{c2} \sin\theta_{22} + m_{vp} l_{11} \dot{\theta}^2{}_{11} l_2 \sin\theta_{22} \quad (6)$$

$$f_{r1} = b_{1vp} \dot{\theta}_{11} \quad (7)$$

$$f_{r2} = b_{2vp} \dot{\theta}_{22} \quad (8)$$

$$fn_{1p} = m_{11} l_{c1} g \cos\theta_{11} + m_{22} g (l_{c2} \cos(\theta_{11} + \theta_{22}) + l_{11} \cos\theta_{11} + m_{vp} g (l_{22} \cos(\theta_{11} + \theta_{22}) + l_{11} \cos\theta_{11} \quad (9)$$

$$fn_{2p} = m_{22} l_{c2} g \cos(\theta_{11} + \theta_{22}) + m_{vp} l_2 g \cos(\theta_{11} + \theta_{22}) \quad (10)$$

Final equations

$$\ddot{\theta}_{11} = \frac{\tau f_{1p} - f_{n1p} - f_{r1} - P_{11} - S_{12} * \ddot{\theta}_{22}}{S_{11}} \quad (11)$$

$$\ddot{\theta}_{22} = \frac{(\tau f_{2p} - f_{n2p} - f_{r2} - P_{21} - S_{12} * \ddot{\theta}_{11})}{S_{22}} \quad (12)$$

Table 1.Paramaeters of manipulator

Parameters for Link1	Values for Link 1	Parameters for Link 2	Values for Link 2
m_{11}	0.392924 kg	m_{22}	0.094403 kg
l_{c1}	0.104648 m	l_{c2}	0.081788 m
l_{11}	0.2032 m	l_{22}	0.1524 m
I_{1P}	0.0011411 kg m^2	I_{2P}	0.0020247 kg m^2
b_{1vp}	0.141231 N-m/radian/s	b_{2vp}	0.3530776 N-m/radian/s
m_{vp}	0.56699 Kg		
g	9.81 m/		

4 Nonlinear Proportional Integral Controller (NPIC)

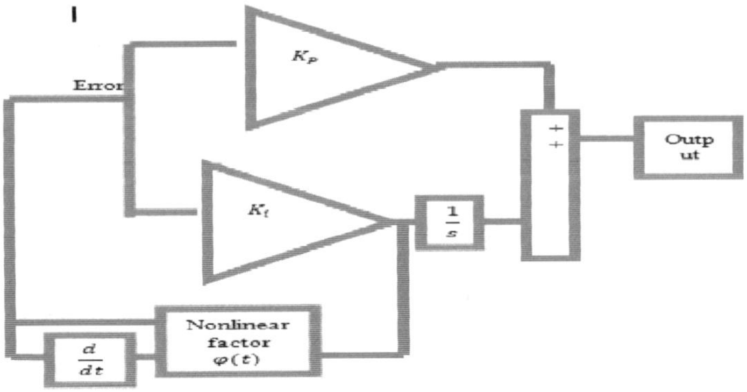

Fig. 1. figure representing NPIC Controller

IOP Conf. Series: Materials Science and Engineering 1104 (2021) 012001 doi:10.1088/1757-899X/1104/1/012001

Currently Linear controllers are being using very frequently for controlling of robotic manipulator. or very few nonlinear controllers might be designed This controller offers a completely different approach for dealing with the nonlinear model and also handles very well the slow time varying errors and various uncertain parameters. This Proposed controller have various advantages over the traditional controllers like faster speed, more accuracy it also provides better stability to the system the PI controller combination help to fully removed the steady state error it has good transient response and is robust to handle nonlinearities. The nonlinear controller is designed with the help of a time-varying nonlinear factor ψ(t), The function of this controller is that it is multiplied with integral gain to change it in run time and to compensate all the uncertainty which appears automatically.it has been designed by a mathematical equation.

$$\varphi(t) = \left[1 - \exp\{-\left(ae^2(t) + br^2(t)\right)\}\right]^2$$

E=instantaneous error
R=derivative of error
a,b= constants
ψ(t)=nonlinear factor

sine wave can be given at the input. This waveform is best suited for trajectory tracking planning of manipulator as the manipulator trajectory path is same as that of sine wave and further simulation can be run on MATLAB/SIMULINK.

5 Optimization through Genetic Algorithm

Genetic Algorithm is an optimal technique for finding the minimum and maximum value of error here genetic algorithm has been used to find the minimum value of error and IAE is the objective function which did not add any additional weight to the sysem this optimization tool is inbuilt in the MATLAB by typing the optimtool in the command window this tool open and it consist of various steps

In the tool we just need to specify the number of variable which we are using and also the objective.

IOP Conf. Series: Materials Science and Engineering 1104 (2021) 012001 doi:10.1088/1757-899X/1104/1/012001

6 Results

In the graphs represented below x axis shows the fitness value which corresponds to the objective function here integral absolute error has been taken as an objective function whereas Y axis represent the number of generations (number of times genetic algorithm has run) starting from the first graph the value of best fitness has come 0.04438 which is the highest value of error as compared to the other graphs the table below also reflects the comparison between the values of error and we can say that we have obtained the least value of error finally by optimizing the controller through genetic algorithm each graph represented here has been run for 100 generation

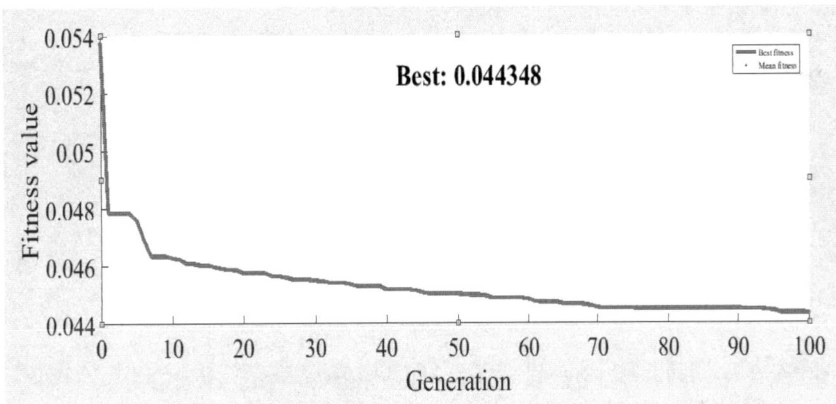

Fig. 2. **Fitness value after first iteration**

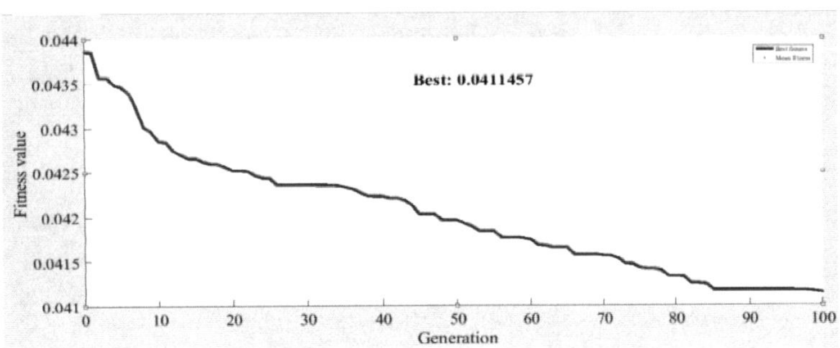

Fig. 3. Fitness value after Second Generation

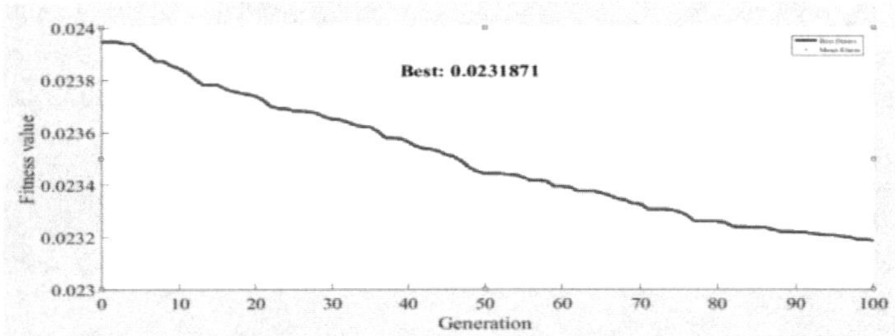

Fig. 4. Fitness value after third generation

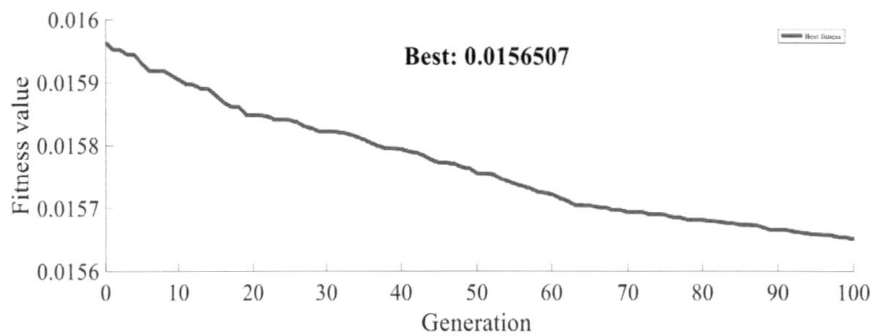

Fig. 5. Fitness Value after fourth generation

7 Conclusion

In this paper we have reduced the value of error by observing the gains value after each generation and apply the new gains value in form of lower and upper bound and observed that we have finally optimized the non linear proportional integral controller and obtained the least value of error.NPIC controller has vast applications it can also be used for controlling. Pneumatic controllers especially for flow control as it provides stability because of integral controller absolute error has been taken as an objective function **whereas Y** axis represent the number of generations (number of times genetic algorithm

IOP Conf. Series: Materials Science and Engineering 1104 (2021) 012001 doi:10.1088/1757-899X/1104/1/012001

References

1. M. W. Spong, M. Vidyasagar, Robot Dynamics and Control, 2nd Edition, John Wiley and Sons,2004.
2. M. Gopal. Digital Control and State Variable Methods, 3rd edition, Tata McGraw Hill,2009.
3. RachidManseur, Robot Modeling and Kinematics, Firewall Media, 96-100,2007.
4. M. W. Spong, Seth Hutchinson and M. Vidyasagar, Robot Modeling and Control, 1st Edition, John Wiley &Sons,Inc.
5. M. Loudini, "Modeling and Intelligence Control of an Elastic link Robot manipulator, " Int. J. Adv. Robot System, vol. 10, pp. 1-18, January 2013
6. A. F. Amer, E.A. Sallam, W. M. Elawady, "Adaptive Fuzzy Sliding Mode Control using-SupervisoryFuzzyControlfor3DOFplanerRobotmanipulators,"Appl.Soft Compt., vol. 11, pp. 4943-4953, December2011.
7. R. Sharma, K.P.S.Rana and V.Kumar, "Performance analysis of fractional order fuzzy PID controllers applied to a robotic Manipulator," *Expert Systems with Applications,* vol73, no. 41(9) pp. 4274–4289, September 2019.
8. V. Kumar, K.P.S. Rana, J. Kumar. Mishra and S.S. Nair, "A robust fractional order fuzzy *P*+ fuzzy *I*+ fuzzy D controller for nonlinear and uncertain system," *International Journal of Automation and Computing* , vol. 14, no. 4, pp. 474-488, August 2017.
9. V. Kumar and K.P.S. Rana, "Nonlinear adaptive fractional order fuzzy PID control of a 2-link planar rigid manipulator with payload, "*Journal of the Franklin Institute vol. 354, no. 2, pp. 993-1022,* 2017
10. J. Kumar, V. Kumar and K. P. S. Rana, "A Fractional Order Fuzzy PD+I Controller for Three Link Electrically Driven Rigid Robotic Manipulator System," *Journal of Intelligent and Fuzzy Systems*, vol. 35, no. 5, pp. 5287-5299, November 2018.
11. D.S. Weile,& E. Michielssen. "Genetic Algorithm Optimization applied to electromagnetics," A review. IEEE Transactions on Antennas and Propagation, Volume: 45 , Issue: 3 , Mar 1997C. I-Chuan; Y. Jeng-Rern, "2 – 13GHz broadband CMOS low voltage mixer with active balun designed for UWB systems," *IEEE International Conference on Electron Devices and Solid-State Circuits 2010*, pp.1-4, 15-17 Dec. 2010.
12. Z. Kong, JiaWenhua, Zhang Guodong& Wang Lifu. "Normal parameter reduction in soft set based on particle swarm optimization algorithm". Applied Mathematical Modelling, Volume 39, Issue 16, August 2015
13. A Ghanbari,SMRKazemi, F Mehmanpazir.,& MM Nakhostin. "A cooperative ant colony optimization-genetic algorithm approach for construction of energy demand forecasting knowledge based expert systems". Knowledge-Based Systems, Volume 39, February 2013
14. K. Jagatheesan, B. Anand, Sourav Samanta., Nilanjan Dey, Amira S. Ashour&Valentina E. Balas "Design of a proportional-integral-derivative controller for an automatic generation control of multi-area power thermal systems using firefly algorithm." IEEE/CAA Journal of AutomaticaSinica, Volume: 6 , Issue: 2 , March 2019.
15. X.S. Yang, & S. Deb, "Cuckoo Search via Lévy Flights. Proceedings World Congress on Nature & Biologically Inspired Computing," India, 978-1-4244-5053-4,january 2009
16. X.S. Yang, & A.H. Gandomi, Bat-algorithm: "a novel approach for global engineering optimization. Engineering Computations,"Vol. 29 No. 5, pp. 464-483,July 2012.
17. Y. Ohtani, & T. Yoshimura, "Fuzzy control of a manipulator using the concept of sliding mode,"International Journal of Systems Science, Vol 27, Issue 2, May 1996.

18. A. Hazzab, I.K. Bousserhane, M. Zerbo, & P. Sicard, "Real-time implementation of fuzzy gain scheduling of PI controller for induction motor machine control," Neural Processing Letters, 24 (2006) 203-215.
19. R.Sharma,S.Bhasin,P.Gaur, D.Joshi, "A switching-based collaborative fractional order fuzzy logic controllers for robotic manipulators,"Applied Mathematical Modelling, Volume 73, September 2019.R.Sharma,S.Bhasin,P.Gaur, D.Joshi, "A switching-based collaborative fractional order fuzzy logic controllers for robotic manipulators,"Applied Mathematical Modelling, Volume 73, September 2019.
20. F. Lin. (2007) "Robust control design: An optimal control approach, "John Wiley & Sons Ltd., England.
21. George Thuruthel, Thomas, et al. "Control strategies for soft robotic manipulators: A survey." *Soft robotics* 5.2 (2018): 149-163.
22. Zhang, Shuang, et al. "Adaptive neural control for robotic manipulators with output constraints and uncertainties." *IEEE transactions on neural networks and learning systems* 29.11 (2018): 5554-5564.
23. Xiao, Bing, Shen Yin, and OkyayKaynak. "Tracking control of robotic manipulators with uncertain kinematics and dynamics." *IEEE Transactions on Industrial Electronics* 63.10 (2016): 6439-6449.
24. Khairudin, Mohammad, Zaharuddin Mohamed, and Abdul Rashid Husain. "Dynamic model and robust control of flexible link robot manipulator." *Telkomnika* 9.2 (2011): 279.
25. Rus, Daniela, and Michael T. Tolley. "Design, fabrication and control of soft robots." *Nature* 521.7553 (2015): 467-475

Design and analysis of nozzles for pneumatic windshield wiper

Anirban Sur [1] Nitin Solke[2] Sharnil Pandiya[3] Ketan Kotecha [4]

1,2 Symbiosis Institute of Technology, Symbiosis International (Deemed) University, Pune 412115, Maharashtra, India,
3,4 Symbiosis Centre for Applied Artificial Intelligence, Symbiosis International (Deemed) University, Pune 412115, Maharashtra, India;

anirban.sur@sitpune.edu.in

Abstract. Almost all the vehicles use mechanical wipers to clean the windscreen when rain or other elements such as dirt hit the surface of the windshield. This helps maintain visibility so the driver can see through the windshield. Also, mechanical wipers can damage the windscreen over time with friction and partially hinder the visibility as they only clear the screen partially. There are no alternative models currently available for solving the same problem. This paper proposes an air jet windshield wiper system consists of a high-speed air outlet mounted on the edge of the windshield through which a high-speed mass of air, called as air-beam, is released over the windshield. This can be achieved with the help of an air nozzle. The air-beam ejected by the outlet of the nozzle is designed to cover most of the surface of the windshield. The beam thickness is used to push rain drops over the windshield to the roof of the vehicle and beyond. In this paper, necessary calculations were done for selecting suitable nozzles for the project. To verify the results, FEM analysis and experimental analysis has been done.

Keywords: Windshield, Air jet wiper System, Nozzle

1. Introduction

A windshield wiper is a device which is used to eliminate rain, debris, snow from a vehicle's front windshield. For cars, trucks, buses, train locomotives, watercraft with a cabin, and some aircraft wipers are legal necessities. A wiper generally consists of a metal arm, one end pivots. The other end has a long rubber blade attached to it. The arm is powered by a motor, often an electric motor, although pneumatic power is also used for some vehicles[1-3]. The blade is swung back and forth over the glass, pushing water or any other obstructions from its surface[4]. The credit for inventing the windscreen wiper goes to an American inventor Mary Anderson (United States Patent No. 748,801, 1903)[5]. She invented the first operational wiper. In her patent, she termed wiper as a 'window cleaning device.' Before that, drivers used to get out of the car and clean the windshield manually. On vehicles made after 1969, the speed is usually adjustable, with several continuous speeds, and often intermittent settings. Most personal vehicles use two synchronized radial-type arms, while many commercial vehicles use one or more pantograph arms. The basic principle of the wiper has continued to this day.

Even after 115 years, nothing has changed. But it has been observed that the uneven wiping action results in streaking of water and reduced driver visibility. Especially during heavy rain and snowfall, the rate of cleaning of the windshield is slow, which reduces the visibility of the driver. As wipers have a direct impact on visibility, this is usually rather negative, and there are enough reasons to replace the principle.

So improvement is required of the conventional wiper to increase the visibility of the driver, to provide more efficient cleaning and eradicate the problems of the traditional wiper such as continuous replacement, scratches on the windshield, and also reduce the maintenance cost.

By looking for a solution to all these problems, a question is raised: Is there already enclosed means of transport with a windshield without wipers? Some scientists have already invented different methods for improving the visibility of the driver during using of the wiper. At 1989, A. Demir et al. investigated T100 Hema air Jet texturing nozzle with varying geometrical configurations [6]. They reported that the exit shape has virtually no effect on the airflow while other parameters contribute to the formation of the emerging jet which performs the texturing. An air wiper unit was patented (Japan Patent No. 5,419,005, 1995)[7] by Japan in 1995. It reported that the air wiper includes a first air ejection unit and a second air ejection unit. The first ejection unit gives out compressed air upwards near windscreen glass and along the outer surface of the windscreen glass to form an air curtain. The second air ejection unit is arranged. An air curtain is generated being spaced away from the windscreen glass more than the first that oscillates and forms a second air curtain through the oscillating movement. The second air ejection unit is formed of multiple nozzles, which forms the second air curtain. A warm air supplying device is arranged in front of and near the lower end of the windscreen glass to feed warm air upwards along the outer surface of the windscreen glass.

Ho-Young Kim [8] reported the study of the sliding velocity of Liquid Drops Down on an Inclined Solid Surface. High-velocity impact of a liquid droplet on a rigid surface. Haller Knezevic, 2002 [9] Was reported by Haller Knezevic. They simulated water droplets impacting at the surface at 500 m/s (1.5 Mach number), and found that the spattered water droplets had the velocity over 6000 m/s from the point of impact (18 Mach number). Alidoost Dafsari 2016 [10] experimented with the atomization and internal flow field mechanism of a hollow nozzle under different inlet pressures. Variations in the flow rate, spray cone angle, and radial spray density was also obtained under different working conditions. The results show that with the increase of inlet pressure, the growth of flow rate tends to be slower, and the airflow field in the nozzle swirl chamber is squeezed but with a more stable shape. Jagmit et al. investigated the effect of Nozzle Geometry on Critical-Subcritical Flow Transitions. They identified the optimum nozzle geometry that maximizes critical pressure ratio while minimizing pressure drop across the nozzle.

After studying the literature mentioned above, it can be concluded that scientists are trying to replace the windshield wipers in the car by an air jet. The jet of air should be strong enough and aligned with the glass to deflect the raindrops that come into contact with the windshield.

2. Design Of Wiping System

This system is designed here for trucks and trailers because, in trucks, compressed air is readily available, used in the braking system. A fraction of this air will be used for operating the pneumatic windshield wiper.

As the air-beam blows over the surface of the windshield, it carries away elements deposited on the windshield with it. If the device is activated during rainfall, it can carry droplets of rain away from the windshield.

With appropriate airspeed and air volume, the device can ensure a completely dry windshield even during rainfall. This can allow a tremendous increase in visibility for the driver and occupants of the vehicle. Essential components of wiping systems are compressed air, pressure regulator, pneumatic pipes, and nozzles.

The air jet windshield wiper system consists of a high-speed air outlet mounted on the car at the windshield through which a high-speed mass of air, henceforth an air-beam, is released over the windshield. The air-beam is characterized by the velocity of its flow, its direction of motion and its shape. The velocity of the air-beam is substantially the average velocity of the air particles in the beam. Similarly, the direction of the air-beam is the direction in which the majority of the air particles in the air-beam are traveling.

Air from the reservoir is passed through a valve to start/stop the system. In the stop position, the valve is closed, and no air is passed ahead. When the valve is open, the air enters the piping and enters into the nozzle. In the nozzle, the pressure energy from the incoming air is converted into kinetic energy, which enables air beam impinging on the windscreen at higher kinetic energy. The air discharged from the nozzle at high velocity will be used for wiping the water droplets from the windshield.

For calculation of the mass of raindrop, raindrops are considered as a sphere. The maximum diameter of a raindrop is 5 mm [11]. If the raindrop gets any larger than 5 mm, it usually splits into two separate drops. Small raindrops are more spherical than larger ones[12]. The frictional drag from the air makes the more massive drops to have a hamburger bun shape. The volume of a small raindrop can be approximated by using the sphere formula[13].

$$m = \boldsymbol{\rho} \times \frac{4}{3} \times \boldsymbol{\pi} \times \boldsymbol{r}^2 \tag{1}$$

where, m= mass of the rain droplet, $\boldsymbol{\rho}$ = density of water, r = radius of rainwater droplet

According to Spilhaus[14], V_t = Terminal Velocity (The constant speed that a freely falling object eventually reaches when the resistance of the medium through which it is falling prevents further acceleration) is 9 m/sec.

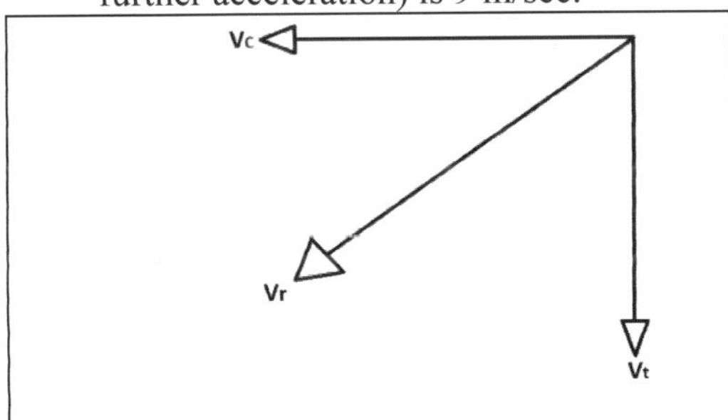

Fig. 1 Resultant Vector

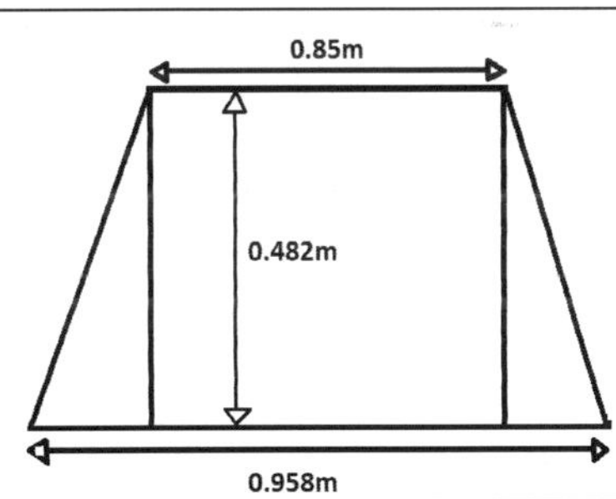

Fig. 2 Windshield diagram

V_c = Velocity of car,

Relative Velocity (v_r) $= \sqrt{V_t^2 + V_c^2}$ (2)

The momentum of a single raindrop is calculated by Eq. (3)

$$p = m \times v_r \tag{3}$$

From the rainfall intensity surveyed at various locations in the country, the mean number of raindrops that fall on the windshield can be estimated to be 500 [15] and 1mm of rainfall equals to 1 liter of water on $1m^2$ area in an hour [16-20]. Fig 2 shows dimensions of a flat windshield (Consider windshield of Maruti 800 car, and slanted angle of 5°), area 0.749 m^2 [21-22]. For deflecting rainwater, the mass flow rate of air should be higher than the mass flow rate of water ($\dot{m}_{air} > \dot{m}_{water}$).

For calculating the outlet velocity at the end of the nozzle, the momentum energy is equated with the kinetic energy of the steady flow of water

$$p \times a \times = \frac{1}{2} \times \dot{m} \times v^2 \tag{4}$$

With the help of the above formulae, the velocity of raindrop and angle of raindrop has been calculated. Fig 3 and Fig 4 represents the same.

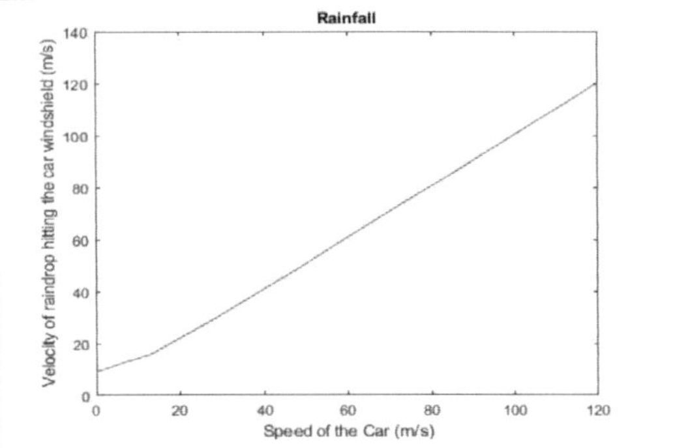

Fig. 3. The velocity of raindrop vs. speed of the car

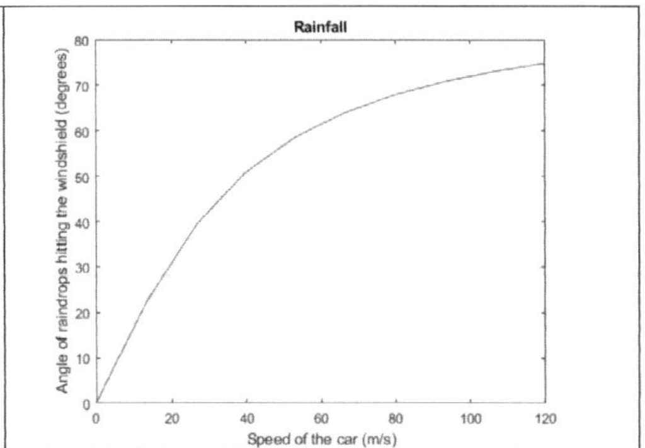

Fig. 4. The angle of raindrops vs. speed of the car

After developing the CAD model in CATIA, ANSYS FLUENT software was used for analyzing the nozzles, which were eventually finalized. The purpose of using this software was to test the flow and outlet velocity of the nozzle.

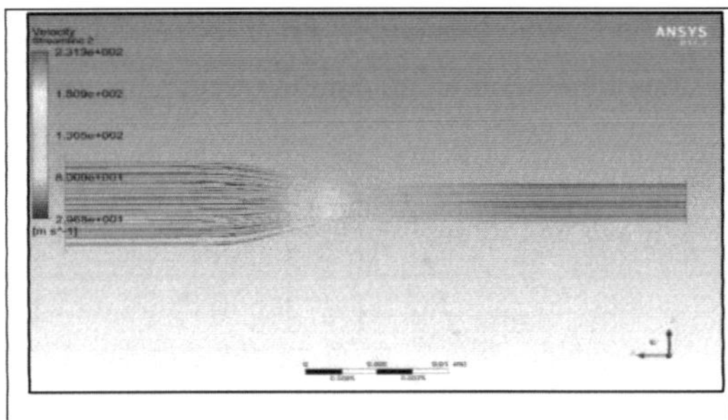

Fig. 5: Long Jet Nozzle

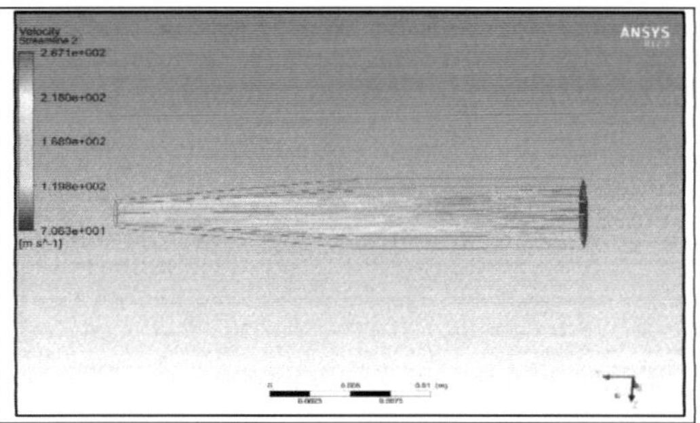

Fig. 6: Cone Jet Nozzle

Fig. 5 represents the streamline model of the long jet nozzle where inlet pressure is kept 2 bar and outlet pressure at 1 bar. The outlet velocity obtained from the analysis was 231 m/s. The velocity is high, and the flow also has a descent spread angle. Fig. 6, shows the analysis of the cone-jet nozzle for the same operating conditions. It can be seen that it has the highest velocity of 267 m/s. Fig. 7 represents the streamline model of the flat jet nozzle for the same operating conditions. The outlet velocity obtained is 136 m/s, which is less than the long jet and cone nozzle.

E

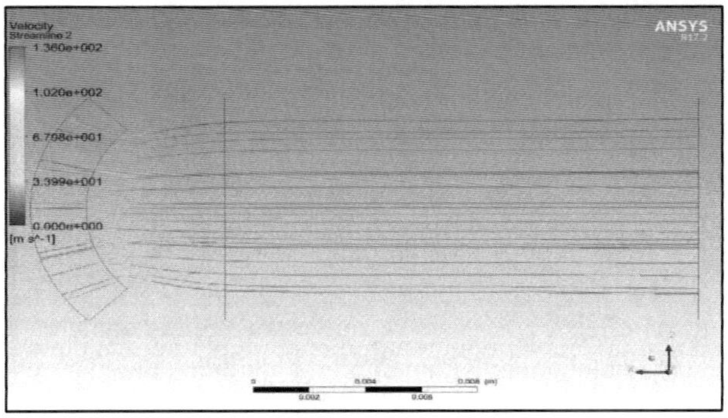

Fig. 7 Flat Jet Nozzle

3. Experimental Setup:

The purpose of this testing is to check the area covered by the nozzle and check the actual results.

To begin with the experiment, first, collect some sand in a bowl. Check for any impurities in the sand particles. With the help of a sieve spread the sand on a flat surface and make sure it is evenly distributed. Meanwhile, turn on the compressor, let the pressure build-up to 4 bar. Place the nozzle in a position as shown in Fig. 13. Check all the fitting in pipes. Make sure there is no loose-fitting, as it will create pressure loss. After everything is checked, turn on the knob. The compressed air will flow through the nozzle and erode the sand. Take a long ruler and a protractor. Measure the range of a nozzle with the help of a long ruler and a protractor. The displacement is the distance between the nozzle and the farthest point of sand eroded. The angle is the area it covers discharging from the nozzle outlet.

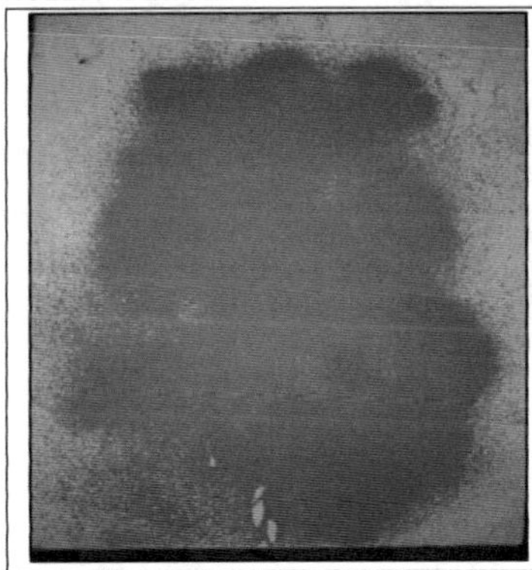	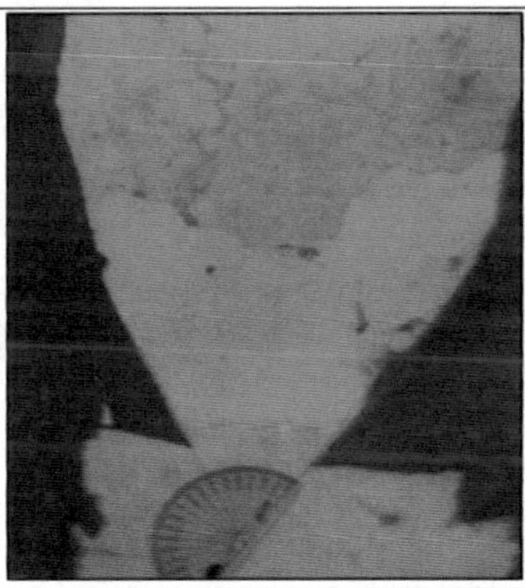	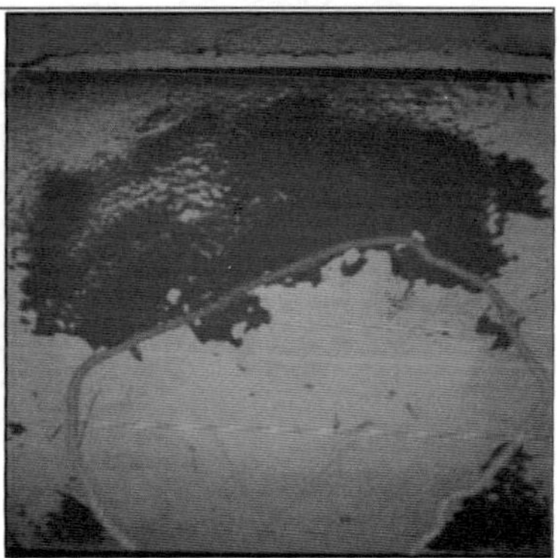
Fig. 8 Sand	Fig. 9 Spray Area of Long Jet Nozzle	Fig. 10 Spray Area of Flat Jet Nozzle

The spray pattern of the long jet nozzle has a well-defined spray pattern with a spray angle of $60^{\circ}$and has an effective range of around 600 mm, which is shown in the given Fig. 8. The flat jet nozzle has an undefined and irregular spray pattern with a spray angle of around 90^{0} but has a very low effective range measuring roughly 300mm.

Graphs are shown (Fig 11 and 12) are based on experimental results drawn. It shows the nozzle displacement and sprays angle of the long jet nozzle (blue) and flat jet nozzle (brown).

IOP Conf. Series: Materials Science and Engineering 1104 (2021) 012002 doi:10.1088/1757-899X/1104/1/012002

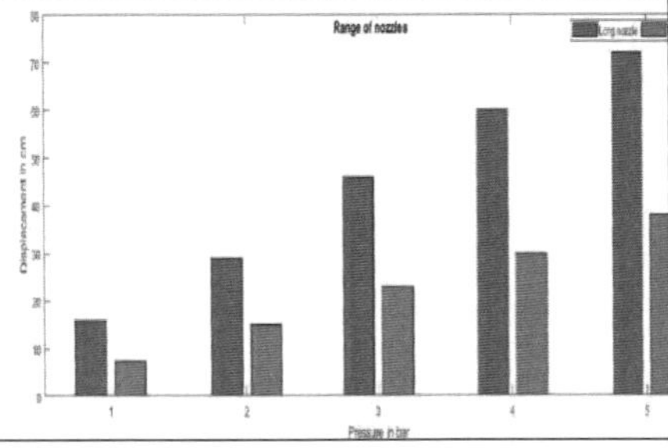

Fig. 11 Displacement vs pressure

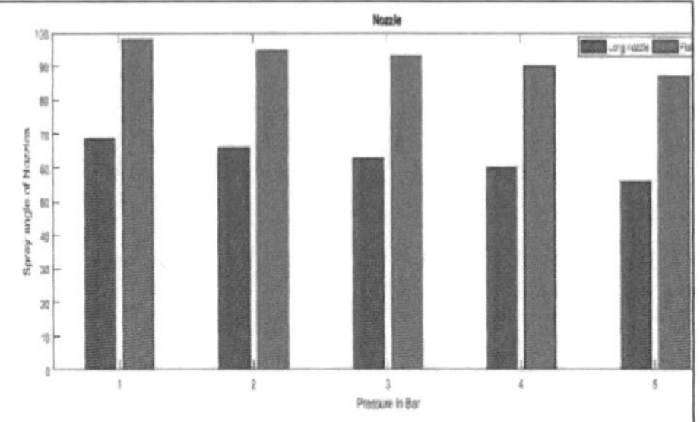

Fig. 12 Spray angle vs. pressure

Different arrangements of nozzles were tried to get the best results. Some of the arrangements are shown below. In the given Fig 11, six nozzles were arranged on the windshield. High-pressure air from the compressor is passed through the pneumatic pipes and is equally distributed among the six nozzles. During this arrangement, the area covered is more, but the outlet velocity is reduced due to the pressure drop, which shows this arrangement insufficient. In the given Fig. 13, 14, two different nozzles were arranged on the windshield (long jet and cone nozzle). This arrangement was efficient in cleaning the nozzle, but the area covered was less. Among the two, the long jet nozzle was more efficient.

4. **Results**

For the cleaning of the windshield, four nozzles were tested. All the tests done give the following results.

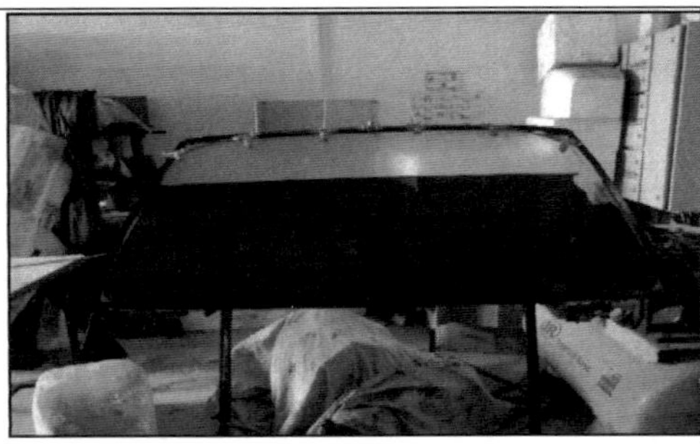
Fig. 13 Nozzle Arrangement (a)

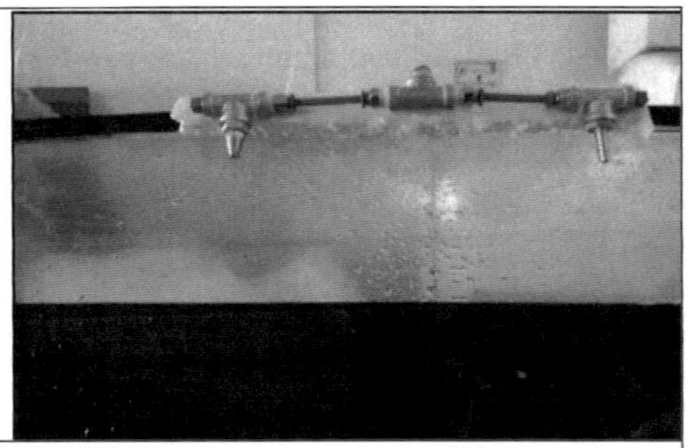
Fig. 14 Nozzle Arrangement (b)

Flat Jet nozzle:
The outlet velocity of this nozzle is 136 m/s, but the spray angle is 90°. The distance covered by this nozzle is around 300 mm, which is less compared to other nozzles. This is due to the widespread angle of this nozzle. As the need for the project is for high velocity, so choosing this nozzle is not a good option.

Long Jet nozzle:
The outlet velocity of this nozzle is 231 m/s. The distance covered is up to 600mm and the spray angle is 60^0. The results obtained by calculating exit pressure, exit velocity, and spray pattern of different nozzle manually and on Ansys shows that Long Jet Nozzle is suitable for the system. The spray angle problem can be encountered by mounting a rotating mechanism for this system.

Cone nozzle:
The outlet velocity of this nozzle is 267 m/s. The distance covered is almost the same as the long jet nozzle and even the spray angle is the same. This nozzle can also be used for the system but the main drawback is that they are heavy than the long jet nozzle. Lighter nozzles are more comfortable

to arrange on the windshield.

Theoretical results were different from the actual result considering pipe loss, pressure drop, friction losses, and other external forces were not considered. The air pressure decreased when using 5/6 nozzles but was considerably enough while working with two nozzles.

Truck minimum compressor pressure is 5 bar and this system is made for all the conditions. This system pressure is also kept up to 5 bar to check if it can work under all the conditions. The results obtained by this pressure were sufficient to wipe off the windshield.

5. Conclusion

This paper presents an alternative method to clean the windshield of an automobile. It uses compressed air at

A certain pressure and velocity to wipe out the raindrops and other particles from the windshield. From the above results, it can be inferred that the pneumatic wiper system is cost-effective, needs less maintenance compared to the traditional method of cleaning the windshield with wipers.

The pressurized air could not cover some parts of the windshield area. But this problem can be fixed in the future by designing custom nozzles. Overall, this system is feasible as per assumptions and calculations. Further, this system can also be used in conjunction with the newly developed nanotechnology-based water-repellant coatings, which will significantly increase the effectiveness of the system.

The system can be made compact without compromising the performance to be used in cars, which may replace the present wiper system completely. It can have different speed settings which can be set according to the rain intensity. The above system can be automated, and it can be automatically switched ON, on detecting rain and stops when the rain stops. This will reduce manual intervention.

Rain sensors, along with 555 timers and the transistor, can be used to switch on the compressor. The rain sensor will detect the rain, and this signal will be processed by 555 timers and take the desired action. Rain sensor works on the principle of using water for completing its circuit, so when rain falls on it, its circuit gets completed and sends out the signal to 555 timer IC

References

[1] Yang, Z. G., X. M. Ju and Q. L. Li . Numerical Analysis on Aerodynamic Forces on Wiper System. *Proceedings of the Sixth International Conference on Fluid Mechanics, AIP Proceedings* 1376, 213-217 2011.

[2] Lee, S. H. and et al. Numerical Study on the Aerodynamic Lift on the Windshield Wiper of High-Speed Passenger Vehicles, In *The Proceedings of Asian Symposium on Computational Heat Transfer and Fluid Flow* 3, 185-190 2009

[3] Clarke, J. and R. Lumley Problems Associated with Windscreen Wiping. *SAE Technical Paper.* 1960.

[4] Gaylard A. P., A. C. Wilson and G. S. J. Bambrook . A Quasi-Unsteady Description of Windscreen Wiper Induced Flow Structures. In *6th MIRA International Vehicle Aerodynamics Conference*. 2006

[5] Anderson, M. *United States Patent No. 748,801,* 10th November 1903.

[6] A Demir. A Study of Air-Jet Texturing Nozzles: The Effects of Nozzle Configuration on the Air Flow. *Journal of Engineering for Industry, 112*(1), 97-104, May 1989.

[7] Mori, T. *Japan Patent No. 5,419,005* May 30, 1995.

[8] Ho-Young Kim, H. J. (, December 10). Sliding of Liquid Drops Down an Inclined Solid Surface. *Journal of Colloid and Interface Scence*, 372-380, 2002.

[9] Haller Knezevic, Y. V. A Computational Study of High-Speed Liquid Droplet Impact. *Journal of Applied Physics, 92*(5), 2821-28, September 2002.

[10] Alidoost Dafsari, R. &. An Experimental Study on Flow and Spray Characteristics of Pressure Swirl Nozzle with Variation of Swirl Chamber Length. *Proceedings of the Autumn Conference of Korea Society of Automotive Engineers*, 118-123, 2016.

[11] Cobert, J. H. *Physical Geography Manual.* U.S.: Kendall Hunt, 2003.

[12] Jallet, S. and et al. Numerical Simulation of Wiper System Aerodynamic Behavior. *SAE Technical Paper.* 2001

[13] Jeong, J. and F. Hussain On the identification of a vortex. *Journal of Fluid Mechanics* 285, 69-94. 1995

[14] Spilhaus, A. (1947, November 29). Raindrop Size, Shape, and Falling Speed. *Journal of Meteorology, 5*, 108-110.

[15] V. Sasi Kumar, S. S. Drop size distribution of rainfall of different intensitie — *Indian Journal of Radio and Space Physics, 32*, 217-220, 2003.

[16] Brouwer, C. *Introduction to Irrigation.* Rome: Food and Agriculture Organization of the United Nations, 1985.

[17] Billot, P., S. Jallet and F. Marmonier Simulation of Aerodynamic Uplift Consequences on Pressure Repartition - Application on an Innovative Wiper Blade Design. *SAE Technical* 01-1043. 2001

[18] Cadirci, S., S. E. Ak, B. Selenbas and H. Gunes Geometric Modifications to Minimize Lift Acting on a Simplified Front Windshield Wiper Blade. *Journal of Thermal Science and Technology* 36(2), 103-109. 2016

[19] Cadirci, S. and et al. Numerical Investigation of Turbulent Flow over a Windscreen Wiper Blade. In *ASMEInternational Mechanical Engineering Congress and Exposition.* 2016

[20] Dawley, M. Aerodynamic Effects on Automotive Components. *SAE Technical Paper.* 1965

[21] Jagmit Singh, L. E. Effect of nozzle geometry on critical-subcritical flow transitions. *Heliyon, 5*(2), 1-19, February 2019.

[22] R.Saidur, N. R. A review on compressed-air energy use and energy savings. *Renewable and Sustainable Energy, 14*(4), 1135-1153, May 2010.

Assessment of Biofuel Resource Potential, Prospects, Challenges and Utilization in Ethiopia: Sourcing Strategies for Renewable Energies- A Review

Michael G. Bidir[1], *N.K. Millerjothi*[2], *Muyiwa S. Adaramola*[3], *Ftwi Y. Hagos*[4] *and R.C. Singh*[5]

[1, 2] *School of Mechanical and Industrial Engineering, EiT-M, Mekelle University, Mekelle, P.O. Box 231, Ethiopia, email:* [1]*michael.gebreyesus@mu.edu.et,* [2]*millerjothi.kalamegam@mu.edu.et*

[3]*Department of Environmental Science and Natural Resource Management, Norwegian University of Life Sciences, 1432 Ås, P.O. Box 5003, Norway, email:* [3]*muyiwa.adaramola@nmbu.no*

[4]Department of Mechanical Engineering, University Malaysia Pahang, 25000 Kuantan, P.O. Box 12, Malaysia, email: [4]*ftwi@ump.edu.my*

[5]Department of Mechanical Engineering, Delhi Technological University, Delhi - 110042, India, email: [5]*rcsingh@dce.ac.in*

Abstract

Lately there has been renewed interest in biomass based renewable energies. This is due to energy security concern, growing environmental benign and spiraling price of fossil fuel. In Ethiopia quality of life and energy consumption is tidily conjoined. This is mainly attributed to fast economic growth, expansion of industry and change of lifestyle of its population. This review paper assesses biofuel resource potential, prospects, challenges of Ethiopia and also investigated strategies for its transport utilization, with particular emphasis on sourcing options for cleaner energies from biodiesel. Possessing huge biomass potential and all types of topographies and climatic conditions that can accommodate the growth of various potential biofuel feed stocks; biofuels are believed to be best renewable alternatives to replace transport fuels. Attracted from its potential, the country has started production and utilization of biofuels, mainly biodiesel and ethanol since 2007. Although the biodiesel production is marred by many challenges. Generally, the economic and environmental benefits of biofuel development have been advocated in many literatures; however, the Ethiopian biofuel development policy and action plan lacks systematic study of strategies for efficient use of biofuel resource beyond the saving of hard currency by replacing petroleum-based fuels with biofuels, pastoral agricultural development and lessen the exhaust emissions. In this study it is shown that, updating the policy consolidating genuine biofuel development strategies depending on different evaluation adjustments, encouraging global expertise, involvement in specialized technical coordination, using up-to-date technologies and the introduction of advanced biofuel production may help in resolving the ongoing energy problems in the country.

Keywords: Biodiesel, Biofuel, Biomass, Energy, Ethanol, Transportation

1. Introduction

The world is in the transition from fossil fuels dominated energy systems to low-carbon and sustainable energy systems with significant attention on energy efficiency and renewable/sustainable energy resources that safeguard the environment as per the Kyoto goals [1]. Energy resources play substantial roles in the economic development of every country of which, most of the world energy needs are delivered by the fossil based energy sources [2]. Among these sources, oil plays a vital part in the industrial development, transportation, farming sectors, and to meet numerous other essential human needs [2], [3]. Despite the danger of running out these non-renewable energy resources, global demand has been extensively expanded in the last few decades [4] and currently, it is stretched by 10^5 times more than its natural formation [5]. Therefore, the world is facing many energy-related problems such as rapid and unstable fuel price, uncontrollable emissions and anxieties of national energy security [6].

The fast population growth, industrialization, and change in human living styles are the fundamental reasons behind the increasing demand and the present energy crisis [4], [7]. These problems have led researchers to seek alternative energy sources to our existing carbon-based energy sources [8]. These alternatives incorporate hydropower, solar energy, wind energy geothermal energy, nuclear energy, and bioenergy (biofuels) resources. Olatomiwa Bifarin (2015) reported that biofuels are the best alternatives, especially for Africa, due to the continent's enormous biomass potential of all the enumerated alternative energy resources [9]. Biofuels are usually liquid and gaseous fuels, which can be derived from agricultural yields [10], animal wastes and municipal (domestic and industrial) wastes (both solid and wastewater). Similarly, human geography studies show that robust economic distributional influences can happen due to biofuel utilization [11]. Thus, the enthusiasm for biofuel development has been growing and has established remarkable attention in present-day as the best alternative replacement of fossil fuels, especially in transportation and power generation sectors [12]–[14]. Correspondingly, due to their environmental recognition, and scope of seen benefits, biodiesel and ethanol are considered as viable alternatives to conventional diesel and petrol fuels [15], [16]. Due to the above mentioned and other several supposed advantages, many countries are promoting their sustainable use. Similarly, Ethiopia, as a net oil importer country and having an enormous agricultural base, has started developing biofuels with an aspiring goal.

Ethiopia's fuel import expense has intensified the strain of the country's equilibrium of payments significantly. For instance, the Ethiopian Petroleum Supply Enterprise (EPSE) purchased 3.3 million tons of oil at an expense of US$1.37 billion in 2017[17]. The imported volume surpasses the sum from the previous year by 10% and the expense is 248 million USD higher than that of 2016 [18], and it keeps increasing as shown in Figure1. This is affecting the balance of expenses negatively as it is increasing with time. In reaction to the price upswing of oil, and partially in reaction to climate fluctuations, Ethiopia has been looking into its energy strategies to move from imported oil fuels to domestically produced biofuels [19].

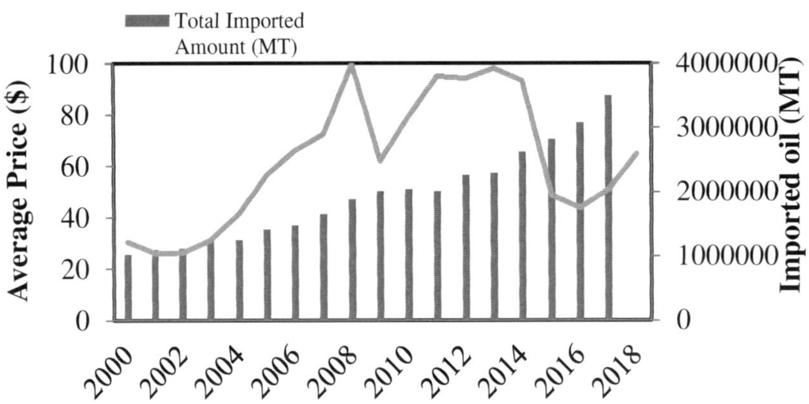

Figure -1: Net energy imports of Ethiopia up to 2016, data adapted from [17].

Furthermore, Kebede et al (2010) reported that 1% growth in GDP requires a 0.55% rise in energy use in developing countries [20]. This indicates that any increment in the current per capita GDP in these countries depends on expanded use of electricity (energy, in general). Hence, to accommodate the exponentially growing energy needs of the country, biofuel development is a priority. Most of the perceived gains from biofuel development include inexhaustibility and biodegradability, increase in the economy by declining imported fuels by exploiting indigenous assets for biofuel production, energy access enhancement, rural society growth, deprivation lessening, and greenhouse gas discharges reductions[16], [21].

Therefore, it is imperative to overview the revealing issues focusing on Ethiopia's potential and status of biofuel development and forecast the future of its growth, depending on the possibilities for biofuel development, policies implemented, motivational mechanisms in place and utilization schemes. In this review, assessment of type biofuels that are being used today in Ethiopia along with their development progresses and utilization pattern, their economic, environmental, and social impacts (considering their advantages and disadvantages during the course of their generation up to utilization), are presented and discussed. Few possible obstacles to the biofuel development in Ethiopia are discussed in the last section before concluding remarks and recommendations are suggested.

2. Biofuel Resource Potential in Ethiopia

Ethiopia has huge biomass potential and many non-edible and edible oil crops, which can be used for the production of biofuel. Furthermore, it has all types of topographies and climatic conditions that can accommodate the growth of various potential biofuel feed stocks. Africa is viewed as one of the real extension regions for the development of biofuel feedstock, and research has demonstrated that Ethiopia has an appealing potential for the establishment of biofuels in Africa [22]. Ethiopia could deliver up to 60 Peta-joules (PJ) of energy from sugarcane, which is comparable to 1.7 billion liters of bio-jet fuels [22]. It has roughly about 30% of its land distinguished as feasibly accessible for purposes other than food-based plant growth and ecological preventions [22], about 25 x 10^6-hectare accessible land and favorable natural conditions for biofuel development [23]. There is about 7 x 10^5 hectare recognized suitable land only for sugar cane cultivation [24]. Furthermore, different oil-bearing seeds and potential biomasses including castor bean, jatropha, cassava [25], croton, palm, Pongamia, and moringa [26] grows in Ethiopia readily. Besides, castor and jatropha plants are the most known biodiesel feedstock's which are considered as native seeds [26]. As shown in *Figure-2* about 327,094-hectare land has been identified as favorable land for biofuel feedstock plantation and was given to investors in 2008 in four regions of the country mainly for farming of castor bean, jatropha, and palm trees [24].

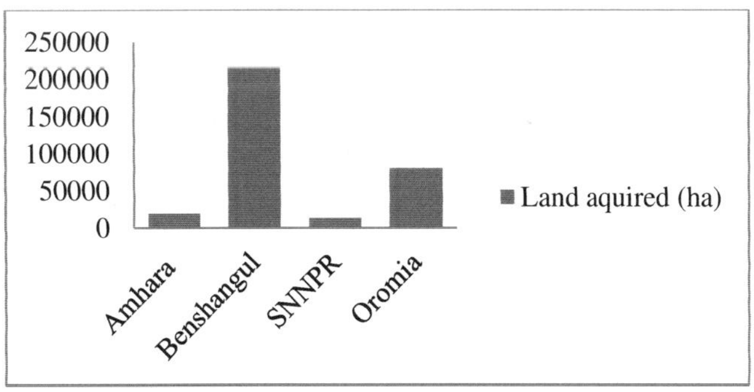

Figure-2: Land allotted for biofuel development in various districts of Ethiopia[24].

3. Prospects of Biofuel Production

Numerous African nations including Ethiopia are motivated by biofuel energies. For examples, Zimbabwe and Malawi began various biofuel activities during the 1980s [27]. In

Tanzania, biofuel advancements exist on various scales [28]. In Nigeria, there are biofuel production endeavors from 1st generation biomass feedstock inputs[29], sustainable biofuel development in Botswana without sizable increase in food price since 2005 is reported by Kgathi D et al. [30], production of biodiesel from jatropha in West African countries exists [31], however, the majority of these efforts are in their earliest stages. Moreover, the consolidated impacts of environmental change, the unpredictability of fuel costs, the ongoing food crisis and the worldwide financial downturn has positioned a feeling of anxiety among planner's, ventures and experts to discover economical and reasonable arrangements in biofuel developments.

In such respect, Ethiopia is an important case for examining the advancement and capability of biofuels in Africa. Excited by the different typically depicted open doors lying in front of the development of biofuels, the Ethiopian government boosted a broad bioenergy development program and appointed project announcement in 2007 for castor and jatropha. These two products were particularly promoted for their flexibility and ability to develop on non-arable land and dry climatic conditions [32]. Castor, Jatropha, and Pongamia have been recognized as primary biodiesel feedstock by the government [33]. As indicated by [34], from all African nations, Ethiopia was reported to have the biggest planted jatropha (about 20,000 ha) in the year 2012. Yet, the majority of these plants are not cultivated for fuel production purposes instead either growth was for housing fences or utilized for ecological restoration purposes.

To reduce the amount of imported crude oil, Ethiopia has been producing a considerable amount of bioethanol from sugarcane mainly in a sugar factory, which is being used for petrol blending, merely in a lower proportion. Currently, Fincha and Metehara sugar plants in Ethiopia are delivering around 30 million liters of ethanol from molasses per annum. Even though Ethiopia had planned to produce 1.7 billion liters of bioethanol per year from sugarcane molasses with effect from 2007, [35], however, the production amount reached about 400 barrels per day in 2016 (that is 146 000 liters) (*Figure-3*)[36]. The plan was very ambitious, and currently, only two facilities from the above-mentioned plants are producing ethanol much below the anticipated amount. However, developments to produce additional ethanol are in progress at large portions in many other new under construction national sugar factories. There is a prospect of additional ethanol generation when sugar industrial facilities under development come to finish. At present there are about 13 sugar factories, 8 of them are operational and 5 of them are under construction, of which one of them is expected to start ethanol production soon [37].

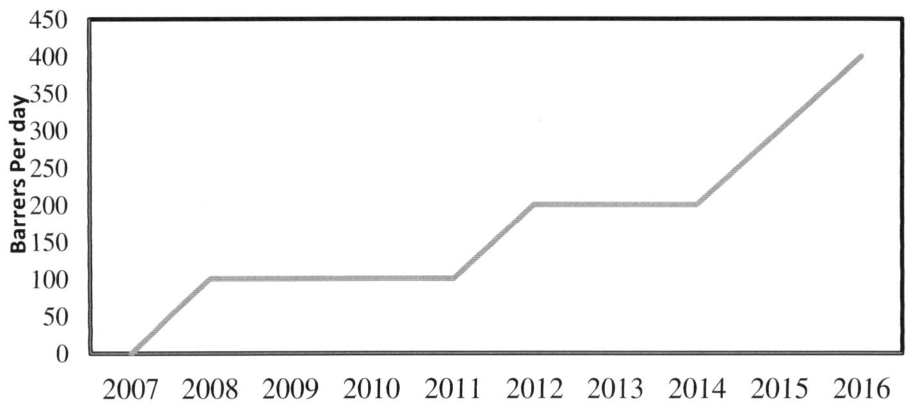

Figure-3: Ethiopia ethanol production, data adapted from[36].

Ethiopia had a plan to achieve the E20 goal by 2015. But, the nation's plan was hindered because of the absence of ethanol supplies following postponements in redesigning sugar plants and poor executing methods [38]. For example, the highly anticipated Tendaho sugar processing plant (one of the 13 sugar factories) was set to start production in January 2013, with a production capacity of 19,000 metric tons of sugar; 120 MW of power and generation of 63 million liters of ethanol, still the ethanol production is yet to start[39]. Similarly, the new Kessem plant has been expected to come to function, delivering 12,500 m^3 of ethanol every year alongside to 153,000 metric tons of sugar in 2015 and yet not operational [40]. However, additional projects are in the pipeline that can increase the biofuel production amount of the country. For example, Natifa, a construction company from Israeli has made secured a $100 million credit to produce 20,827 cubic meters of ethanol from Wolkayit Sugar industrial facility in Tigrai Regional State[41]. Additionally, Ethiopia and Eugen Schmitt (a German distillation technology company mainly involved in the design, manufacturing, and assembly of alcohol production facility) are developing a $51 million ethanol plant having a generation capacity of 60,000 liters of ethanol per day at the Wonji Shoa Sugar Factory since October 2017. Eugen Schmitt will have 83% of the share, while the Ethiopian government and three different investors will take the remaining 14% and 3% shares, respectively[42].

In Ethiopia, ethanol blending with gasoline was started in 2009 [35]. Nile Oil (private company) was the first enterprise to start mixing 5% ethanol mixing with gasoline (E5). Currently, companies called Oil Libya and National Oil are doing the blending process [43]. Subsequently, in five years between 2011 and 2015, the country has a mixed aggregate of 59.6 million liters of ethanol equivalent to 46.9 million US dollars of fuel imports based on the E10 approach [43]. The mixing of ethanol is exclusively done in Addis Ababa and it is meeting 10-15 percent of transport fuel utilization in the capital [44]. The country has a target of increasing the national ethanol mixing facilities in different places of the nation in the near future.

For biodiesel production, the Ethiopian government is spending 2.8 million dollars since July 2014 to deliver 5 X 10^8 liters of biodiesel from jatropha annually with the support of the Norwegian government fund. Undertaking in 18 districts in five regions which are expected to help about 14 million agriculturists and rural communities[45]. Whereas, biodiesel production and blending it with conventional diesel fuel for a diesel engine is yet to start.

4. Utilization Schemes of Biofuel in Ethiopia

Cross-sectional time sequence evidence of 25 years for 20 Sub-Sahara African countries reveals that wood fuel represents 70% of energy utilization, trailed by oil [46]. Similarly, Ethiopia is perceived as the poorest country beside it is a net energy importer [47]. Ethiopia population lives below the energy poverty line with minimal access to electricity while its main energy supply is largely from biomass [48]. Waste and biomass are the nation's essential energy sources accounting 92.4% of Ethiopia's primary energy supply, trailed by oil (5.7%) and hydropower (1.6%)[49]. The Ethiopian Ministry of Water, Irrigation, and Electricity most recent report demonstrate that the electric access of the nation has achieved 57% while the number of connected households to the grid are just 2.8 million[50]. Ethiopia relies on imported crude oil mainly for the transportation and industrial sectors. Furthermore, these fuels are consumed by backup power generation systems in most industries, business, and residential sectors.

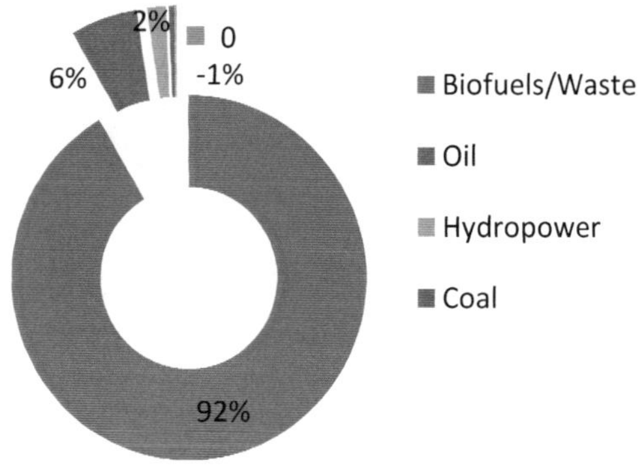

Figure-4: Share of total primary energy supply in Ethiopia in 2015 (equivalent to 49,990 Ktoe)[49].

The current total energy consumption of Ethiopia is around 40,000 GWh, out of this about 92% is utilized by residential apparatuses, 4% by the transportation segment and 3% by manufacturing. The majority of the energy source is secured from biomass. The country's electricity production is about 9000 GWh/year fundamentally produced from hydropower (about 92%), and the remaining 4% from wind energy. Out of the total generated electricity about 11% of it is exported to neighboring countries. Conversely, the country imports petroleum for its transportation sector [17]. Though Ethiopia is blessed with enormous alternative energy resources such as hydro, solar, wind, geothermal, and biomass energies, yet the harnessed amount is very little as shown in *Figure-5.* The Total Primary Energy Supply (TPES) of the country is increasing with time. Besides, it is heavily dominated by biomass, which is usually utilized by low efficient devices.

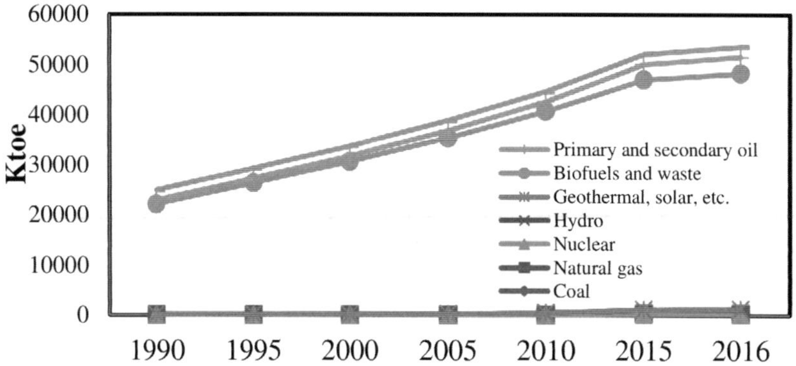

Figure-5: Ethiopia's TPES by source in 2018 [17]

Ethiopia's Total Final Energy Consumption (TFEC) energy is dominated by biomass, followed by oil, electricity, and coal respectively as shown in *Figure-6.* The biomass energy is consumed mainly in residential energy applications, particularly for cooking and heating. The oil energy is consumed largely in the transportation sector and in some backup generators and industries as the main energy fuel. The electricity is most energy supplied to the urbanized areas and its principal consumption is for lighting, cooking and slightly for heating purposes. A portion of it is used in industries as the main energy input. However, almost all the coal consumption is in manufacturing industries. As can be seen from figure 15 and 16, the petroleum fuel consumption is huge and all of it is imported consuming a greater portion of hard currency collected from export commodities of the country. Beside, most of the petroleum fuels are consumed mainly by the transportation sector, thus the need for biofuel as a substitute in this sector makes it anticipatory.

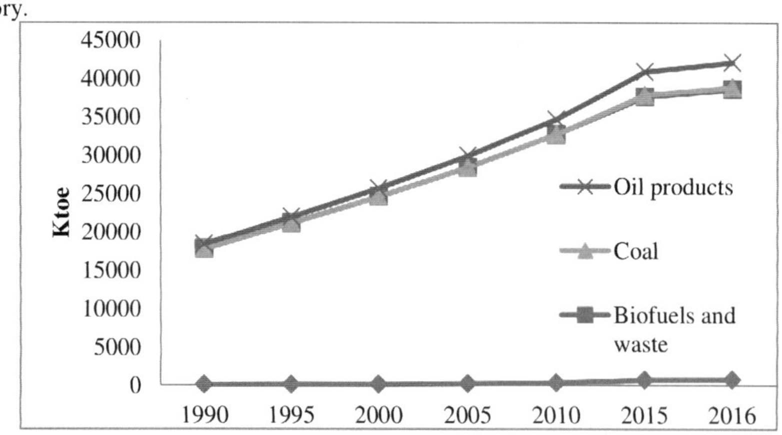

Figure-6: Ethiopia's TFEC by source –in 2016, [17]

5. Possible Challenges of Biofuel Production

Even-though the biofuel development in Ethiopia has got a priority by the government as well as the huge interest by developers to participate on the business because of its perceived economic and environmental benefits, however, its realization to the expectation can be hindered by some barriers if not given a due focus. The possible obstacles that could slow down the biofuel development in Ethiopia are: similar to many developing countries the unstable energy policy, cultural constraint, educational establishment, unsteady economy, absence of core technology and poor capacity building programs[46] could be among the most challenges that the biofuel development program is expected to face.

The dispute on food versus fuel is another obstruction to the development of biofuel[25]. Unless a consensus is reached with the public and stockholders by discussions and awareness creations to convince the public about the pros and cons of the biofuel development agendas, it could be one of the difficulties to be encountered. However, it can be overwhelmed by utilizing non-edible biomass sources[51]. Similarly, besides the food versus energy issue, the association of biofuels with various natural and economic effects such as deforestation, biodiversity, water accessibility, energy-security, food-security and loss of land access [28] are additional issues that need in-depth considerations.

Furthermore, consideration and recognition of the participating stakeholder's values and attitudes are necessary to accomplish sustainable developments[52]. However, the Ethiopian biofuel development policy seems somehow inclined to favor investors than the small scale farmers due to their larger capital and competence. This reason may create susceptibilities in small and large scale developers anticipating interests in feedstock production and processing plants. Moreover, the issues articulated in most public discussions emphasizing land rights and food security are distinctive. Hence, although they miss the technical part of agricultural, economic and market principles, their opinions shall be included in the decision-making considerations. Unless it will be an additional obstacle to the biofuel development endeavors.

6. Conclusion

Despite the huge potential for biofuel development in Ethiopia, the level of production is very small. The Ethiopian biofuel development was very fascinating until it reaches the E10 blending approach in 2011. Afterward, the production amount remained constant and the current production capacity is below the anticipated target (400 barrels per day in 2016). Regardless of government's huge investments in sugar factories to produce ethanol as a second product, the country's goal to start E20 ethanol blending approach in 2015 was not achieved due to delays in most of the sugar factory project completions. Furthermore, the expansion of ethanol blending facilities beyond Addis Ababa to other parts of the country and start biodiesel blending with conventional diesel is yet to be achieved. The upcoming biofuel production expectations are only ethanol from about 13 sugar factories under development. Biodiesel production seems not given due attention.

Due to the lack of basic setups and policy gaps in biodiesel production, big scale biodiesel development seems doubtful. Hence focus shall be given to revitalize small scale local business to produce biodiesel from non-edible oil crops that can be utilized directly for cooking and lighting. The governmental motivational and support schemes should be towards prospering these small scale local businesses as they could have noticeable economic and environmental significances.

The biofuel development policy needs to be reviewed and shall include approaches based on a clear understanding of the current scenarios and future development targets considering its tradeoffs, land use design and economic benefits of biofuel compared to petroleum fuels. Biofuel development can only be feasible for Ethiopia if it is financially feasible and all threats related to its development are alleviated. Thus, in order to avoid undesired outcomes consideration of best existing practices in similar fields and incorporating public opinions in decision making is vital.

Finally, while governmental encouragement is essential for the overall feasibility of biofuel development, the motivational schemes shall focus to make biofuel businesses affordable and cost-effective. As it can be seen from the ongoing many large sugar factory constructions with the intention of increasing ethanol production amount, massive spending on projects intending only to minimize the expenses on petroleum fuel imports is not necessary. Hence, well studied and precisely selected motivational instruments must be implemented.

Reference

[1] *D. K. Jamuwa, D. Sharma, and S. L. Soni, "Experimental investigation of performance, exhaust emission and combustion parameters of compression ignition engine with varying ethanol energy fractions," Energy, vol. 127, pp. 544–557, May 2017.*

[2] *D. Y. C. Leung, X. Wu, and M. K. H. Leung, "A review on biodiesel production using catalyzed transesterification," Appl. Energy, vol. 87, no. 4, pp. 1083–1095, Apr. 2010.*

[3] *P.-J. Shiu, S. Gunawan, W.-H. Hsieh, N. S. Kasim, and Y.-H. Ju, "Biodiesel production from rice bran by a two-step in-situ process," Bioresour. Technol., vol. 101, no. 3, pp. 984–989, Feb. 2010.*

[4] *M. K. Lam, K. T. Lee, and A. R. Mohamed, "Homogeneous, heterogeneous and enzymatic catalysis for transesterification of high free fatty acid oil (waste cooking oil) to biodiesel: A review," Biotechnol. Adv., vol. 28, no. 4, pp. 500–518, Jul. 2010.*

[5] *S. K. Mandotra, R. Kumar, S. K. Upadhyay, and P. W. Ramteke, "Nanotechnology: A New Tool for Biofuel Production," Springer, Cham, 2018, pp. 17–28.*

[6] *A. Demirbas, "Potential applications of renewable energy sources, biomass combustion problems in boiler power systems and combustion related environmental issues," Prog. Energy Combust. Sci., vol. 31, no. 2, pp. 171–192, Jan. 2005.*

[7] *V. Chandrasekaran, M. Arthanarisamy, P. Nachiappan, S. Dhanakotti, and B. Moorthy, "The role of nano additives for biodiesel and diesel blended transportation fuels," Transp. Res. Part D Transp. Environ., vol. 46, pp. 145–156, Jul. 2016.*

[8] *Ladanai SVinterbäck J, "Global Potential of Sustainable Biomass," Technology, vol. 32, 2009.*

[9] *"THE CASE FOR BIOFUEL: The Magic Bullet for Energy Security and Rural Development in Africa – Advanced BioFuels USA," 2015. [Online]. Available: https://advancedbiofuelsusa.info/the-case-for-biofuel-the-magic-bullet-for-energy-security-and-rural-development-in-africa/. [Accessed: 20-Nov-2018].*

[10] *R. A. Voloshin, M. V. Rodionova, S. K. Zharmukhamedov, T. Nejat Veziroglu, and S. I. Allakhverdiev, "Review: Biofuel production from plant and algal biomass," Int. J. Hydrogen Energy, vol. 41, no. 39, pp. 17257–17273, Oct. 2016.*

[11] *G. Oladosu and S. Msangi, "Biofuel-Food Market Interactions: A Review of Modeling Approaches and Findings," Agriculture, vol. 3, no. 1, pp. 53–71, 2013.*

[12] *J. A. Moncada, Z. Lukszo, M. Junginger, A. Faaij, and M. Weijnen, "A conceptual framework for the analysis of the effect of institutions on biofuel supply chains," Appl. Energy, vol. 185, pp. 895–915, Jan. 2017.*

[13] *F. Creutzig et al., "Bioenergy and climate change mitigation: An assessment," GCB Bioenergy, vol. 7, no. 5. pp. 916–944, 2015.*

[14] *A. Elfasakhany, "Investigations on the effects of ethanol–methanol–gasoline blends in a spark-ignition engine: Performance and emissions analysis," Eng. Sci. Technol. an Int. J., vol. 18, no. 4, pp. 713–719, 2015.*

[15] *E. Purdy and P. G. Corpuz, "Philippine Biofuels Situation and Outlook," 2013.*

[16] *S. H. Gheewala, B. Damen, and X. Shi, "Biofuels : economic , environmental and social benefits and costs for developing countries in Asia . : Article type : Advanced Review Article title :," Wiley Interdiscip. Rev., vol. 4, no. 6, pp. 497–511, 2013.*

[17] *"Statistics | Ethiopia - Total Primary Energy Supply (TPES) by source (chart)." [Online]. Available: https://www.iea.org/statistics/?country=ETHIOPIA&year=2016&category=Energy supply&indicator=TPESbySource&mode=chart&dataTable=BALANCES. [Accessed: 23-Jan-2019].*

[18] "Ethiopia: Enterprise Imports Fuel Worth 37 Billion Br - allAfrica.com," 2017. [Online]. Available: https://allafrica.com/stories/201708090849.html. [Accessed: 07-Mar-2019].

[19] T. Ferede, Z. Gebreegziabher, A. Mekonnen, Fantu Guta, J. Levin, and G. Köhlin, "Biofuels, Economic Growth, and the External Sector in Ethiopia," no. July, 2013.

[20] E. Kebede, J. Kagochi, and C. M. Jolly, "Energy consumption and economic development in Sub-Sahara Africa," Energy Econ., vol. 32, no. 3, pp. 532–537, May 2010.

[21] J. Hill, E. Nelson, D. Tilman, S. Polasky, and D. Tiffany, "Environmental, economic, and energetic costs and benefits of biodiesel and ethanol biofuels," Proc. Natl. Acad. Sci., vol. 103, no. 30, pp. 11206–11210, Jul. 2006.

[22] T. Bole-Rental, "RSB Aviation Biofuel Summit," WWF South Africa, 2018. [Online]. Available: www.rsb.org. [Accessed: 16-Nov-2018].

[23] T. Ferede, Z. Gebreegziabher, A. Mekonnen, F. Guta, J. Levin, and G. Köhlin, "Biofuels, Economic Growth, and the External Sector in Ethiopia: A Computable General Equilibrium Analysis."

[24] A. Birhanu and S. Ayalew, "Abadi Birhanu, Shimels Ayalew. A Review on Potential and Status of Biofuel Production in Ethiopia," J. Plant Sci., vol. 5, no. 2, pp. 82–89, 2017.

[25] V. Okudoh, C. Trois, T. Workneh, and S. Schmidt, "The potential of cassava biomass and applicable technologies for sustainable biogas production in South Africa: A review," Renew. Sustain. Energy Rev., vol. 39, pp. 1035–1052, Nov. 2014.

[26] A. Ejigu, A. Asfaw, N. Asfaw, and P. Licence, "Moringa stenopetala seed oil as a potential feedstock for biodiesel production in Ethiopia," Green Chem., vol. 12, no. 2, p. 316, Feb. 2010.

[27] B. Amigun, J. K. Musango, and W. Stafford, "Biofuels and sustainability in Africa," Renew. Sustain. Energy Rev., vol. 15, no. 2, pp. 1360–1372, Feb. 2011.

[28] A. Gasparatos et al., "Biofuels in sub-Sahara Africa: Drivers, impacts and priority policy areas," Renew. Sustain. Energy Rev., vol. 45, pp. 879–901, May 2015.

[29] J. Ben-Iwo, V. Manovic, and P. Longhurst, "Biomass resources and biofuels potential for the production of transportation fuels in Nigeria," Renew. Sustain. Energy Rev., vol. 63, pp. 172–192, Sep. 2016.

[30] D. L. Kgathi, K. B. Mfundisi, G. Mmopelwa, and K. Mosepele, "Potential impacts of biofuel development on food security in Botswana: A contribution to energy policy," Energy Policy, vol. 43, pp. 70–79, Apr. 2012.

[31] R. NDONG et al., "Life cycle assessment of biofuels from Jatropha curcas in West Africa: a field study," GCB Bioenergy, vol. 1, no. 3, pp. 197–210, Jun. 2009.

[32] MoARD, "Biofuel profil.," Minist. Agric. Rural Dev., 2011.

[33] "Environment for Development Biofuel Investments and Implications for the Environment in Ethiopia," in An Economy-wide Analysis, 2015, no. June.

[34] N. Wahl, T. Hildebrandt, and C. Moser, "Insights into jatropha projects worldwide: key facts & figures from a global survey," 2012.

[35] K. Tutlam, "RSB Aviation Biofuel Summit," Event Summery, 2018. [Online]. Available: www.rsb.org. [Accessed: 16-Nov-2018].

[36] "Ethiopia Fuel ethanol production, 1973-2018 - knoema.com," 2019. [Online]. Available: https://knoema.com/atlas/Ethiopia/topics/Energy/Renewables/Fuel-ethanol-production. [Accessed: 13-Feb-2019].

[37] E. Abate, "RSB Aviation Biofuel Summit," Ethiopian Sugar Corporation, 2018. [Online]. Available: www.rbs.org. [Accessed: 16-Nov-2018].

[38] M. Sapp, "Lack of ethanol supply keeping Ethiopia from implementing E20," 2015. [Online]. Available: www.biofueldigest.com. [Accessed: 18-Nov-2018].

[39] "Ethiopia's Tendaho mill to start ethanol production in January : Biofuels Digest," 2013. [Online]. Available: http://www.biofuelsdigest.com/bdigest/2013/11/12/ethiopias-tendaho-mill-to-start-ethanol-production-in-january/. [Accessed: 20-Nov-2018].

[40] "Ethiopia's Kessem mill to produce 12.5 million liters of ethanol production : Biofuels Digest," 2015. [Online]. Available: https://www.biofuelsdigest.com/bdigest/2015/05/06/ethiopias-kessem-mill-to-produce-12-

5-million-liters-of-ethanol-production/. [Accessed: 20-Nov-2018].

[41] *"Israeli company secures $100 million loan for Ethiopian sugar project : Biofuels Digest," 2014. [Online]. Available: http://www.biofuelsdigest.com/bdigest/2014/02/03/israeli-company-secures-100-million-loan-for-ethiopian-sugar-project/. [Accessed: 20-Nov-2018].*

[42] *"German company to invest in US$51 million Ethiopian ethanol plant : Biofuels Digest," 2017. [Online]. Available: https://www.biofuelsdigest.com/bdigest/2017/08/29/german-company-to-invest-in-us51-million-ethiopian-ethanol-plant/. [Accessed: 20-Nov-2018].*

[43] *"Ethiopia saves $46.9 million in gasoline imports over 5 years thanks to E10," biofuel digest, 2015. [Online]. Available: www.biofueldigest.com. [Accessed: 18-Nov-2018].*

[44] *"The Ethiopian Herald," 2018. [Online]. Available: http://www.ethpress.gov.et/herald. [Accessed: 18-Nov-2018].*

[45] *"Ethiopian government investing US$2.8 million in jatropha project : Biofuels Digest," 2014. [Online]. Available: http://www.biofuelsdigest.com/bdigest/2014/07/21/ethiopian-government-investing-us2-8-million-in-jatropha-project/. [Accessed: 20-Nov-2018].*

[46] *Y. S. Mohammed, M. W. Mustafa, and N. Bashir, "Status of renewable energy consumption and developmental challenges in Sub-Sahara Africa," Renew. Sustain. Energy Rev., vol. 27, pp. 453–463, Nov. 2013.*

[47] *P. Nussbaumer, M. Bazilian, and V. Modi, "Measuring energy poverty: Focusing on what matters," Renew. Sustain. Energy Rev., vol. 16, no. 1, pp. 231–243, Jan. 2012.*

[48] *"| Human Development Reports." [Online]. Available: http://hdr.undp.org/en/countries/profiles/ETH. [Accessed: 23-Jan-2019].*

[49] *"Share of total primary energy supply* in 2015."*

[50] *"Ethiopia's electric power coverage surpasses 57 pct - Xinhua | English.news.cn," 2018. [Online]. Available: http://www.xinhuanet.com/english/2018-04/20/c_137125761.htm. [Accessed: 11-Mar-2019].*

[51] *P. C. Munasinghe and S. K. Khanal, "Biomass-derived syngas fermentation into biofuels: Opportunities and challenges," Bioresour. Technol., vol. 101, no. 13, pp. 5013–5022, Jul. 2010.*

[52] *undefined W. (Wissenschaftlicher B. der B. G. Umweltveränderungen), "Welt im Wandel – Zukunftfähige Bioenergie und nachhaltige Landnutzung," 2009.*

Synthesis and Characterization of K_2SiF_6 Hexafluorosilicate

Vartika S.Singh* and S.V.Moharil

Department of Physics, RTM Nagpur University, Nagpur, India

*vartikasingh1710@gmail.com

Abstract. Hexafluorosilicates such as K_2SiF_6 (Hieratite) exists in nature as minerals. The common formula is M_2SiF_6 where M represents alkali ion. All the alkali fluorosilicates including K_2SiF_6 decomposes in the temperature range of 300 to 700 °C with SiF_4 as residual product. SiF_4 has an important application in the solar cells fabrication industry. Due to low thermal stability, it is difficult to prepare these complex fluorides by the conventional methods. Hence, a one step simple wet-chemical route was adopted. Confirmation of successful precipitation of K_2SiF_6 during our process of synthesis was done by using various analysis techniques. Results obtained through XRD, EDAX and SEM measurements are reported and discussed here in brief.

Keywords: Hexafluorosilicate, Synthesis, Precipitation, XRD, EDAX, SEM.

1. Introduction

K_2SiF_6 is a complex fluoride which belongs to a type of compounds collectively called as hexafluorosilicate. Main formula for these fluorides is M_2SiF_6 where M represents an alkali ion. Alkali fluorosilicate K_2SiF_6 decomposes in the temperature range of 300 to 700 °C with SiF_4 as residual product. The reaction commences as: $K_2SiF_6 \rightarrow 2KF + SiF_4$. Its low decomposition temperature makes it major source of fluorine and silicon in chemical processing factories. SiF_4 has an important application in the solar cells fabrication industry. Also more recently, K_2SiF_6 was used in producing tubular SiO_2 by hydrothermal reactions [1]. Hexafluorosilicates are significant in geological research since, K_2SiF_6 also known by the names Demartinite and Hieratite exists in nature as minerals. K_2SiF_6 "Hieratite" is properly used in the manufacture of special glasses, enamels and ceramic materials [2].
More recently, Mn^{4+} activated K_2SiF_6 phosphor finds huge applications as a red component in warm-white LED's used in smart phones, tablets and televisions [3-6].
Week thermal stability of hexafluorosilicates makes difficult to prepare these fluorides via conventional solid state diffusion method. Solid state reaction methods needs very high temperatures for diffusion of initial reactants and require huge apparatuses. Hence, another simple method of synthesis was attempted as K_2SiF_6 is insoluble in water and therefore can be precipitated.

2. Synthesis

Reagents used were of the Analytical grade. Desired amount of potassium carbonate (K_2CO_3) was thoroughly mixed with silicic acid ($SiO_2.H_2O$). Hydrofluoric acid HF(48%) was poured drop wise in the aq. solution of potassium carbonate and silicic acid, till effervescence due to decomposition of carbonate and evolution of CO_2 stops. This process was done at room temperature. Water formed in the reaction and excess HF was slowly evaporated off on a hot plate. The whole procedure was done in a PTFE beaker. Resulting mass was dried and used in further experiments.

3. Results and Discussion

"X-ray diffraction patterns were recorded on a Rigaku MiniFlex II diffractometer. Elemental and morphological studies were performed on scanning electron microscope (SEM, VEGA MV2300T/40) operated at 20.0kV voltage."

3.1. X-Ray Diffraction (XRD) Analysis

Fig.1 shows the XRD pattern of K_2SiF_6 (as prepared sample without any thermal treatment). Stick pattern of ICDD 85-1382 is also shown for the comparison. Excellent match is seen. Cubic form of K_2SiF_6 is thus obtained by the simple wet chemical synthesis.

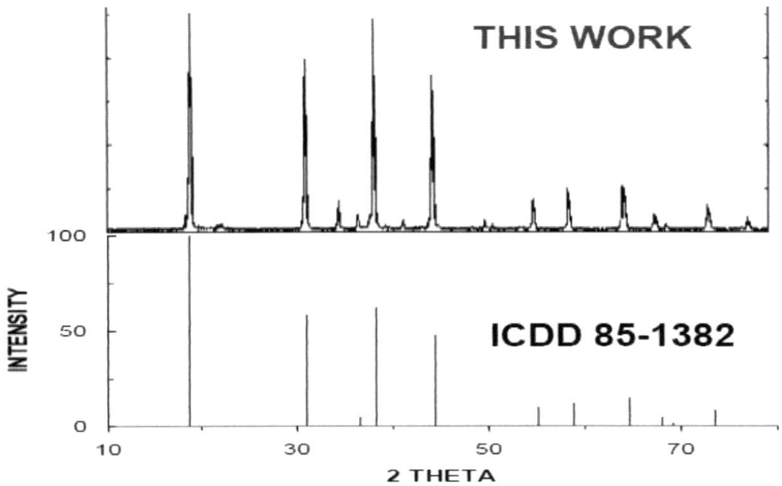

Figure 1.XRD stick pattern of K_2SiF_6 formed in our sample, compared with ICDD 85-1382.

3.2. Crystal Structure

Fig. 2 shows the crystal structure for K_2SiF_6. It crystallizes in cubic system which is its more stable form with space group O_h^5- (Fm3m) with a lattice constant of a = 8.13 Å. The K^+ ions are coordinated by12-fold F^- ions. The Si^{4+} ions are surrounded by 6 F^- ions. The K – F distance is about 2.897 Å and that of Si – F is 1.683 Å. The SiF_6 octahedra lie on the cell edges and on the face-centers.

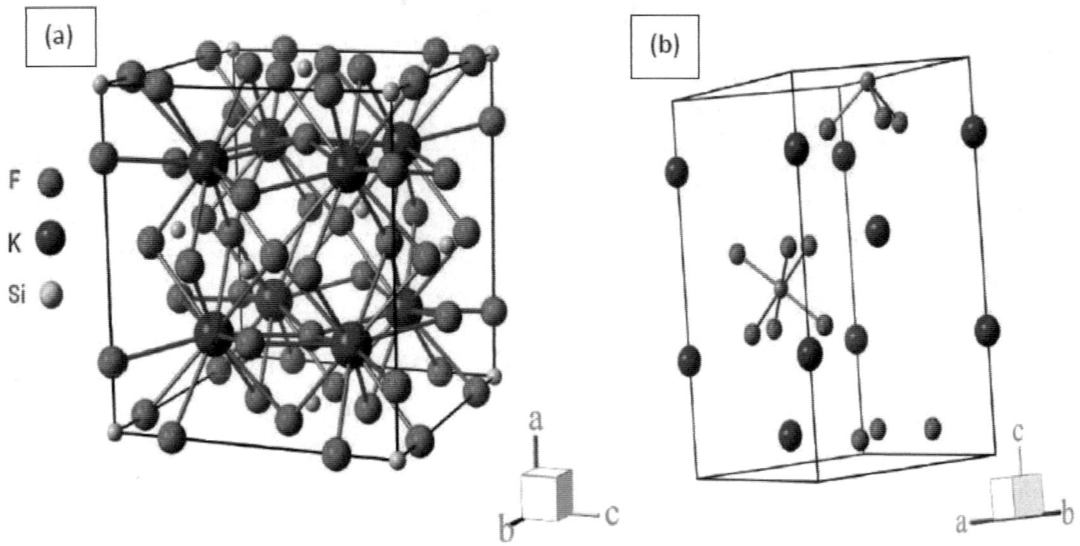

Figure 2. Crystal structure of K_2SiF_6 showing (a) cubic K coordinated unit cell, (b) hexagonal-SiF_6, Si coordinated by $6F^-$ ions.

3.3. Energy Dispersive X-Ray (EDAX) And Scanning Electron Microscope (SEM) Measurements

Previous investigations of the thermal stability of K_2SiF_6 in dry air by Deadmore (1960) [7] revealed that at temperatures above 700 °C, a progressive conversion from K_2SiF_6 to K_3SiF_7 initiates. During this phase transformation, it loses SiF_4 until the composition K_3SiF_7 completes.

Recently, Barve et al [8] carried out TGA/DTA and XRD studies on K_2SiF_6. Two DTA peaks are observed in the temperature range 100-500 °C. The first exothermic peak is observed at 270 °C with the onset at 150 °C, indicates the onset of phase change to K_3SiF_7. Also careful study of their XRD results with increasing temperature, showed that small amount of K_3SiF_7 phase exists even in the sample heated at 350 °C and increases further with the increase in temperature. These observations show that decomposition of K_2SiF_6 starts at 150 °C, itself. Hence due to its very low thermal stability, incorporation of divalent rare earth ions is also very difficult in order to study other physical properties such as luminescence.

Energy dispersive x-ray analysis (EDAX) has been performed on K_2SiF_6 in order to know the actual composition of various elements present in our samples. Two phases, K_2SiF_6 and K_3SiF_7 are known to exist. By measuring relative ratios of different constituent atoms, it may be possible to deduce which of these phases is formed in order to confirm the phase pure formation of K_2SiF_6 by precipitation. **Fig. 3** shows EDAX spectrum of K_2SiF_6 together with experimentally observed values.

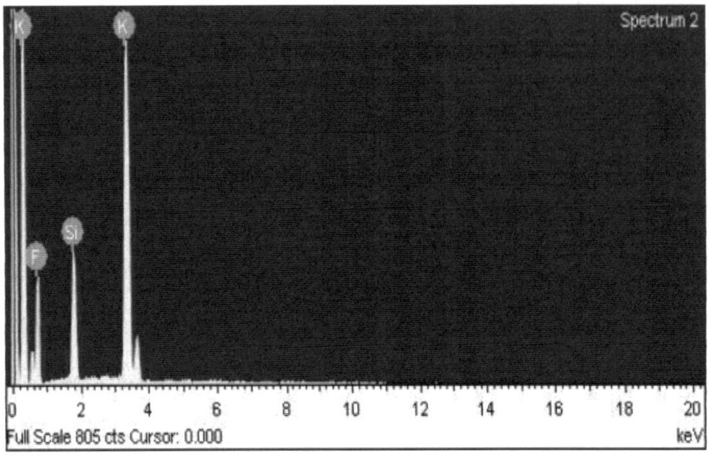

Element	Weight%	Atomic%
F	44.20	60.32
Si	10.29	9.50
K	45.51	30.18
Totals	100.00	

Figure 3. EDAX spectrum of K_2SiF_6

The relative ratios of various atoms for two phases and experimentally observed values are given in the **Table 1.**

Table 1: Results of EDAX analysis of K_2SiF_6 compound.

Atom percent:

	K	**Si**	**F**
K_2SiF_6	22.22	11.11	66.7
K_3SiF_7	27.2	9.1	63.6
observed	30.18	9.5	60.32

Atomic ratio:

	Si	**K**	**F**
K_2SiF_6	1	2	6
K_3SiF_7	1	3	7
observed	1	3.17	6.34

EDAX results show that the composition is closure to K_3SiF_7. However, XRD pattern shows good match with K_2SiF_6. The discrepancy could be due to excess KF adsorbed on K_2SiF_6 particles.
Fig. 4 shows SEM images where (a) shows the image taken at a resolution of 200μm, (b) at 50μm and (c) at 100μm. It can be concluded that the precipitates are particles of irregular shape and are of about 50 μm size.

(a)
SEM MAG: 500 x
HV: 5.0 kV
VAC: HiVac
DET: SE Detector
DATE: 02/06/14
Device: VEGA MV2300T/40
200 μm
Vega ©Tescan
Digital Microscopy Imaging
(b)
SEM MAG: 2.93 kx
HV: 5.0 kV
VAC: HiVac
DET: SE Detector
DATE: 02/06/14
Device: VEGA MV2300T/40
50 μm
Vega ©Tescan
Digital Microscopy Imaging

Figure 4. SEM images of K_2SiF_6, (a) shows the image taken at a resolution of 200µm, (b) at 50µm and (c) at 100µm.

4. Conclusions

XRD results confirmed the formation of K_2SiF_6 hexafluorosilicate in our sample. However, EDAX measurements showed that the relative composition of different constituent elements appears more close to that of its high temperature phase K_3SiF_7. But XRD pattern shows no lines of such impurity present in XRD pattern of our sample. Hence, it could be concluded that this discrepancy could be due to excess KF adsorbed on K_2SiF_6 particles. Also, SEM images show that precipitated particles are of irregular shape and about 50 µm in size.

References

1. Zhong H-X, Huang Q-L, Maa Y-L, Hong J-M, Chen X-T, Xue Z-L 2009 *J.Solid Stat. Chem.* **182** 1679.
2. Brika M G, Srivastava A M 2012 *J. Electrochem. Soc.* **159** J212.
3. Adachia S, Takahashi T 2008 *J. Appl. Phys.* **104** 023512-3.
4. Adachia S, Takahashi T 2009 *Electrochem Solid St.* **12** (2) J20.
5. Kasa R, Adachi S 2012 *J. Electrochem. Soc.* **159** (4) J89.
6. Sijbom H F, Verstraete R, Joos J J, Poelman D, Smet P F 2017 *Opt. Mater. Express.* **7**(9), 3332.
7. Deadmore D L 1960 *Thermal Decomposition of Some Inorganic Fluorides* Ph.D. Thesis, University of Illinois, Feb.
8. Barve R A 2012 *Photoluminescence, Thermoluminescence And Optically Stimulated Luminescence (OSL) In Doped Lithium And Silicon Compounds* Ph.D. Thesis, RTM Nagpur University, April.

Joining of aluminium to polymer by friction stir welding: An overview

S Shankar[1], A Kaushal[1], S Chattopadhyaya[1], P Vilaça[2], F Bennis[3]

[1]Department of mechanical engineering, IIT(ISM), Dhanbad, Jharkhand, India
[2]Department of mechanical engineering, ,Aalto University,Espoo,Finland
[3]Ecole Centrale de Nantes, Nantes, france

sachindrashankar@gmail.com

Abstract. Friction stir welding (FSW) is a potential fabrication method to join the aluminium and polymer dissimilar materials as FSW joins the metals below their melting point. Since the automobile and aviation industries replacing the heavier metals with the lighter one like aluminium and polymer therefore joining of these two materials is the requirement for the industries in the current time. A lot of studies have been already done to join these two materials in the lap, butt and t-joint configuration. Particular attention has been focused to macrostructural and microstructural evolution after welding, microhardness and tensile property. As the temperature is an important factor for both the polymer and aluminium due to difference in chemical and mechanical property, so FSW is a suitable technique compared to other fusion welding processes therefore improved joint property is expected. In the present study, process capability of FSW has been enlightened for joining aluminium-polymer materials for future prospective…

1. Introduction

Automobile industry always looks for materials that reduces the weight of component and environment friendly and so polymers utilization has increased in these industries[1]. Polymers have high specific strength, low thermal expansion, excellent fatigue and fracture strength[2]. Polymers are light weighted and easily formable substances and these are much needed property for automobile and aerospace industry. The technologies such as roll drawing and hydrostatic extrusion has produced polymers with the strength and module comparable to that of metals[3]. Polymers lack in ductility so there usage along with metals such as aluminium, steel, magnesium and etc are done. Automobile industries have replaced the steel components with aluminium as the later gives strength even at reduced mass improving power to weight ratio[4]. Aluminium is a less dense material with good ductility therefore it is joined with polymer to extract its advantageous properties. Aluminium and polymers are used for making vehicle panels and battery housing[5].

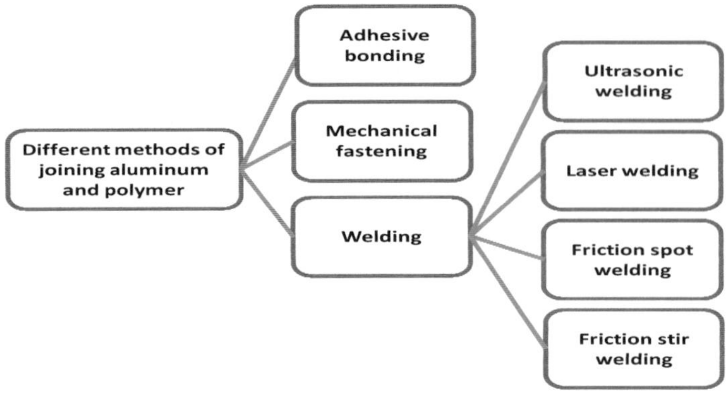

Figure 1 Different methods for joining Aluminium and Polymers[1]

There are 3 main methods for joining aluminium and polymers; these are (1) Adhesive bonding (2) Mechanical fastening and (3) Welding as shown in fig. 1. Resins used for joining Al are epoxies, acrylics, phenolics and temperature resistant polyimides. These resins gets hardened on polymerization or condensation[3]. Surface treatment such as etching and anodizing is done when joints are formed. Adhesive bonding is susceptible to environmental degradation even little amount of moisture degrades the strength as well as a small penetration can destroy the joint [6]. In mechanical fastening method components such as rivets, bolts, etc are used for joining metals and polymers. The problem encountered in this process is that it requires holes in the material which gives rise to stress concentration[1,7]. In aerospace industry adhesive bonding along with mechanical fasteners are used. Sadowski et al [8] reported that when adhesive are used with rivets then energy absorption before facture has increased to 64% compared to when adhesive bonding was used alone. Tensile strength was increased to about 11% to that of analogues bonded joint and 130% to that of simple riveted joint [7,8]. Direct welding of Al and polymer is not possible as they are structurally dissimilar. Polymer have a open structure comprising of chain of micro molecular monomers on the other hand aluminium is a densely packed crystalline structure[1,9,10]. Welding can only be done for thermoplastic as welding requires melting and thermosetting plastics cannot be melted[10]. Welding of al and polymers can be done in various ways but ultrasonic, laser, friction spot welding and friction stir welding gives improved results[1]. Belle et al [11] performed ultrasonic spot welding of aluminium and carbon fiber reinforced polymer (CFRP) and revealed that a tensile shear strength of 30 MPa can be achieved within a welding time of 5 seconds. Laser welding of aluminium and CFRP can be produced using disc type laser[12]. Friction stir spot welding (FSSW) is similar to FSW with a difference that there is no linear movement of the tool. Rotating tool produces heat which is utilized for welding process. Paider et al. [13] performed friction FSSW of aluminium and PP-C30S polypropylene polymer having thickness of about 1.7 to 2mm thickness using a pre-threaded hole in the aluminium plate for mechanical interlocking. It was revealed that with the increase of tool rotational speed thickness of reaction layer increases from 70 to 100 μ also the shear-tensile and cross-tensile load of the hybrid joint increases[13].

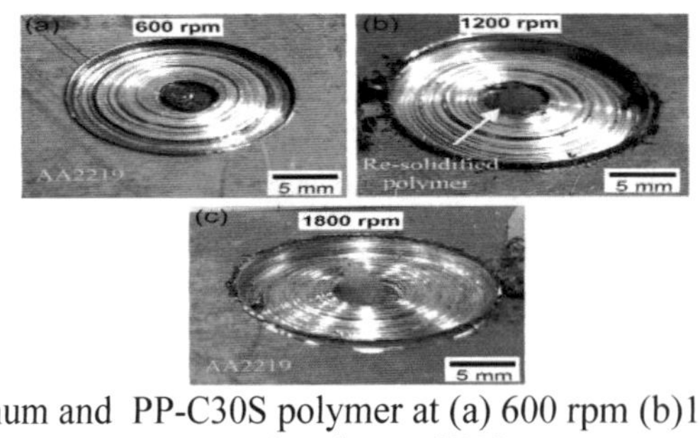

Figure 2 FSSW of aluminum and PP-C30S polymer at (a) 600 rpm (b)1200 rpm and (c) 1800 rpm rotational speed[13]

FSW is a solid state welding method which utilizes the friction between rotating tool with a probe and the parent material for producing heat and when sufficient heat is achieved then tool is moved in linear direction. Khodabakhshi et al [14] performed FSW of aluminum and densed polyethylene. It was revealed that joining occurs due to mechanical interlocking and adhesion between metallic Al chips and consolidated polymeric layers by interface chemical reaction. Formation of thin Al2O3 nano-scale (30 nm) layer was observed at the Aluminum and polymer interface[14].

Figure 3 FSW of aluminum alloy and high polycarbonate[15]

2. Macrostructure

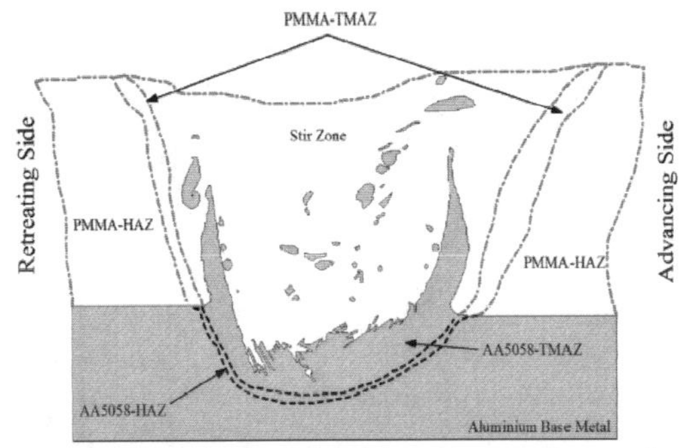

Figure 4 Macroscopic view of the Aluminum polymer lap joint [16]

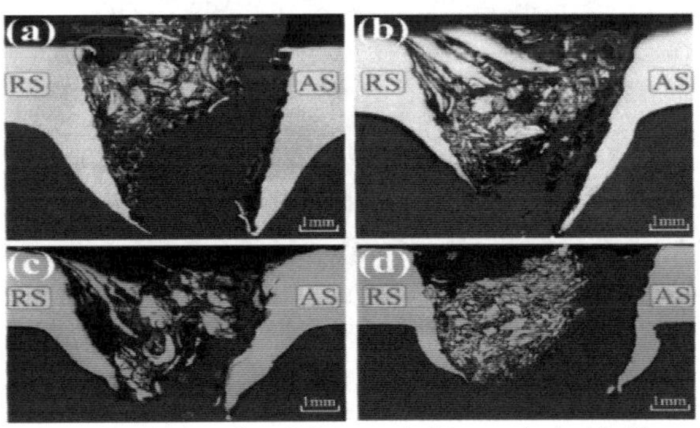

Figure 5 Macrostructure of the Al-polymer lap joint at a) 30 mm/min b) 50mm/min c) 70 mm/min d) 90 mm/min welding speeds [17]

Figure 4 is representing the welding zone of the aluminum polymer lap joint. Identifying the different weld zone is the most difficult part visually, specially the HAZ. Derazkola et al. [16]used FSW technique to join the AA5058 and PMMA polymer, and distinguish the different region in the welded

part as shown in fig. 4. A wider heat affected zone along the advancing side represents the higher heat input. Fragments of aluminum were observed in the melted and re-solidified PMMA matrix. A U-shape structure with two aluminum ramus was observed at the middle of the stir zone. Ramus were observed towards both sides advancing side and retreating side respectively along the joint line. Since the shoulder is in immediate contact with the PMMA therefore most part of the heat resides in the PMMA which restricts the plasticization of aluminum along the joint line due to the low thermal conductivity of PMMA. However ramus are helpful in enhancing the macro and micro mechanical interlocking between the materials which improves the tensile property of the joint.
The shear and stirring action helps in the improved material mixing in the weld stir zone. Huang et al. [17] used FSW technique to join the PEEK polymer and AA6061. Aluminum was placed over the PEEK polymer. The most part of heat generated inside the aluminum alloy thus restricts the movement of polymer outside the weld zone. Apart from the stir zone aluminum arrow were observed along thermo- mechanically affected zone, retreating side and advancing side as shown in fig. 5. The aluminum arrow at joint interface develops a mechanical interlocking and thus improves the tensile property of the joint. The aluminum anchor along the advancing side is bigger than the retreating side due to the high heat input along the advancing side thus results in proper softening of the aluminum alloy. The length of this anchor increases and decreases with the variation of heat input. The length of the anchor sharply decreases with further increase the welding speed to 90 mm/min

3. Microstructure

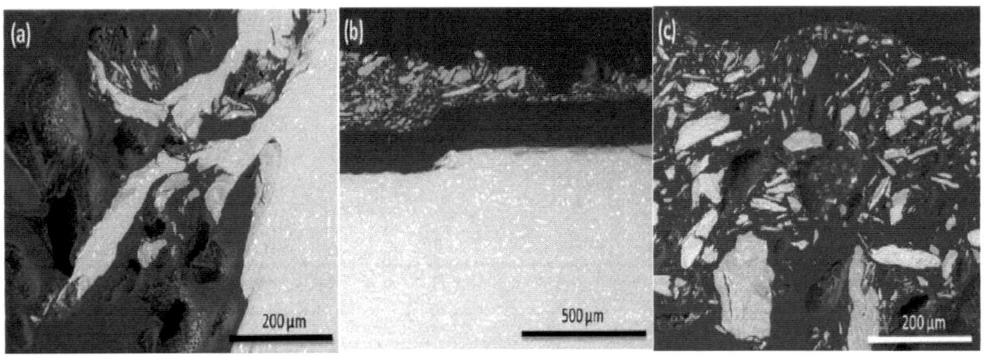

Figure 6 SEM image of different stir zone region a) fragmented aluminum particle b), c) Distribution of aluminum in the stir zone

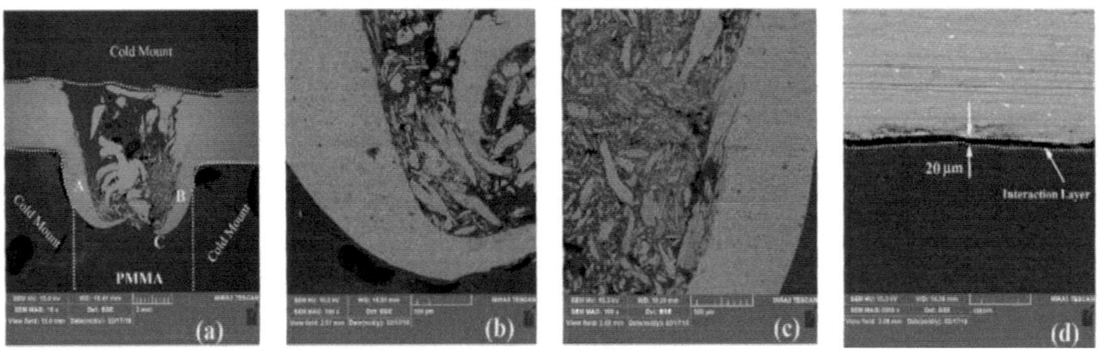

Figure 7 Cross sectional SEM image of Al-PMMA T-joint a) whole cross section b) advancing side C) retreating side d) interfacial layer

The weld stir zone of Aluminum-Polymer joint is characterized by the interlock phenomena and transport of aluminum into the polymer[18]. On the basis of aluminum distribution, stir zone is divided into three distinct region: parent material (Fig.6a), aluminum mix composite structure (Fig.6b

& 6c) and interfacial zone (Fig. 7d). The aluminum particle transported and locked inside the polymer. Movement of these aluminum plays a big role in defining the tensile strength. Severe defects such as void are noticed in the nugget zone as shown in fig. 6a. Derazkola et al. [19] joined AA5754 and PMMA in T-joint configuration and studied the microstructure of the joint as represented in fig. 7 Fewer aluminum fragments are observed along retreating side as compared to the advancing side. Stretched aluminum ramus observed near the interfacial area of Al-PMMA. The weavy lines acts as mechanical interlock between the aluminum-PMMA. Interaction layer is observed between the lower area of aluminum ramus and PMMA without any defects and voids (Fig. 7d). So the joining mechanism depends on the mechanical interlocking, adhesion bonding between the Al/PMMA at interface and distribution of Al fragments in the stir zone.

4. Micro-hardness

Butt and lap joint are the two most common methods to join the aluminum and polymer. So the hardness is measured along the different lines to cover the entire zone of the weld zone. Sahu et al. [20]measured the microhadness along the center line of friction stir welded Al6063 and polypropylene in butt joint configuration as shown in fig. 8. The microhardness decreases continuously from the aluminum side to the weld centre line and increases from the centerline to the interface along the polypropylene side. The highest microhardness of 132HV was observed at 2.5 mm from the weld center line along the aluminum side due to the formation of cabon based brittle compounds at the interface. Measuring of microhardness along the center line does not include aluminum HAZ in lap joint if polymer is placed over the aluminum alloy. Derazkola and Elyasi [21] measured the microhardness along shoreD line and HV line in lap welded AA5058 and polycarbonate as shown in fig. 9. Fig 9 b showing the hardness of polycarbonate reduced after the welding compared to the base polycarbonate material. With the increase in rotational speed the hardness of the resolidified polycarbonate decreases due to the frictional heat at the time of process lead to the moleculer weight reduction. Derazkola and Khodabakhshi [22] claimed that the frictional heating increases the rate of the dynamic recovery and consequent dynamic recrystallization of the aluminum alloy at the time of mixing with polymer in weld zone, reduces the hardness . Derazkola and Simchi observed that the any experimental parameter which affects the size and volume of the aluminum in weld zone affects the hardness distribution.

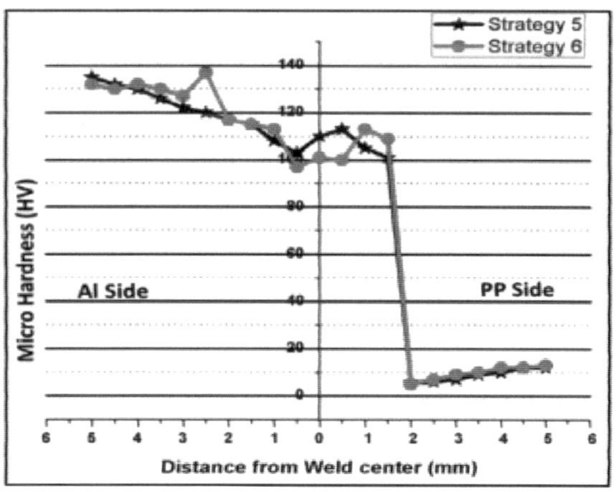

Figure 8 Microhardness of Al6063 and polypropylene along the center line

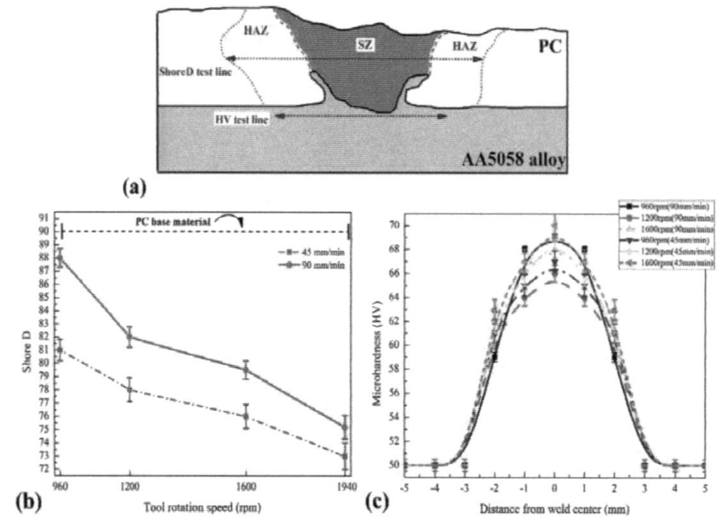

Figure 9 Shore D test and HV test of AA5058 and polycarbonate

5. Bonding mechanism

Mechanical interlocking and chemical adhesion are the main bonding mechanism for development of dissimilar welding between aluminum and polymers[23]. Chemical bonding occurs due to transfer of charge from metal atoms to polymer molecules[24]. Transfer of charge between metal and polymer results in the formation of metal-polymer complexes[24] and these results in chemical adhesion between metal and polymer. Metal surfaces are generally rough when observed on microscopic level. So when liquid polymer is applied to the rough surface then the liquid flows into the holes, dips and microgrooves and conforms to them thus resulting in the formation of interlocking between aluminum alloy and polymers[25]. Yusof et al [25] performed FSSW of aluminum alloy and polyethylene terephthalate (PET) and revealed that the bubbles formed during the welding help in interlocking. Friction stir welding melts the PET and leads to bubble formation near the joining interface due to the heating. Bubbles push the melted polymer in liquid phase into the roughened metal surface where mechanical interlocking occurs and joint strength gets improved[25]. More bonding area is formed when high heat is produced during welding process and it gives better bonding strength[26].

6. Tensile properties

Tensile property of aluminum alloy and polymers fabricated by FSW depends upon various process parameters. These process parameters include positioning of material, speed of rotation, tilt angle, penetration depth and welding speed. Dalwadi et al. [27] friction stir welding of aluminum alloy and acrylic and revealed that on increasing the tool rotational speed joint strength of the welded joint increases but when the welding speed or tool feed is increased then tensile strength first increases up to 40mm/min and there after it goes on decreasing[27]. A maximum joint strength of 7.06 MPa was achieved at1000 rpm and 40 mm/min which is 20% less then PMMA (acrylic) as shown in fig. 10a.

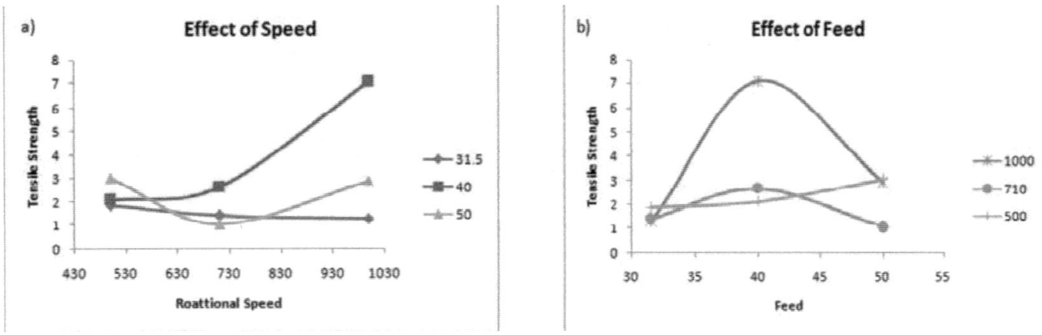

Figure 10 Influence of (a) Rotational speed and (b) welding speed on tensile strength of friction stir welded acrylic and aluminum alloy

Whereas increasing the tool feed first increases and then decreases the strength of the joint (figure 10 (b))[27]. Derazkola et al[16] reported that the tilt angle and penetration depth also effect the strength of the welded joint.

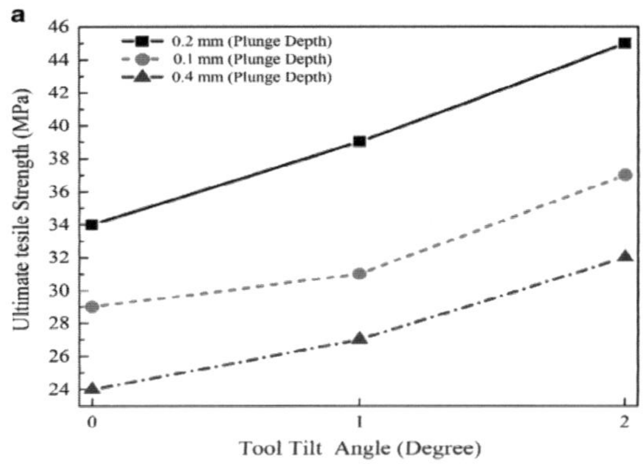

Figure 11 Effect of tool tilt angle and penetration depth on the joint strength of the FSW aluminum alloy and PMMA [16]

With the increment in tool tilt angle, the ultimate tensile strength increases as shown in fig. 11. Similarly Increment in ultimate tensile strength was registered from the plunge depth of 0.1 mm and 0.2 mm and it decreases at the plunge depth of 0.4 mm. This happens because at high plunge depth excessive heating and inappropriate micro and macro interlocking occurs. The highest tensile strength of 45 MPa was obtained at 2° tilt angle and 0.2 mm penetration depth[16]. Derazkola et al[28] employed FSW to join the Al-Mg alloy and poly methyl methacrylate in lap joint configuration. The maximum joint strength obtained was 45.5 MPa obtained at 1250 rpm and 50 mm/min.

7. Conclusion

A brief review on the joining of polymer to aluminum is presented. The main purpose of the article was to study the effect and distribution of material flow on the mechanical properties like tensile strength and microhardness. Article also covers the bonding mechanism between both the materials which is an important factor for the joining

1. Three different joint configurations are possible such as butt joint, lap joint and T joint. The lap joint can be performed by keeping either aluminum on the top or polymer on the top.
2. Ramus or mechanical arrows are formed at the interface in both the lap joint configurations which enhances the mechanical interlocking and thus the tensile property.

3. Frictional heating affects the microhardness profile in the stir zone due to the dynamic recovery and consequent dynamic recrystallisation.

References

[1] Kah, P., Suoranta, R., Martikainen, J. and Magnus, C., 2014. TECHNIQUES FOR JOINING DISSIMILAR MATERIALS: METALS AND POLYMERS. *Reviews on Advanced Materials Science*, *36*(2).

[2] Ghori, S.W., Siakeng, R., Rasheed, M., Saba, N. and Jawaid, M., 2018. The role of advanced polymer materials in aerospace. In *Sustainable composites for aerospace applications* (pp. 19-34). Woodhead Publishing.

[3] Ramarathnam, G., Libertucci, M., Sadowski, M.M. and North, T.H., 1992. Joining of polymers to metal. *WELDING JOURNAL-NEW YORK-*, *71*, pp.483-s.

[4] Cavezza, F., Boehm, M., Terryn, H. and Hauffman, T., 2020. A Review on Adhesively Bonded Aluminium Joints in the Automotive Industry. *Metals*, *10*(6), p.730.

[5] Weber, A., 2014. New Techniques for Joining Plastics to Metal. *Assembly Magazine*.

[6] Campbell, F.C., 2004. A Way to Reduce Assembly Costs through Parts Integration. *Manufacturing Processes for Advanced Composites*,

[7] Moldovan, E., Tierean, M.H. and Stanciu, E.M., 2017. Overview of joining dissimilar materials: Metals and polymers. *Bulletin of the Transilvania University of Brasov. Engineering Sciences. Series I*, *10*(1).

[8] Sadowski, T., Golewski, P. and Zarzeka-Raczkowska, E., 2011. Damage and failure processes of hybrid joints: adhesive bonded aluminium plates reinforced by rivets. *Computational Materials Science*, *50*(4), pp.1256-1262.

[9] Liu, C.T., Zhu, J.H., Brady, M.P., McKamey, C.G. and Pike, L.M., 2000. Physical metallurgy and mechanical properties of transition-metal Laves phase alloys. *Intermetallics*, *8*(9-11), pp.1119-1129.

[10] Amancio-Filho, S.T. and Dos Santos, J.F., 2009. Joining of polymers and polymer–metal hybrid structures: recent developments and trends. *Polymer engineering & science*, *49*(8), pp.1461-1476.

[11] Balle, F., Wagner, G. and Eifler, D., 2007. Ultrasonic spot welding of aluminum sheet/carbon fiber reinforced polymer–joints. *Materialwissenschaft und Werkstofftechnik: Entwicklung, Fertigung, Prüfung, Eigenschaften und Anwendungen Technischer Werkstoffe*, *38*(11), pp.934-938.

[12] Katayama, S., Kawahito, Y. and Mizutani, M., 2012. Latest progress in performance and understanding of laser welding. *Physics Procedia*, *39*, pp.8-16.

[13] Paidar, M., Ojo, O.O., Moghanian, A., Pabandi, H.K. and Elsa, M., 2019. Pre-threaded hole friction stir spot welding of AA2219/PP-C30S sheets. *Journal of Materials Processing Technology*, *273*, p.116272.

[14] Khodabakhshi, F., Haghshenas, M., Chen, J., Shalchi Amirkhiz, B., Li, J. and Gerlich, A.P., 2017. Bonding mechanism and interface characterisation during dissimilar friction stir welding of an aluminium/polymer bi-material joint. *Science and Technology of Welding and Joining*, *22*(3), pp.182-190.

[15] Patel, A.R., Kotadiya, D.J., Kapopara, J.M., Dalwadi, C.G., Patel, N.P. and Rana, H.G., 2018. Investigation of mechanical properties for hybrid joint of aluminium to polymer using friction stir welding (FSW). *Materials Today: Proceedings*, *5*(2), pp.4242-4249.

[16] Derazkola, H.A., Fard, R.K. and Khodabakhshi, F., 2018. Effects of processing parameters on the characteristics of dissimilar friction-stir-welded joints between AA5058 aluminum alloy and PMMA polymer. *Welding in the World*, *62*(1), pp.117-130.

[17] Huang, Y., Meng, X., Wang, Y., Xie, Y. and Zhou, L., 2018. Joining of aluminum alloy and polymer via friction stir lap welding. *Journal of materials Processing technology*, *257*, pp.148-154.

[18] Moshwan, R., Rahmat, S.M., Yusof, F., Hassan, M.A., Hamdi, M. and Fadzil, M., 2015. Dissimilar friction stir welding between polycarbonate and AA 7075 aluminum alloy. *International Journal of Materials Research*, *106*(3), pp.258-266.

[19] Derazkola, H.A. and Simchi, A., 2019. An investigation on the dissimilar friction stir welding of T-joints between AA5754 aluminum alloy and poly (methyl methacrylate). *Thin-Walled Structures*, *135*, pp.376-384.

[20] Sahu, S.K., Pal, K., Mahto, R.P. and Dash, P., 2019. Monitoring of friction stir welding for dissimilar Al 6063 alloy to polypropylene using sensor signals. *The International Journal of Advanced Manufacturing Technology*, *104*(1-4), pp.159-177.

[21] Derazkola, H.A. and Elyasi, M., 2018. The influence of process parameters in friction stir welding of Al-Mg alloy and polycarbonate. *Journal of Manufacturing Processes*, *35*, pp.88-98.

[22] Derazkola, H.A. and Khodabakhshi, F., 2020. Development of fed friction-stir (FFS) process for dissimilar nanocomposite welding between AA2024 aluminum alloy and polycarbonate (PC). *Journal of Manufacturing Processes*, *54*, pp.262-27

[23] Sahu, S.K., Panda, M., Mahto, R.P., Pal, S.K., Pal, K. and Dash, P., 2017. Experimental investigation to join Al 6063 alloy to polypropylene using friction stir welding. In *10th international conference on precision, Meso, micro and Nano Engineering (COPEN 10). Indian Institue of Technology Madras, Chennai-600036 India* (pp. 727-731).

[24] Ho, P.S., 1990. Chemistry and adhesion of metal-polymer interfaces. *Applied surface science*, *41*, pp.559-566.

[25] Yusof, F., Muhamad, M.R.B., Moshwan, R., Jamaludin, M.F.B. and Miyashita, Y., 2016. Effect of surface states on joining mechanisms and mechanical properties of aluminum alloy (A5052) and polyethylene terephthalate (PET) by dissimilar friction spot welding. *Metals*, *6*(5), p.101.

[26] Goushegir, S.M., Dos Santos, J.F. and Amancio-Filho, S.T., 2015. Influence of process parameters on mechanical performance and bonding area of AA2024/carbon-fiber-reinforced poly (phenylene sulfide) friction spot single lap joints. *Materials & Design*, *83*, pp.431-442.

[27] Dalwadi, C.G., Patel, A.R., Kapopara, J.M., Kotadiya, D.J., Patel, N.D. and Rana, H.G., 2018. Examination of mechanical properties for dissimilar friction stir welded joint of Al alloy (AA-6061) to PMMA (Acrylic). *Materials Today: Proceedings*, *5*(2), pp.4761-4765.

[28] Derazkola, H.A., Khodabakhshi, F. and Simchi, A., 2018. Friction-stir lap-joining of aluminium-magnesium/poly-methyl-methacrylate hybrid structures: thermo-mechanical modelling and experimental feasibility study. *Science and Technology of Welding and Joining*, *23*(1), pp.35-49.

IOP Conf. Series: Materials Science and Engineering 1104 (2021) 012006 doi:10.1088/1757-899X/1104/1/012006

Prediction of breakdown hours of load haul dumper by long short term memory network

Vaibhav Bisht[1], Sarthak Kumar[1], Anil Kumar Agrawal[1], Mohd A H Siddiqui[1,2], Somnath Chattopadhyaya[1]

[1]Indian Institute of Technology (Indian School of Mines), Dhanbad, Jharkhand 834001, India.

[2]Coal India Limited, Western Coal Field Limited, Nagpur, 440001,

anil.102024@gmail.com

Abstract. The profitability and feasibility of a system depends on the optimisation of the work of the subsystem as a unit. Here, the breakdown hours of an LHD machine is predicted by modelling a time series on a Neural Network. By accurately predicting the breakdown hours one can plan for efficient mining operations, maintenance and parts supply chain for replacements.

1. Introduction

Mine productions are sophisticated systems comprising of many subsystems. The profitability and feasibility of the system depend on the optimization of the work of the subsystem simultaneously as a unit. LHD machines are employed to haul ore or waste rock from the site of digging and unload it into ore passes or trucks. Being able to predict the time for which the LHD will breakdown help be able to plan for the mining operations and preparing for maintenance activity of the LHD. The breakdown hours of the LHD depend on complex factors and can be formulated as a time series. Traditionally time series have been modelled using Autoregressive Moving Average (ARMA) Model. However, with the advances in neural networks and computation power time series have been modelled using various deep neural network. In this paper, try to model the breakdown data for four LHD machines using Long Short Term Memory networks.

2. Literature survey

Effectiveness of an LHD is mainly affected by the availability, reliability and maintainability of the machine, and its capability to perform as expected. Traditionally, the breakdown of the LHD machines is studied as a stochastic process and modelled using Markov models. Several case studies have been done to perform the reliability analysis of LHD machines [1,2,3]. These studies use graphical methods and statistical tests to find the best fit distribution for characterisation of the failure data. These methods are based on some a priori assumptions which may not be easy to validate for real-life problems.

Treating the reliability analysis as a time series problem helps to deal with these biases. Ho and Xie (1998) [4] have used the Autoregressive Integrated Moving Average (ARIMA) model [5] for reliability forecasting and analysis. However, comparable or even better results have been achieved using artificial neural networks [6,7]. Neural networks' ability to approximate any distribution has provided exceptional results. This study aims to take the application of neural networks to reliability analysis further by using a long short term memory(LSTM) network. LSTMs are a type of (recurrent neural network) RNN that have been shown to be very useful for training temporal and sequential data. LSTMs have solved the problems of vanishing gradient, and long term dependencies faced with standard RNNs and have found use in varied problems. Many studies have been done using them on financial market data [8,9]. In this analysis, we are going to fit an LSTM on breakdown data of four LHD machines and analyse its performance.

3. Machine details

Load Haul Dumper (LHD) machines are loaders which are similar to conventional front end loaders but developed for the toughest of hard rock mining applications, with overall production economy, safety and reliability in mind. They are rugged, highly manoeuvrable and exceptionally productive. More than 75% of the world's underground metal mines use LHD for handling the muck of their excavations.

LHD has powerful prime movers, advanced drive train technology, heavy planetary axles, four-wheel drive, articulated steering and ergonomic controls. Their narrower, longer and lower profile make them most suitable for underground application where height and width are limited. As the length is not a limitation in a tunnel and decline LHD are designed with sufficient length. The range improves axial weight distribution and bucket capacity can be enhanced. The two-part construction with central articulation helps in tracking and manoeuvrability. In mining, there is a limitation for shifting heavy equipment. Sometimes, an LHD has to be moved through a shaft while dismantled.

Western coalfields limited, Nagpur uses LHD machines manufactured by EIMCO ELECON. Data of four LHD machines were collected, the serial numbers of which were- 811-682, 811-784, 811-785 and 811-685.

4. ANN model

The neural network is a learning model that achieves efficient results in a wide range of both supervised and unsupervised machine learning tasks. Machine perception tasks, where the raw underlying features are not individually interpretable, is where they are uniquely suited. Unlike traditional methods that rely upon hand-engineered features, their success is attributed to their ability to learn hierarchical representations. Despite their ability to understand complex underlying relations, standard neural networks have limitations. Most notably, they rely on the assumption of independence among the training and test examples. The entire state of the system is lost after each data point is processed. If each model is generated independently, this presents no problem. But in case of a time series where data points are related, this is unacceptable. Additionally, standard networks take fixed-length vectors as input. Thus new network structures have been proposed to extend these powerful learning tools to model data with temporal or sequential structure and varying length inputs and outputs.

RNNs are connectionist models with the ability to selectively pass information across sequence steps while processing sequential data one element at a time. Thus they can be used to model data where input and/or output consists of sequences of elements that are autocorrelated. Further, RNNs can simultaneously model sequential and time dependencies on multiple scales. Traditionally, recurrent neural networks have been difficult to train, and often contain millions of parameters. However, recent advances in network architectures, optimization techniques, and parallel computation have enabled solving many machine learning problems with them. In particular, systems based on long short-term memory (LSTM) architectures have demonstrated exceptional results on tasks as varied as language translation, image captioning, time series modelling, and handwriting recognition.

To overcome the problem of vanishing gradients, Hochreiter and Schmidhuber [1997] introduced the LSTM model. Standard RNN has a chain-like structure with a repeating neural network module, as shown in figure 1.

IOP Conf. Series: Materials Science and Engineering 1104 (2021) 012006 doi:10.1088/1757-899X/1104/1/012006

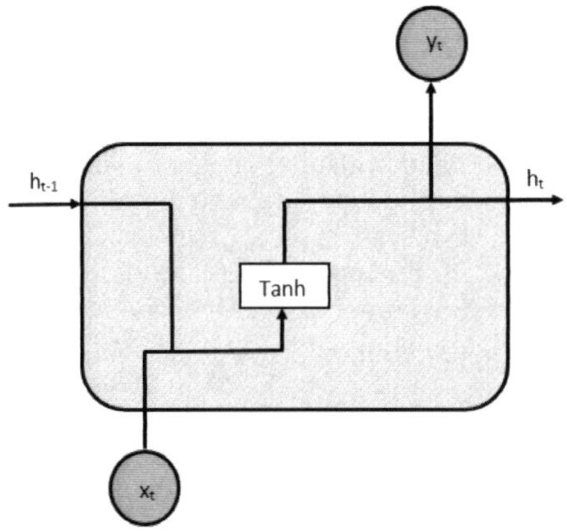

Fig 1: The repeating module of a RNN

The module takes in previous cell state h_{t-1} and input x_t and the tanh neural layer performs the computation. The updated cell state h_t is a function of previous cell state h_{t-1} and input x_t.

To overcome the problem of long term dependencies the repeating module of a standard RNN is modified to produce the repeating module of an LSTM, as shown in figure 2.

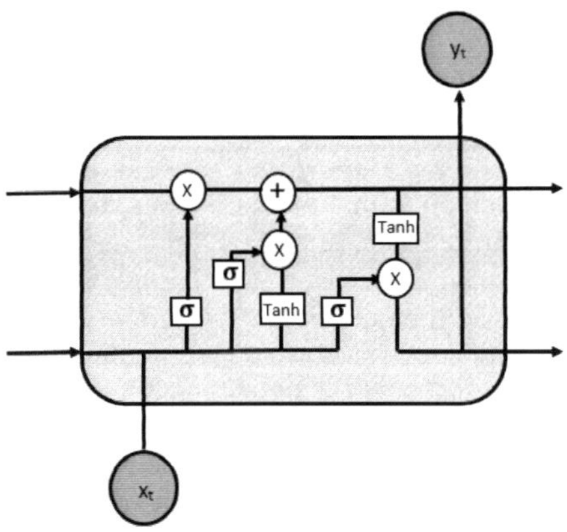

Fig 2: The repeating module of a LSTM

The different interacting layers in an LSTM selectively control the flow of information. The key building block is a structure called gate which enables the LSTM to selectively add or remove information from the cell state. And barriers consist of a neural net layer like a sigmoid and a pointwise multiplication. Enabled by the interacting neural net layers and pointwise operations LSTM process information in four simple steps:

- Forget – It forgets the irrelevant parts of the previous state.

- Store – It stores relevant new information into the cell state.
- Update – It selectively updates the current cell state values.
- Output – The output gate controls what information is sent to the next time step.

For this analysis, the LSTM code has been implemented using Keras, as shown in figure 3.

```
In [48]: regressor = Sequential()

         regressor.add(Dense(units=10))

         regressor.add(LSTM(units=100,return_sequences=True,input_shape=(X_train_rnn.shape[1],1)))
         regressor.add(Dropout(0.2))

         regressor.add(LSTM(units=100))
         regressor.add(Dropout(0.2))

         regressor.add(Dense(units=1))

         regressor.compile(optimizer='adam', loss='mean_squared_error')
```

Fig 3: Keras implementation of the LSTM network

5. Data collection and modelling

The data for this experiment was provided by the Western Coalfields Limited, Saoner Mine. The breakdown data of each of the four LHD machines was recorded, starting from July 2017 up to December 2019.

To prepare the data to be fed into the neural network, the following steps were taken:

- The breakdown data were scaled to the range of [0,1] using the MinMaxScaler() function provided by the sci-kit learn library.
- The sequential scaled data were then arranged into an input dataset and an output dataset. For each time step, the corresponding output data is the next time step in the series.
- The input and output dataset was then split into training, validation and test sets with an approximate split of 70/15/15 per cent.

Finally, the training, validation and test sets were reshaped to be fed into the long short term memory network.

6. Analysis

We have used the Mean Squared Error(MSE) as the cost function to be optimised by the adam optimiser. MSE is the mean of the squares of the error calculated for each data point and unlike Mean Absolute Error(MAE) punishes harshly output values that stray far away from the actual value. The derivatives calculated on the cost function are backpropagated in the network to update the weights. Root Mean Squared Error(RMSE) has been used to report the model error for the four LHD machines. RMSE is the square root of MSE and hence makes it easy to interpret the result in comparison to the input values.

$$RMSE = \sqrt{\sum_{i=1}^{n} \frac{(\hat{y}_i - y_i)^2}{n}}$$

7. Result

The sequential breakdown data for the four LHD machines was fit into the LSTM model. Table 1 and figures 4,5,6, and 7 show the comparison of the actual vs predicted breakdown hours for the four machines using a line plot.

The average RMSE on the validation set for the four machines comes out to be 27.461075 and the average RMSE on the test set for the four machines comes out to be 27.7685, showing the model does well on both validation and testing set across all machines.

Table 1. Comparison of the actual vs predicted breakdown hours

LHD No	Validation RMSE	Test RMSE
811-682	14.59941	13.98681
811-784	33.42955	32.85165
811-785	39.53129	28.09883
811-607	22.28421	36.13755
Average	27.461075	27.7685

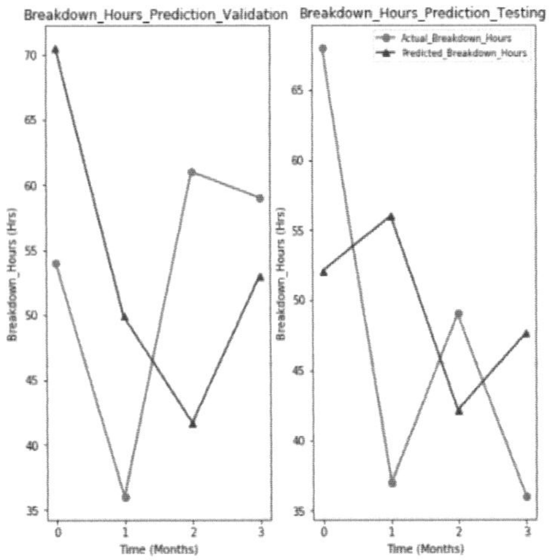

Fig 4: Plot for machine 811-682

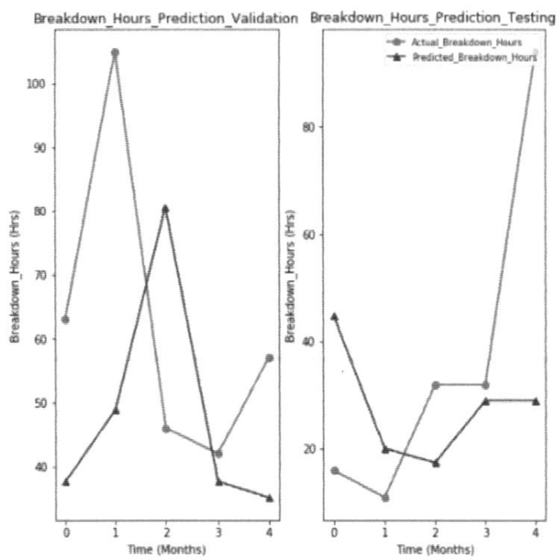

Fig 5: Plot for machine 811-784

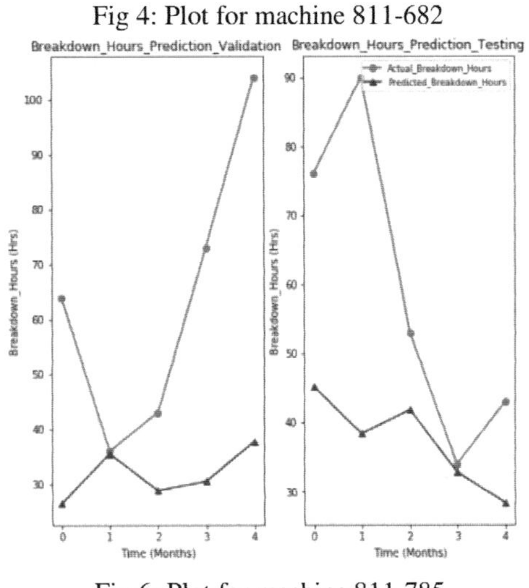

Fig 6: Plot for machine 811-785

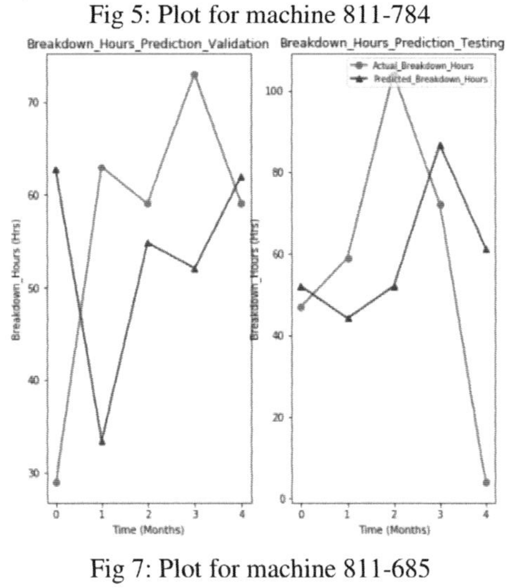

Fig 7: Plot for machine 811-685

8. Conclusion

A priori knowledge of the breakdown hours of an LHD machine in an underground mining operation allows for effective resource management, forecasting production and is also helpful for the scheduling preventive maintenance of the machine. With the proposed model, we are able to predict the breakdown hours for the next month given the data for previous months with an acceptable margin of error. The LSTM model has been successful in capturing the complexity and dependence of the series on time with good accuracy.

Acknowledgement

Authors are thankful to Coal India Limited for field investigation as well as technical supports.

References

1. Barabady J and Kumar U 2008. Reliability analysis of mining equipment: A case study of a crushing plant at Jajarm Bauxite Mine in Iran. *Reliability engineering & system safety*, *93*(4), pp.647-653.
2. Barabady J 2005 January. Reliability and maintainability analysis of crushing plants in Jajarm Bauxite Mine of Iran. In *Annual Reliability and Maintainability Symposium, 2005. Proceedings.* (pp. 109-115). IEEE.
3. Vashistha S, Agrawal A K, Siddiqui M A and Chattopadhyaya S 2019, November. Reliability and Maintainability Analysis of LHD Loader at Saoner Mines, Nagpur, India. In *IOP Conference Series: Materials Science and Engineering* (Vol. 691, No. 1, p. 012013). IOP Publishing
4. Ho, S L and Xie M 1998. The use of ARIMA models for reliability forecasting and analysis. *Computers & industrial engineering*, *35*(1-2), pp.213-216.
5. Box G E, Jenkins G M and Reinsel G C 2011. *Time series analysis: forecasting and control* (Vol. 734). John Wiley & Sons.
6. Chatterjee S and Bandopadhyay S 2012. Reliability estimation using a genetic algorithm-based artificial neural network: An application to a load-haul-dump machine. *Expert Systems with Applications*, *39*(12), pp.10943-10951.
7. Ho S L, Xie M and Goh T N 2002. A comparative study of neural network and Box-Jenkins ARIMA modeling in time series prediction. *Computers & Industrial Engineering*, *42*(2-4), pp.371-375.
8. Roondiwala M, Patel H and Varma S 2017. Predicting stock prices using LSTM. *International Journal of Science and Research (IJSR)*, *6*(4), pp.1754-1756.
9. Tsang G, Deng J and Xie X 2018, August. Recurrent Neural Networks for Financial Time-Series Modelling. In *2018 24th International Conference on Pattern Recognition (ICPR)* (pp. 892-897). IEEE.

IOP Conf. Series: Materials Science and Engineering 1104 (2021) 012007 doi:10.1088/1757-899X/1104/1/012007

Designing and Prototyping for Conservation and Effective Utilization of Waste Heat from Air Conditioner

Amit Choudhari[1] Aditya Rane[2] Shamir Talkar[3] Pawan Rayar[4] Dhananjay Shukla[5]

[1] Design Engineer, Worley, Mumbai, India
[2] Project Engineer, Radcorps fire and safety Pvt Ltd, Mumbai, India
[3] Design Engineer, Arkk Consulting, Mumbai, India
[4] Professor, Dwarkadas J Sanghvi College of Engineering, Mumbai, India
[5] Professor, Dwarkadas J Sanghvi College of Engineering, Mumbai, India

Email:- amitchoudhari.r@gmail.com

Abstract. Nowadays Energy is becoming a crucial part of human needs and it's essential for the day to day activities. We use an enormous amount of energy either from fossil fuels, electricity, or solar to heat the water to fulfil the need of residential or industrial sectors, and unfortunately, we consciously lose out a significant amount of energy in the form of heat from the condensers of air conditioning. Hence conservation of energy and its effective utilization has become imperative. The main objective of this research paper is to demonstrate how the waste heat which is dissipated from the condenser coil of a 1-ton Air conditioner is being utilized for heating the water and used to increase the Coefficient of Performance (COP) of the system. Therefore, by redesigning the condenser part of AC we would be able to appease the necessity of the heated water for bathing, cleaning, space heating, etc. in hostels, restaurants, for domestic and industrial use as a primary unit or with a hybrid water heater system.

1. Introduction

The use of air conditioners has increased significantly in the past few years including commercial as well as residential areas. Its effect is directly on the consumption of electricity and other resources. From an advancement perspective, especially the HVAC industry has grown incredibly well in terms of energy consumption, compact-ness, reduction in carbon footprints, use of environment-friendly refrigerants with a reliable and sustainable product. [1] Even though we are approaching to reach the peak in the advancement but unfortunately every year a huge amount of energy in the form of heat from the condenser of air conditioners is consciously released into the atmosphere in the form of waste which is pathetic. This research paper is about the methods or techniques by which a large portion of waste heat of air conditioner can be recovered and utilized for our day to day applications. In this research paper, two experiments are conducted to demonstrate how the modification in some portion of the air conditioner results in the recovery of a huge amount of heat which is otherwise gets wasted. The cost and efficiency are the two-parameter that governs the designing of any new system or doing some modification in the existing one by giving utmost importance to the primary function and quality. Therefore, considering all these parameters a system is designed and prototyped. From a future perspective, this experiment may help the researcher to rethink and design a system by which, such a great amount of energy can be utilized for human needs.

IOP Conf. Series: Materials Science and Engineering 1104 (2021) 012007 doi:10.1088/1757-899X/1104/1/012007

2. Constructional Details:

Annotations used in schematic representation as figure 1 are as follows,

a. Window Air Conditioner
b. Condenser coil
c. High density poly Ethylene (HDPE) Tank
d. Water Sprinkler
e. Supporting frame
f. Water tap
g. Temperature sensor
h. External pump

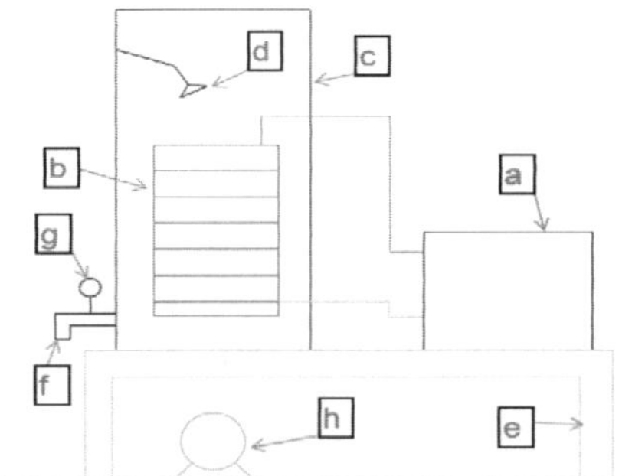

Figure 1. Schematic representation

Figure 2. Constructional setup

2.1. Window Air Conditioner & it's Condenser Coil

In this experimental setup an old window air conditioning of 1 TR capacity is used which is as shown in figure 3. [2] In this A.C, the condenser portion is reconstructed by adding a copper coil to it which is then placed inside the tank as shown in figure 4 to carry out the experiment.

Figure 3. Window Air Conditioner

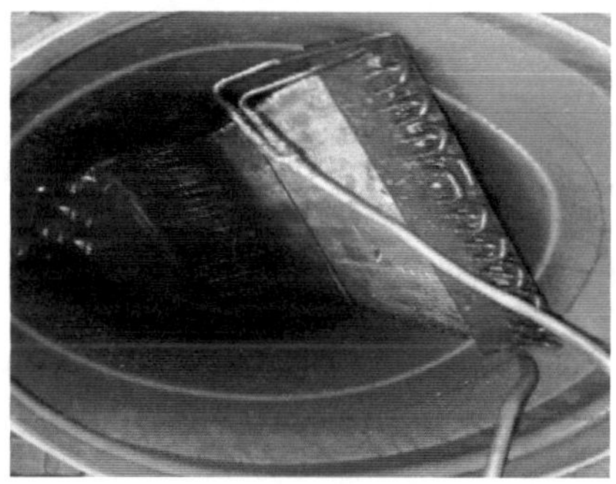

Figure 4. Condenser coil immersed in tank

2.2. Water tank and Shower head

Figure 5. HDPE water tank

Figure 6. Shower head to sprinkling water

The high-density polyethylene tank is shown in figure 5, It is used because of the cost involved in using a metallic tank of the same size. It can withstand a temperature of 85°C without any significant change in physical properties as its melting point is 111.90°C which is obtained from the manufacturer's manual. To carry out experiment a water sprinkler which is a showerhead used and connected with the tank as shown in figure 6.

2.3. *Supporting Table and Water tap: -*

As the window air conditioners are bulky and 150 liters of the tank with 100 liters of water, is used for the experiment, so to support these a robust base frame is designed and fabricated which is as shown in figure 7. Its dimensions are 50 cm×25 cm to hold the Air conditioner and 45 cm × 20 cm to hold the water tank & made up of mild steel. A water tap is also installed with the tank to drain out the water from the tank which is shown in figure 8.

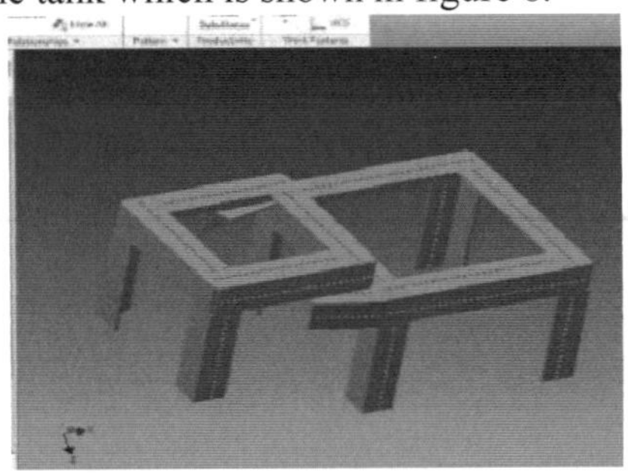

Figure 7. Supporting Table

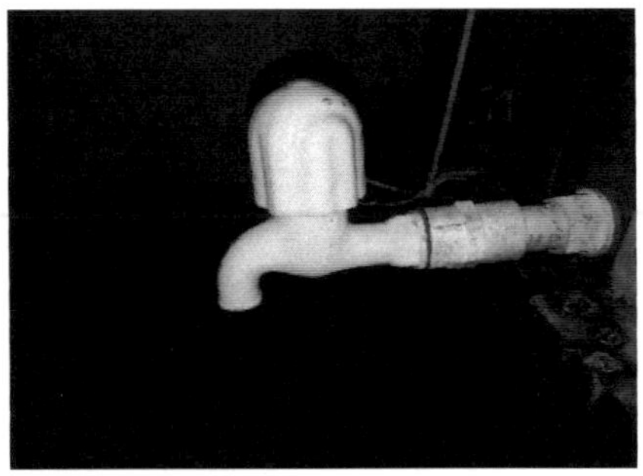

Figure 8. Water tap

3. Working:

As shown in figure 2, the entire assembly is placed over the frame. For analysis purposes after the theoretical conceptualization, it becomes imperative to validate it by using different experiments hence we conducted two experiments. In 1st experiment, as the Air conditioning is turned on the refrigeration cycle starts. [3] When the liquid refrigerant reaches at evaporator inlet after the expansion coil it absorbs the heat of the system and gets converted into vapor which then passed through the compressor where it pressurizes, and high-pressure high-temperature vapor enters into the condenser. Since the coil placed inside the water there is a significant amount of heat released to the water and as a result, this waste heat gets utilized for heating the water. In 2nd experiment, an external motor is used to force the water over the coil, and with the help of a water sprinkler, it gets converted into small droplets. [4] Here convection is taking place with air as well as the water droplets as a result more amount of heat gets released and sub-cooling takes place. Whereas in conventional A.C there is forced convection takes place over the coils by blowing air over it with the help of a fan.

4. Experimental Result

4.1. *Sample Calculation: -*

Following are the specification of the experimental setup: -
The Window A.C is 3-star rating, so the energy efficiency ratio (EER) is 2.7 [6]

Capacity of A.C	= 1 TON
Usage of Water	= 100 Liters
Running hours	= 1 Hour
Rated Power Input	= 1407 Watt
Rated Cooling Capacity	= 3517 Watt
Type of Refrigerant	= R-22

All the heat will be absorbed by the water inside the tank. The total amount of heat absorbed by the water is given by the equation,

$$Q = m * Cp * (T_1 - T_2)$$

Where,
T_1 and T_2 are Inlet and outlet temperature of the water.
In an ideal condition, the entire rejected heat will be absorbed by the water and is equal to the sum of the power input and amount of heat absorbed from the room which is equal to cooling capacity. So, we can write,

$$1407 + 3517 = 100 * 4187 * (T_1 - T_2)$$

$(T_1 - T_2) = 0.01176°C/\text{Sec}$ equivalent to increase in temp. of 7 °C in 10 minutes.

4.2. Observation table

Table 1. Increase in temperature of water observed during experiment

Sr. No.	1	2	3	4	5	6	7
Time (In minutes)	0	10	20	30	40	50	60
Expected Temp. increment (°C)*	28	35	42	49	56	63	70
Observed Temp. increment (°C)	28	33.5	37.4	41	43	45.7	47.8

*Expected time is calculated in topic 4.1

Table 1 shows the result obtained from the 1st experiment. In this experiment, the condenser coil is immersed inside the water. The amount of water used in this experiment is 100 liters in a tank having a capacity of 150 liters. On the day of the experiment, the atmospheric temperature is 28°C and the tank is open to the atmosphere. [5] When the air-conditioner is turned on the temperature of the water goes on increasing and it is recorded with the help of a digital thermometer in 10 minutes intervals. The ideal consideration was our temperature will increase linearly as time goes on as per the calculation in topic 4.1 but when we analyzed the result, we found that as the temperature goes on increasing, for the same period, the temperature gradient is decreasing which is as shown in figure 9.

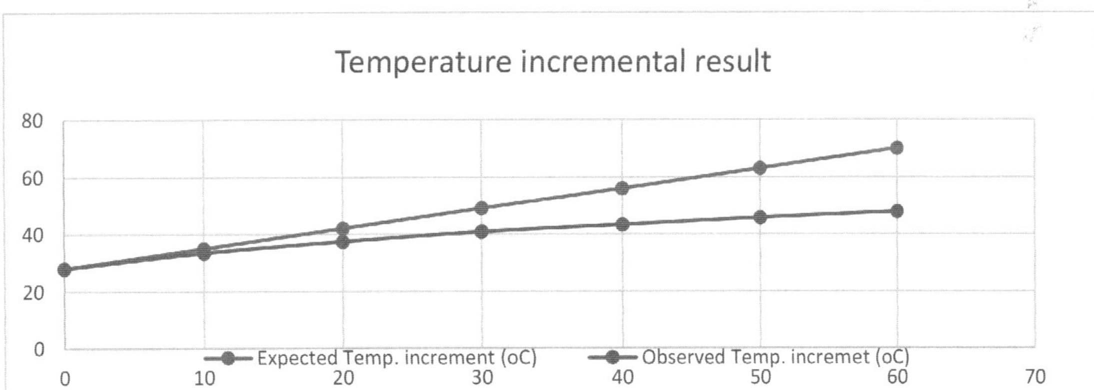

Figure 9. Comparison of expected temperature increment with experimental data

In figure 9, the blue line indicates the increment in temperature of the water, and the red line shows the increment in temperature result obtained from 1st experiment.

In the 2nd experiment, we use a pump and a shower head to spray the water over the coil in fine droplets which is installed just above the condenser coil inside the tank. In this case, the water is stored in a separate tank and then at different volume flow rates, it is sprayed over the coil. The pump discharge is 2.5 lpm and the outlet temperature of the water observed is 42.8°C which is optimum. Even though the outlet temperature of the water is lower than experiment 1 but it is the best method to recover the waste heat from the air conditioner effectively and efficiently. A sample calculation for each experiment is given in the below section. The amount of power consumed for running the pump is negligible in comparison to recovering the amount of heat, so its calculation is not included.

5. Sample Calculation: -

5.1. Experimental results obtained for calculation: -

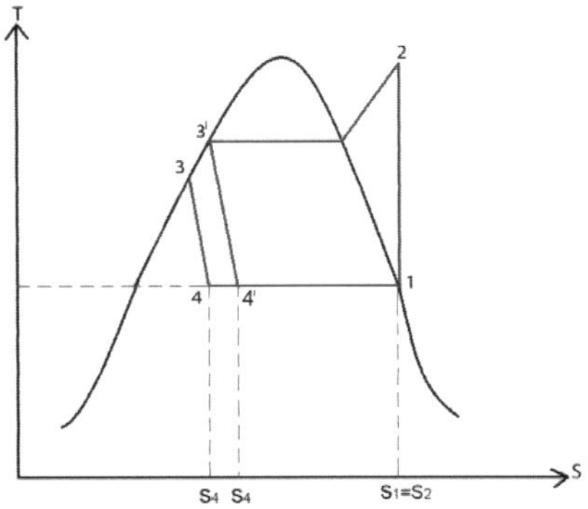

Figure 10. Temperature (T) – Entropy (s) diagram

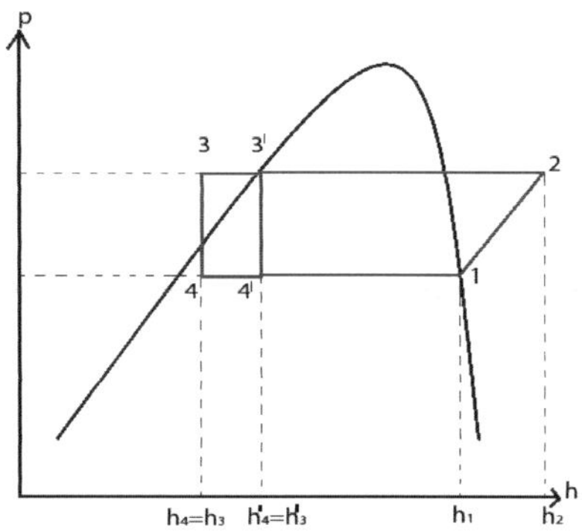

Figure 11. Pressure (P) - Enthalpy (h) diagram

Table 2. Enthalpy and entropy at different points as per figure 10 and figure 11

Properties	1	2	3	4	3'	4'
h (Enthalpy)	397.9	426.3	302.5	302.5	318.3	318.3
s (Entropy)	1.779	1.788	1.187	1.311	1.191	1.323

Here,
h_1 , h_2, h_3, and h_4 = Enthalpy of refrigerant at entry of compressor, condenser, expansion device and evaporator respectively with dismantling the coil of condenser and immersing inside the water as shown in figure 4.
$h_{3'}$ and $h_{4'}$ = Enthalpy of refrigerant at exit of condenser and expansion device respectively without dismantling the coil of condenser.

$$\text{COP of Conventional Air conditioning} = \frac{(h_1 - h_{4'})}{(h_2 - h_1)} = \frac{(397.90-318.3)}{(426.30-397.90)} = 2.8028$$

From figure 11, $h_4' = h_4 + (h_4'-h_4)$

$$\text{COP of Modified Air conditioning system} = \left(\frac{(h_1 - h_4)}{(h_2 - h_1)}\right) = \left(\frac{397.90-302.5}{426.30-397.60}\right) = 3.324$$

The above sample result is as per experiment 2nd Experiment which shows that there is a significant rise in COP of the system from 2.80 to 3.32 which could only be possible after modifying the condenser assembly.

5.2. Amount of waste heat recovered:

5.2.1 In 1st experiment: -

As the result obtained from the experiment, we can say that only a certain amount of the heat rejected from the condenser coil can be reutilized for the purpose of any specific application i.e. for heating of residential water storage system for day to day use, for industrial purpose, etc.
As a result, the water temperature increases from 28°C to 47.8°C in a period of 1 hr. So, the total amount of waste heat recovered in 1 hr. is,

$$Q_W = m * C_p * \Delta t$$

$$Q_W = 100 * 4.187 * (47.80 - 28) = 8290.26\ KJ$$

5.2.2 In 2nd experiment: -

Total amount of heat recovered (Q_W) in the period of 1 hr. is, (Outlet temp. is 42.8°C)

$$(Q_W) = m * C_p * \Delta t$$

$$Q_W = 2.50*60*4.187*(42.8\text{-}28) = 9295.14\ \text{KJ}$$

6. Conclusion: -

In summary, by doing modification in the existing system by keeping in mind the future obligations and waste energy in the form of heat which is dissipated in the atmosphere from air conditioners periodically, must be adequately utilized. Modification in condenser design not only makes it possible to use the waste heat appropriately but also unquestionably there is a marginal increment (from 2.80 to 3.32) in the coefficient of performance of the entire system. By this experiment, we could also be able to extract a considerably larger amount of heat (i.e. ~ 9300 KJ) which can be utilized for the sterilization in hospitals and medicals, for bathing purpose in hotels, hostels and may also be used for various domestic applications, in the restaurants, in various chemical industries for preheating the salts, ashes and different chemical solutions as a primary or secondary heating source. In the future, it may be one of the best source of energy at the economical rate, which must be utilized effectively for fulfilling the need of the human being, especially when the modern conventional air conditioners will be a part of solar modules and will use it as a primary source of power supply.

References

[1] S.C. Kaushik, M. Singh., Feasibility and Design studies for heat recovery from a refrigeration system with a heat exchanger, Heat Recovery Systems & CHP, Vol.15(1995)665673
[2] P. Sathiamurthi, PSS. Srinivasan "Design and Development of Waste Heat Recovery System for air Conditioning Unit, European Journal of Scientific. Research, Vol.54 No.1 (2011), pp.102-110.
[3] M. Joseph Stalin, S. Mathana Krishnan, G. Vinoth Kumar, "Efficient usage of waste heat from air conditioner", International journal of advances in engineering &technology, July 2012. ISSN: 2231-1963.
[4] Rakesh Jain, Performance Improvement of a Boiler through Waste Heat Recovery from an Air Conditioning Unit, international Journal of Innovative Research in Science, Engineering and Technology Vol. 2, Issue 2, February 2013
[5] H.I. Abu-Mulaweh, "Design and performance of a thermosiphon heat recovery system", Applied Thermal Engineering, Vol.26, (2006) 471–477
[6] https://www.bijlibachao.com/air-conditioners/air-conditioner-selection-understand-tonnage-eer-cop-and-star-rating.html

Numerical Analysis of Inertance Pulse Tube Refrigerator

Mahmadrafik S. Choudhari[1], B. S. Gawali[1] and Prateek Malwe[1]

[1]Department of Mechanical Engineering, WCE, Sangli, India- 416415

Email: rafikmith@gmail.com

Abstract. The inertance type pulse tube refrigerator (IPTR) has been extensively used in industries due to its small size and efficiency. The physical phenomena of oscillating flow in IPTR is complex and it can be examined by using a commercial computational fluid dynamics (CFD) software. Researchers are not so much interested in the 3D analysis of IPTR due to high computational time but for better visualization, it is very beneficial. A two-dimensional axis-symmetric model and 3D model is considered for simulation. The temperature at the cold heat exchanger (CHX) and phase difference between pressure and mass flow rate at CHX is demonstrated. In this paper, CFD analysis of IPTR has been done and various issues involved in the simulation are presented for the betterment in the desired solution.

1. Introduction

Pulse tube refrigerator (PTR) is one of the promising contenders to the conventional Gifford-McMahon and Stirling cryocoolers, due to absence of moving parts in the cold region. PTR is a highly reliable device with low vibration and high life. A PTR is an oscillating flow loop type device consists of various components in which thermal interactions take place between solid and fluid. Continuous improvements in the performance of PTR is going on since basic PTR invented [1], which lowers the achievable refrigeration temperature. In all PTRs, the correct nature of physical operation occurring in the system is not well understood and the reason behind this is the poor understanding of phase difference and thermal relaxation. Another reason is the processes of pressurization and depressurization not well defined. Most important among all is the phase lag between the mass flow rate and pressure. This phase lag is adjusted by the inertance tube in various IPTC designs.

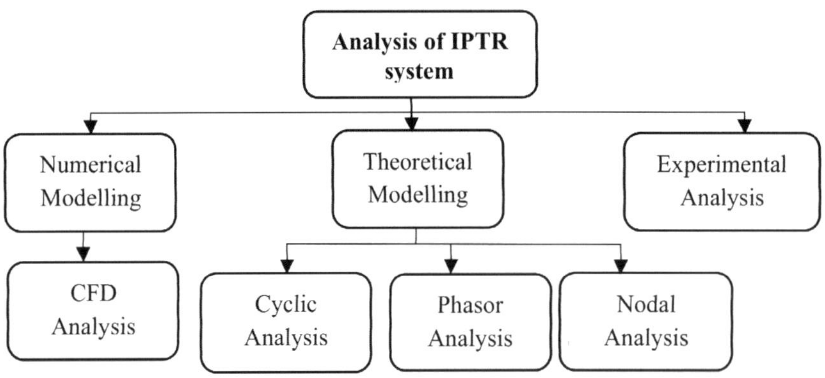

Figure 1. Different analysis techniques of the IPTR system

The different modelling techniques for the analysis of Inertance pulse tube refrigerator (IPTR) system is shown in Figure 1. Theoretical modelling is done by using various approaches like cyclic analysis, nodal analysis, or phasor analysis for the analysis of PTRs [2-4]. These models are convenient and simple but do not have sufficient considerations for the study of fluid flow and heat transfer.

Computational models have the capacity and it considers heat transport and fluid flow details for the analysis of PTR. The tool for the numerical solution of Navier-stokes equations is CFD commercial code, such as Fluent. Using CFD simulation, various difficulties are overcome like heat transfer phenomenon in regenerators and heat exchangers, modelling of oscillating flow in the system.

CFD analysis of simple orifice type of pulse tube cryocooler operating at high frequency is done by Barrett and Arsalan [5] using the axis-symmetric model. The solid material and gas physical properties are taken as constant while doing analysis. They have shown the swirling flow pattern presented in the velocity vectors. Numerical investigation of IPTR is done to observe the significance and extent of multi-dimensional flow effects in the IPTR system operating at high-frequency with two-dimensional axisymmetric model [6]. Chen [7] studied the thermodynamic cycles in an IPTR system by means of CFD simulation method. Numerical study of a 3D IPTR system from a design perspective has been done by Kumar [8]. Various issues like the simulation of the compressor and negative volume error occurred in it, are investigated which are involved in CFD simulation of an IPTR in detail. In this paper, two-dimensional axisymmetric and 3D model CFD simulations are conducted to analyze the IPTR. The IPTR model considered for the analysis is taken from the work reported by Gawali [9].

2. Simulated System

2.1. Model building and meshing

The pulse tube refrigerator contains all cylindrical components. The model is generated using ANSYS workbench DesignModeler.

2.1.1. 2D-Axisymmetric model. The schematic diagram of IPTR is shown in Figure 2. The dimensions of IPTR is given in Table 1. The different components are generated using the function of 'Add

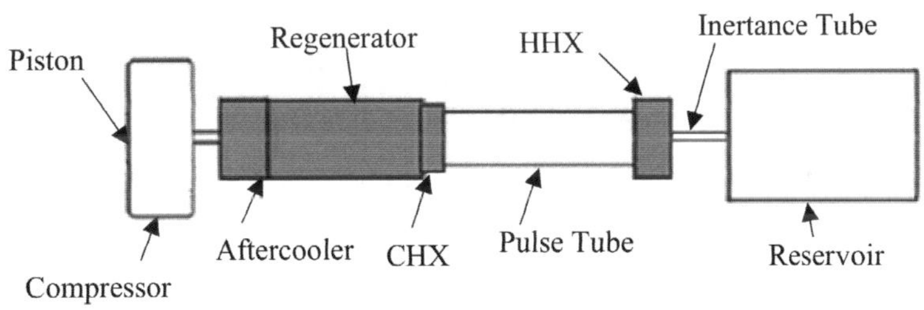

Figure 2. IPTR block diagram (not as per scale)

frozen' instead of 'Add material'. This is done to avoid a single solid as a body because boundary conditions should be given to specific components of IPTR. So after forming all solids then do 'Form new part' which act as one body of different solids components. This geometry is drawn symmetric along the x axis to take advantage of the axisymmetric model. IPTR 2D axisymmetric model is as shown in Figure 3.

ANSYS meshing module is used for the meshing of the object using elements of triangular and rectangular shape. However, in the current case, piston oscillates and generates pressure waves of the high amplitude of approximately ±5 bar. Operating pressure in this case in the system is 16 bar. The compressor is the most critical and essential component from a meshing point of view.

Table 1. Dimensions of IPTR [2]

Component	Radius (m)	Length (m)
A(Compressor)	0.03090	0.01300
B(Transfer line)	0.00300	0.03796
C(Aftercooler)	0.01400	0.0107
D(Regenerator)	0.01400	0.05400
E(CHX)	0.01400	0.00879
F(Pulse tube)	0.00480	0.12850
G(HHX)	0.00500	0.009
H(Inertance tube)	0.00125	2.00000
I (Reservoir)	0.03889	0.10000

It is obvious that to obtain pressure waves dynamic meshing should be done [10]. Compressor meshing is done by rectangular elements only to avoid negative cell volume detected error and the aspect ratio has to be maintained nearly equal to one to achieve better numerical stability.

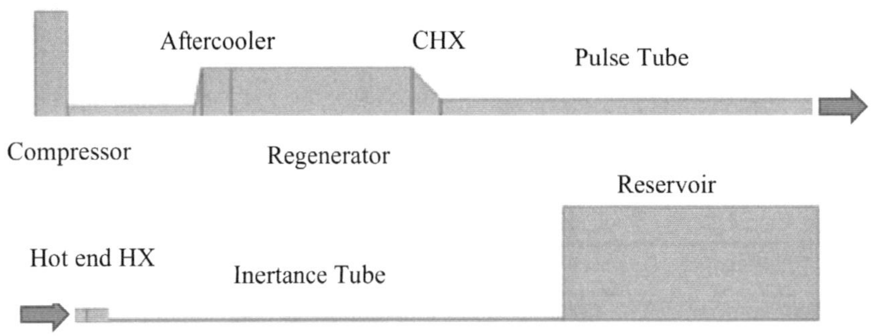

Figure 3. 2-D axisymmetric model

The components of the IPTR system have meshed with finer elements for better numerical stability as shown in Figure 4. Rectangular elements have been used to mesh the inertance tube, pulse tube and regenerator for suitable wave propagation. In the current case, total nodes are created 3916.

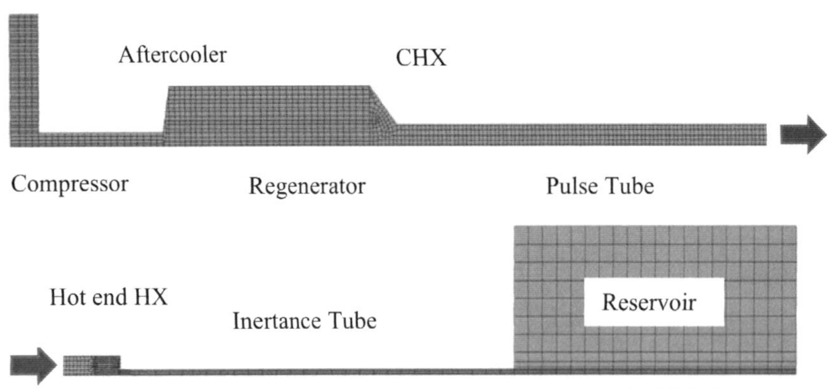

Figure 4. Meshing of the 2D-axisymmetric model of IPTR

2.1.2. 3D model. 3D IPTR model is drawn using geometry design modeller by taking same dimensions as given in Table 1 and it is shown in Figure 5.

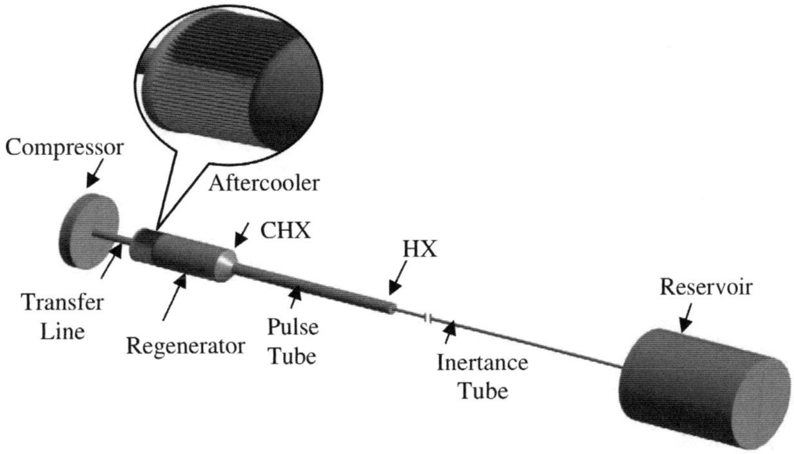

Figure 5. 3D IPTR model

Meshing of 3D IPTR model has been done and it is shown in Figure 6. For optimal meshing 3,96,568 nodes are created.

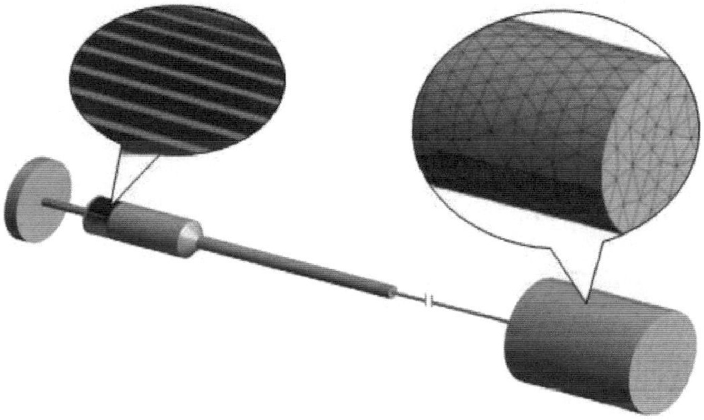

Figure 6. Meshing of 3D model

2.2. Governing equations

The mass, momentum and energy equations are the governing equations for the current 2-D axis-symmetric model [11]. The governing equations for the gas domain are given by

Mass:

$$\frac{\partial \rho}{\partial t} + \nabla \cdot (\rho \vec{v}) = 0 \tag{1}$$

Momentum:

$$\frac{\partial}{\partial t}(\rho \vec{v}) + \nabla \cdot (\rho \vec{v} \vec{v}) = -\nabla p + \nabla \cdot (\bar{\bar{\tau}}) \tag{2}$$

Energy:

$$\frac{\partial}{\partial t}(\rho E) + \nabla \cdot (\vec{v}(\rho E + p)) = \nabla \cdot \left(k_{eff}\nabla T + (\bar{\bar{\tau}}_{eff} \cdot \vec{v})\right) \tag{3}$$

Where ρ, $\vec{v}$, p, T, k_{eff} and E are the fluid density, velocity, static pressure, temperature, thermal conductivity and total energy of the fluid, respectively. Stress tensor is denoted by $\bar{\bar{\tau}}$.

Similarly, the governing equation for the porous zones includes mass, momentum and energy equations are given as However, for porous zones, the properties of the solid matrix should be taken into account.

Mass:

$$\frac{\partial(\varepsilon\rho)}{\partial t} + \nabla \cdot (\varepsilon\rho\vec{v}) = 0 \tag{4}$$

Momentum:

$$\frac{\partial}{\partial t}(\varepsilon\rho\vec{v}) + \nabla \cdot (\varepsilon\rho\vec{v}\vec{v}) = -\varepsilon\nabla p + \nabla \cdot (\varepsilon\bar{\bar{\tau}}) + S_i \tag{5}$$

Where ε is the void number, S_i is the source term added in the momentum equation. Source term made of two parts: the Darcy term and the Forchheimer term [11]:

$$S_i = -\left(\frac{\mu}{\alpha}v_j + C_2\frac{1}{2}\rho v_{mag}v_j\right) \tag{6}$$

Where α is the permeability and C_2 is the inertial resistance factor.

$$\alpha = \frac{{D_p}^2}{150}\frac{\varepsilon^3}{(1-\varepsilon)^2} \tag{7}$$

$$C_2 = \frac{3.5}{D_p}\frac{(1-\varepsilon)}{\varepsilon^3} \tag{8}$$

Energy:

$$\frac{\partial}{\partial t}(\varepsilon\rho_f E_f + (1-\varepsilon)\rho_s E_s) + \nabla \cdot \left(\vec{v}(\rho_f E_f + p)\right) = \nabla \cdot \left(k_{eff}\nabla T + (\bar{\bar{\tau}}_{eff} \cdot \vec{v})\right) \tag{9}$$

Where, E_f = Fluid energy
E_s = Solid medium energy
ε = void number
k_{eff}= effective thermal conductivity

2.3. *Fluent*

The fluent configuration for the current case and boundary conditions applied to different components are tabulated in Table 2 and Table 3 respectively.

Table 2. Setup

Setup	Selection
Solver	Pressure based solver
Computation	Double precision mode
Material	Ideal Gas (Helium)
Energy and Turbulent Model	Enabled and Standard κ − ε model (default mode)
Solution Methods	Pressure velocity coupling, SIMPLE, second-order transient

Table 3. Cell zone and Boundary conditions

Component	Cell zone and Boundary conditions
Compressor, Transfer line, Pulse tube, Reservoir	Cell zone conditions: Fluid: Helium, Solid: Steel Boundary conditions: Walls: Adiabatic
Aftercooler, Hot end heat exchanger	Cell zone conditions: Porous medium (Copper) Viscous-resistance (D): 2.47×10^{9} m^{-2} Inertial resistance (C): 26254 m^{-1}, Porosity (φ): 0.664 Boundary conditions: Wall Temperature: 300K
Regenerator	Cell zone conditions: Porous medium (Steel) Viscous-resistance (D): 4.09×10^{10} m^{-2} Inertial resistance (C): 107306 m^{-1}, Porosity (φ): 0.662, Boundary conditions: Walls: Adiabatic
Cold end heat exchanger (CHX)	Cell zone conditions: Porous medium (Copper) Viscous-resistance (D): 2.10×10^{9} m^{-2}, Inertial resistance (C): 9234 m^{-1}, Porosity (φ): 0.69, Boundary conditions: Walls: Adiabatic
Inertance tube	Cell zone conditions: Fluid: Helium, Solid: Copper Boundary conditions: Walls: Adiabatic

In the dynamic mesh section, layering mesh method is used. Three different sections considered in the compressor. First one is piston, second one is axis and last one is compressor wall. In fluent language, compressor axis and wall is defined as deforming and piston is defined as rigid body. In current case, piston motion is horizontal (x direction). Oscillating motion of piston is obtained by using a user defined function (UDF). This UDF is shown below, which produce desired pressure waves. The velocity is given to piston at 50 Hz frequency.

```
#include "udf.h"
DEFINE_CG_MOTION (velocitycomp, dt, vel, omega, time, dtime)
{
real frequency = 50.0;
real dw = 2.0 * 3.14 * frequency;
real disp = 0.005;
NV_S(vel, = ,0.0);
NV_S(omega, = ,0.0);
vel[0] = dw * disp * cos (dw * time);
}
```

The convergence criteria for the simulation and operating system parameters are considered as per tabulated in Table 4 and Table 5 respectively. Standard initialization technique is used for the initialization. Time step size taken as 3e-4 sec.

Table 4. Convergence criteria

Convergence criteria	
Continuity, Velocity, $\kappa-\varepsilon$	1e-03
Energy	1e-06

Table 5. Operating parameters

Working fluid	Helium
Frequency	50 Hz
Operating pressure	16 Bar

3. Results and Discussions

3.1. 2D-axisymmetric

The simulation of PTR is done using FLUENT software for mainly studying the temperature drop at the cold end of the pulse tube i.e. CHX. The simulation should be carried out until the cyclic steady state is achieved.

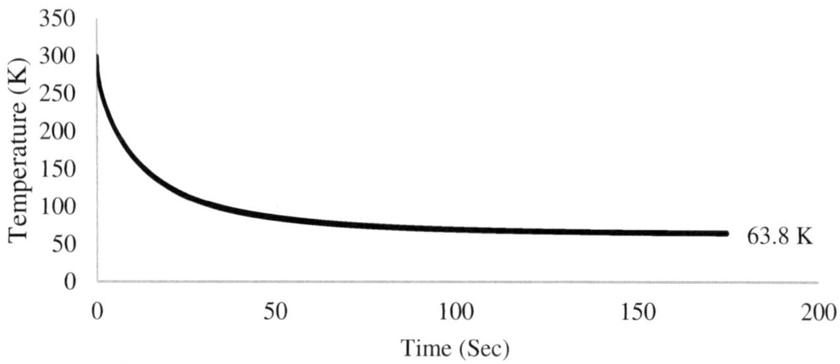

Figure 7. Cool down curve at no load condition

Experimental temperature achieved at CHX by Gawali [2] is 39 K at no load condition. Numerical temperature value achieved at CHX is 63.8 K as shown in Figure 7. Main reason for the difference between numerical and experimental result is due to aftercooler modelling. Physically aftercooler is slotted heat exchanger but in the numerical simulation it is taken as porous section due to difficulty in the 2D axisymmetric modelling. Temperature contour of the PTR is shown in Figure 8.

Figure 8. Temperature contour

3.2. 3D-PTR Model:

Temperature achieved at CHX is 26.5 K as shown in Figure 9. Gawali [2] achieved 39K experimentally. Difference between numerical and experimental result is due to heat transfer losses occurred in the different part of the system.

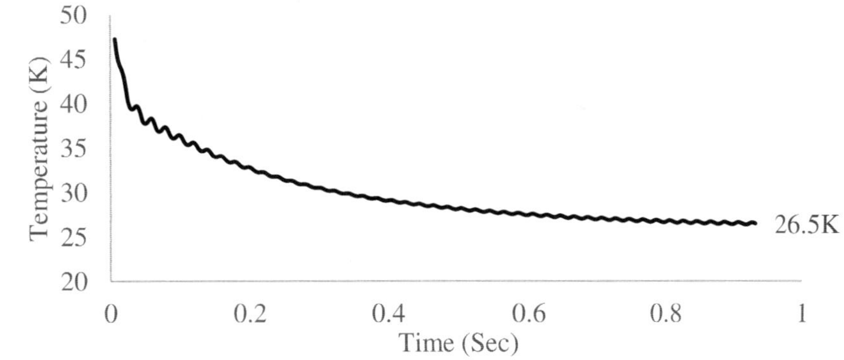

Figure 9. Cooldown curve

Pressure and mass flow rate at cold end of pulse tube are plotted in Figure 10. This plot is used to find out phase angle between pressure and mass flow rate which is crucial for refrigerating effect.

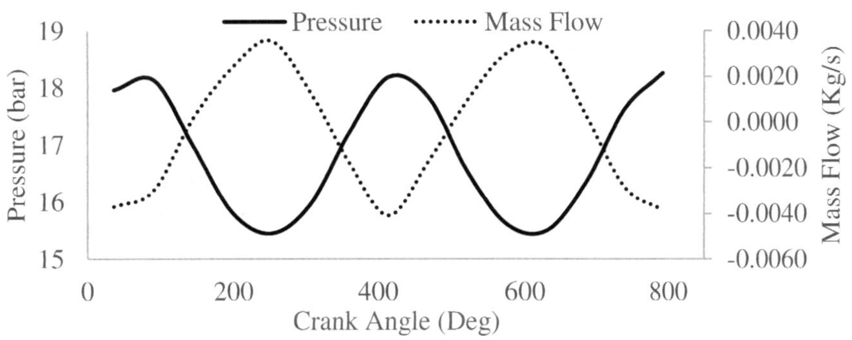

Figure 10. Pressure and mass flow rate at cold end of pulse tube

4. Conclusion

Summarizing, the methodology for analysing an IPTR in CFD, using 2D-axisymmetic and 3D model is proposed. Gawali's IPTR model [2] is validated and results of the presented model are adequately agree with the Gawali model. This simulation demonstrates the time varying temperature at cold heat exchanger. Also the phase difference in pressure and mass flow rate at cold end of pulse tube is also predicated. This methodology can be used for design of IPTR and same methodology can be extended to do the modifications in the system and analyze it.

References

[1] Gifford W E and Longsworth R C 1964 *Advances in Cryogenic Engineering* **10** pp. 69-79.
[2] Gawali B S and Narayankhedkar K G 2006 *AIP Conference Proceedings* **823** pp. 11-18. .
[3] Zhu S W and Chan Z Q 1994 *Cryogenics* **34** pp. 591-595.
[4] Atrey M D, Bapat S L and Narayankhedkar K G 1990 *Cryogenics* **29** pp. 341-347.
[5] Barrett F and Arsalan R 2004 *Adv. Cryog. Eng.* **49** pp. 1493–1499.
[6] Cha J S, Ghiaasiaan S M, Desai P V, Harvey J P and Kirkconnell C S 2006 *Cryogenics* **46** pp. 658-665.
[7] Chen L, Zhang Y, Luo E, Li T and Wei X 2010 *Cryogenics* **50** pp. 743–749.
[8] Kumar P, Gupta A K, Sahoo R K, Jena D P 2019 *Cryogenics* **98** pp. 125-138.
[9] Gawali B S, Atrey M D and Narayankhedkar K G 2003 *ICEC 19* pp. 391-394.
[10] Pfotenhauer J M and Baik J H 2002 MA: Springer (US, Boston) pp. 249–57.
[11] Fluent 6.3 User's Guide, Fluent Inc.

IOP Conf. Series: Materials Science and Engineering 1104 (2021) 012009 doi:10.1088/1757-899X/1104/1/012009

Material selection and design of shell and tube heat exchanger for thermos-syphon steam generator

Nitin S Solke[1 [0000-0002-9008-306X]], Anirban Sur[2 [0000-0002-6256-7470]] and Mandar Sapre[3[0000-0002-5545-8778]]

[1, 2, 3] Symbiosis Institute of Technology, Symbiosis International (Deemed University), Pune, India

nitinsolke@gmail.com

Abstract. Pure steam generated by conventional boiler is not devoid of chemical additives, rust or other undesirable materials and fails to meet essential quality of steam needed for pharmaceutical and health care process industries for WFI(Water for Injection), sterilization of medical equipment and clean room humidification on account of the limitations in achieving desired standards needed by the qualities of ideal pure steam such as pH value, conductivity, bacteria count etc. as prescribed by the International Society of Pharmaceutical Engineers (ISPE). Dry and pure steam is produced by 'pure steam generators' using purification process of the plant steam and doesn't contain any liquid droplets, any anti scaling additives and corrosion inhibitors. - Thermo siphon type Pure Steam Generator has been chosen in the present project for its extremely good heat transfer performance at low temperature differences, less sensitivity to large changes in process condition and more stable operation. In this paper design of the heat exchanger for generation of pure steam has been discussed. Here shell and tube type heat exchanger has been selected due to its flexibility of design, which allows wide range of pressures and temperatures. Standards laid out in ASME BPE for selection of material, CGMP and TEMA for design and manufacturing methods are followed for efficient and safe steam generation. Parameters given by the industry to meet customers' requirements have been considered while designing the HE, its accessories, piping, valves and instruments. Thermal design of the two phase shell and tube heat exchanger has been done manually and validated using HTRI suite. The mean % error between HTRI results and manual results is 12.05%.

Keywords: Thermo siphon, shell and tube heat exchanger design, pure steam generator.

...

1. Introduction

Pharmaceutical and health care process industries where the steam comes in contact with a pharmaceutical or medical product, use pure steam to avoid contamination [1]. Steam from conventional boiler (Plant steam), if utilized in such cases, does not meet requisite specifications because of the undesirable dissolved materials, chemical additives, or rust [2]. Pure Steam Generator produces pure steam known as Pyrogen, which is a free dry steam and in the condensed state meet the requirements of water for injection (WFI) [3]. Pharmaceutical industries use pure steam for the sterilization purposes. Steam Sterilization is mainly required in the manufacture of water used for injectable and use of pure steam for creating a sterile environment, or into autoclaves where vials and stoppers are sterilized [4]. Pure steam is also used for some other function like Clean Room Humidification.

The following tables illustrates the purity required of pure steam and the purity of water used in pharmaceutical industries as per published standards.

Table 1. Purified Water/ Feed Water Specification [5].

Heading level	Example
P.H Value	5 to7 at 25°C
Conductivity	≤ 5 µs /cm at 25°C
Chlorides	Absent
Sulphates	Absent
Metals	Absent
Nitrates	Absent
Micro Organisms	< 1000 cfu/ml
Total Organic Carbon	< 500 ppb
Bacterial endotoxin	< 25 EU/ml
Silica	< 1ppm

Table 2. Pure steam specification [6].

Heading level	Example
P-H Value	5 to 7 at 25°C
Conductivity	≤ 1.3 µs at 25°C
Total Bacteria count	≤ 10 cfu / 100 ml
Total organic carbon	< 500 ppb
Bacterial Endotoxin	< 0.25 EU/ml
Non-Condensable Gases	<3.5%
Moisture Content	< 5%
Pressure	2.5 kg/ cm^2 to 3 kg/cm^2
Temperature	121°C to 136 ° C

Pure steam should meet all the qualities mentioned above. Above Properties are collected after condensing the pure steam.

2. Overview of Pure Steam Generator

It is possible to purify the plant steam to some extent by using filters to remove undesired particles such as rust particles; however, it is not practically possible to remove vaporized condensable contaminants from steam [7]. Therefore, it becomes desirable to generate pure steam in a pure form, rather than purifying the plant steam. For producing dry steam it is ensured that feed water does not contain any corrosion inhibitors and anti-scaling additives[8]. Advantages and disadvantages of different types of pure steam generators which are available in industries has been discussed in table 3[9].

Table 3. Advantages & Disadvantages with respect to type of generator.

Generator type	Advantages	Disadvantages
(1) Kettle Re- boiler type Pure Steam generator.	Low investment cost. Simple design like a boiler. All safety and controls.	Higher water holds up capacity
	Already established as in case of commercial boiler.	May lead to high down rates (High running cost)
(2) Thermo – Syphon type Pure Steam Generator	- Reduction in stress fatigue of tubes due to Short evaporator column - Easy maintenance. - Short external evaporator.	Recirculation rates and pressure drops (For two phase flow) are critical Special design care is to be taken
(3) Falling film evaporator	Good steam quality	Long evaporator columns resulting in
	Least water hold up capacity	(a) Stress fatigue of tubes (b) Difficult maintenance.
	Commonly manufactured and used in India	(c) Large space required for disassembly

Based on above comparison we decided to design Thermo-siphon type Pure Steam Generator. Thermo-siphon type re-boiler gives extremely good heat transfer performances at low temperature differences, less sensitive large changes in process condition and therefore its operation is more stable[10]. As per market survey, approximately 107 new pharmaceutical plants are expected to come up in the country, in which number of pure steam generators required will be about 285[11]. Total market potential for steam generators in next three years is about Rs 45 crores. Fig 1 shows details of thermosyphon pure steam generator.

Like most pure steam generators, thermos-syphon pure steam generator is fitted with feed water heaters and makes use of generator's blow down as the heating media [12]. This naturally results into added obvious advantage of cooling the blow down avoiding discharge of very hot and flashing water. If the feed water supply pressure is inadequate there may be a need of a feed pump. Pressure of feed water should be above the maximum steam pressure by 0.5 bar to 0.75 bar. This will overcome the pressure drop in pipes and valves[13].

Quality of condensate of pure steam is monitored by a sample cooler fitted with a conductivity meter. Conductivity of pure steam measured by conductivity meter is used for checking the final quality of steam [14]. It is necessary to supply the plant steam to the generator at higher pressure than the required pure steam pressure (7 - 8.5 bar).

Higher pressure difference between the plant and the pure steam pressure increases rate of pure steam generation. Plant steam pressure should be maintained higher than the pure steam pressure by 2 to 3 bar for maximum pure steam production. Plant steam consumption will be greater than the 10% to 20% of pure steam production [15].

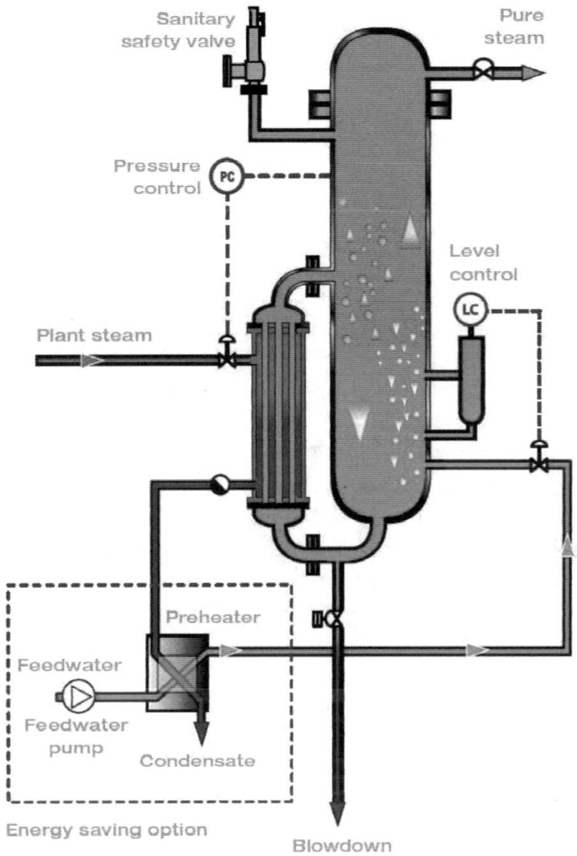

Fig. 1. Schematic Diagram of Thermosyphon Pure Steam Generator

Conventional boilers has tendency to corrode fire tube sheets materials, so clean and pure steam generator boilers do not use gas or oil as the primary heat source. Therefore, in clean and pure steam generator boilers, plant steam or pressurized hot water or thermal oil is most commonly used as a heating source.

Even though steam generators use treated water as feed water, there is tendency to have concentrated contaminants in the low levels of hold up volume as the water is evaporated. Blow down process is often used to discharge these contaminants. The frequency of blow down is determined by the quality of the feed water. For the high quality feed water intermittent blow down may be adequate. However, for lower quality feed water, 15% continuous blow down of the feed water is required. Automatic control of the process use conductivity measurement of the holdup liquid. Blowdown often leads to flashing after release from the generator pressure and need to be cooled down. Blowdown water is used in larger generators to enhance energy efficiency by preheating the feed water through heat exchanger. Design of this heat exchanger has been discussed in this paper.

3. Heat Exchanger Design

Shell and Tube Heat Exchangers offers the flexibility of design allowing its use in wide range of pressures and temperatures making it one of the most popular types of heat exchanger. Two main categories of Shell and Tube exchanger are used in industries [16-17]. One used in the petrochemical industry and pharmaceutical industry and the second type are used in power industry such as power plant condenser and feed water heater. Heat exchangers used in the petrochemical and pharmaceutical industry should follow TEMA (Tubular Exchanger Manufacturers Association) guidelines. So here we have followed TEMA guidelines for heat exchanger design.

Shell and tube heat exchanger consist of shell, containing bunch of tubes, front header and rear header. Through the front header fluid enters the tube side. Sometime front header is also called stationary header. The tube side fluid leaves the exchanger from rear header after passing through multiple passes and is returned to the front header. Tube bundle comprises of the tube sheets, baffles and tie rod for holding the bundle together.

A double tube sheet design is adopted for welded shell and tube heat exchangers to prevent product contamination in the case of a tube joint failure. It's imperative that the product contact surface must not be scored during fabrication when the tubes are to be expanded into the inner and outer tube sheets. It is to be ensured that tubes shall be seal welded to the outer tube sheet. And the distance between inner and outer tube sheets should be sufficient to allow leak detection and examination. Moreover, tube sheets and channels should be drainable [18].

For designing following standards are to be compulsorily taken into consideration while designing any product for pharmaceutical industry. The International Society of Pharmaceutical Engineers (ISPE) states the qualities of ideal pure steam required for pharmaceutical industry for pH value, conductivity, bacteria count etc. As per ASME BPE rules the material of construction is selected, its design methodology is finalized, fabrication techniques inspection etc. are finalized. As per CGMP general equipment requirements are considered so as to have highly efficient as well as best quality product. TEMA guidelines are compulsory to follow so as to design heat exchanger. TEMA standards provide most current, efficient design and manufacturing methods. [19-20].

4. Material Selection

Material selection is of utmost importance in the design of pure steam generator and heat exchanger due to tendency of corrosion of best of the steel materials and possibility of unacceptable impurity levels. Feed water use for producing pure steam tends to corrode carbon steel used in boiler and heat exchanger of the steam generation plant. This causes damage to components and pipe work besides contamination of steam. Corrosion of components tend to rapidly proliferate leading to the formation of 'rouge'.

Corrosion resistant grades of stainless steel is used for manufacturing components of high purity steam system to avoid any possibility of corrosion as there is high possibility that the corrosion products may reach the point of use. Materials of construction are so selected that they are capable of withstanding the temperature, pressure, and chemical corrosiveness ensuring the purity and integrity of the product. Normally, stainless steel materials of 316, 316L grades and higher alloys have proven to be acceptable [16]. When nonmetallic materials are used (elastomers, adhesives plastics), the manufacturer shall specify the materials shall carry a compliance certificate.

Table 4. Stainless steel standards for pure steam generation.

Material	Weight% AISI304	Weight% AISI316L	Weight% AISI316
Iron (Fe)	66.4-74	65	65
Nickel (Ni)	8-10.5	12	10-14
Chromium (Cr)	18-20	17	16-18

Manganese (Mn)	2	2	2
Sulphur (s)	0.03	0.03	0.03
Carbon (C)	0.08	0.03	0.08
Phosphorous (P)	0.045	0.045	0.04
Titanium (Ti)	-	-	Max 0.7
Molybdenum (Mo)	-	2.5	2-3
Silicon(Si)	1	1	1
Iron (Fe)	66.4-74	65	65

As per requirement given by the industry following parameters have been used for heat exchanger design. Following equations have been used for design of heat exchanger.

4.1 Energy balances

Total heat carried by water has been calculated by equation 1

$$Q_w = (m_w \times cP_w \times \Delta T_w) \text{ pure water} + (m_w \times \lambda_w) \text{ pure water} + (m_s \times cP_s \times \Delta T_s) \quad (1)$$

In the above equation,
Q_w = Heat carried by water (kcal/ hr)
m_w = mass of water (kg/hr)
cPw = Specific heat of water (kcal/kg-°C)
ΔTw = Temperature difference between inlet and outlet of temperature (°C)
λ_w = Latent heat of water (kcal/kg)
m_s = mass of steam = mw (kg/hr)
cP_s = Specific heat of steam (kcal/kg-°C)
ΔT_s = Temperature difference of superheated steam (°C)
Log mean temperature of the heat exchanger is calculated by equation 2

$$\text{LMTD} = \frac{(T2 - Ta) - (T1 - Tb)}{\ln\frac{(T2 - Ta)}{(T1 - Tb)}} \quad (2)$$

Here,
Inlet Temp of pure water 30° C (T_a)
Outlet temp of pure water 140 ° C (T_b)
Inlet Temp of Plant Steam 164 °C (T_1)
Outlet temperature of plant steam 160°C (T_2)

4.2 Overall heat transfer coefficient (UD)

Value of overall heat transfer coefficient is assumed for estimating the size of the heat exchanger and after calculation the value has been validated.

$$U_D = 950 \text{ kcal / m}^2 \text{ °c [23]} \quad (3)$$

Area of the heat exchanger and tubes no inside the heat exchangers are calculated using equation 4 and 5.

$$A = \frac{Qw}{U_D \times LMTD} \quad (4)$$

Here
A = Heat transfer area m^2
N_T = no of tubes in heat exchanger
$N_T = A / \pi$ Do L
D_o = Outer dia of tube (m)

L = length of the tube (m) = 0.400m

4.3 Number of passes

Here flow inside the heat exchanger tubes considered as turbulent flow and accordingly no of passes are chosen. From heat transfer data book Reynolds no. (Re) has been chosen Re = 4000

$$R_e = 4 \times m_w \times \frac{\left(\frac{N_P}{N_T}\right)}{(\pi \times \mu \times D_i)} \quad (5)$$

Here,
N_p = number of pass,
D_i = inner diameter of pipe (m)
μ = kinematic viscosity of water (kg/ m-s)
Tube size and actual tube count have been calculated using equation 6 and 7.
The initial design of the heat exchanger is completed here.

4.4 Adequacy of the initial design

Fluid properties dependent on temperature are assumed constant since these are not available. The viscosity correction factors is assumed as one. Overall coefficient is calculated using equation 6.

$$U\ req = \frac{Q_W}{N_T \times \pi \times D_o \times L \times (LMTD)} \quad (6)$$

Convective heat transfer coefficient inside the tube wall (hi) is calculated using equation 7. We use k = Thermal conductivity of SS 316 L

$$hi = \left(\frac{K}{D_i}\right) \times 0.023\, Re \times 0.8 Pr \times \frac{1}{3} \times \left(\frac{\mu}{\mu_w}\right) \times 0.14 \quad (7)$$

Convective heat transfer coefficient for outer surface of tube (h_o) is calculated using equations 8 to 13 as follows. Diameter of Shell D_S = 0.300 m

$$\text{Baffle Spacing } B = 0.2 \times D_S \quad (8)$$

Area of Shell is calculated using equation 9.

$$A_S = \frac{D_S \times \acute{C} \times B}{\left(144 \times \frac{P_T}{1.44}\right)} \quad (9)$$

In above equation, $\acute{C}$ = 0.00635 B=0.06 P_T = 0.03175 [24]

$$Re = De \times \frac{G}{\mu} \quad (10)$$

In above equation D_e = 0.0185 m

$$G = \frac{m_w}{A_s} \quad (11)$$

$$j_H = 0.5 \times \left(1 + \frac{B}{D_S}\right) \times \left(0.08 \times Re^{0.6821} + 0.7 \times Re^{0.1772}\right) \quad (12)$$

$$ho = j_H \times \left(\frac{K}{D_e}\right) \times Pr^{1/3} \times \left(\frac{\mu}{\mu_w}\right) \times 0.14 \quad (13)$$

Clean overall coefficient of heat exchanger has been calculated using equation 14.

$$U_C = \left(\frac{D_O}{h_i \times D_i}\right) + \frac{\left(D_O \times \ln\frac{D_O}{D_i}\right)}{2 \times k} + \left(\frac{1}{h_o} - 1\right) \tag{14}$$

U_C should be greater than U_{req} then only next process will continue.

The fouling factor for the tube side is considered as **0.000102** m^2.K/ kcal

$$R_D = R_{Di} \times \left(\frac{D_o}{D_i}\right) + R_{Do} \tag{15}$$

4.5 Design overall heat transfer coefficient is calculated from equation

$$U_D = \frac{1}{U_C} + R_D - 1 \tag{16}$$

As per heat exchanger design data hand book if UD > Ureq, heat exchanger design should be thermally stable. Over-surface and over-design of heat exchanger is monitored by equation 17 and 18.

For finding over surface and overdesign, overall heat transfer coefficient of heat exchanger is used here for convenience of calculation rather than surface areas.

$$\text{Over-surface} = (U_C / U_{req}) - 1 \tag{17}$$

$$\text{Over-design} = (U_D / U_{req}) - 1 \tag{18}$$

For shell and tube heat exchanger pressure drop is very important factor for calculating rate of heat transfer. Pressure drop inside the shell and tube have been calculated using following equations. The friction factor is calculated using equation 19.

$$f = 0.4137\,(Re^{-0.2585}) \tag{19}$$

$$G = \frac{m_w \times \left(\frac{N_P}{N_T}\right)}{\pi \times \frac{D_i^2}{4}} \tag{20}$$

The friction loss is calculated using following formula. Frictional pressure loss due to flow inside the tube is calculated as

$$\Delta P_f = 1.644 \times f \times N_P \times L \times G^2 / 7.50 \times 10^{12}\, D_i \times S \times \Phi \tag{21}$$

The tube entrance, exit, and return losses are estimated using $\alpha_r = (2\,N_P - 1.5)$

$$\Delta Pr = 1.664 \times 1.334 \times 10^{-13} \times (2Np - 1.5) \times G^2/S \tag{22}$$

Nozzle Diameter = 0.0254 m

$$Re_{nozzle} = \frac{(4 \times m_w)}{\pi \times D_n \times \mu} \tag{23}$$

$$G_n = \frac{m_w}{\pi \times \left(\frac{D_n^2}{4}\right)} \tag{24}$$

After calculation it has been found that Reynolds no is more than 2000 i.e. it is turbulent flow. So we opt following pressure formula for finding pressure drop inside the nozzle.

$$\Delta P_n = 1.644 \times 4 \times 10^{-13} \times Ns \times G_n^{\,2} / S \tag{25}$$

So total Pressure Drop in tube side is calculated as

$$\Delta P = P_f + P_r + P_n \tag{26}$$

Pressure drop on the shell side of heat exchanger is calculated using following formulae. Following two equations are used for calculating the friction factor:

$$f_1 = (0.0076 + 0.000166\ D_S) \times Re^{-0.125} \tag{27}$$

$$f_2 = (0.0016 + 5.8 \text{ x } 10 - 5\ D_S) \times Re^{-0.157} \tag{28}$$

It is assumed that $B / D_S = 0.06$, therefore the friction factor (f) is calculated by:

$$f = 144 \text{ x } f_2 \tag{29}$$

$$Re_{nozzle} = 4m / (\pi \times D_n \times \mu) \tag{30}$$

$$G_n = m / (\pi \times D n^2 /4) \tag{31}$$

$$\Delta P_n = 1.644 \times 4 \times 10^{-13} \text{ x } N_s \text{ x } G_n{}^2 / s \tag{32}$$

$$\Delta P_f = 1.644 \times f \times G^2 \times D_S \text{ x } (N_b + 1) / 7.50 \times 10^{12} D_e \times S \times \Phi \tag{33}$$

The total shell-side pressure drop is:

$$\Delta P_o = \Delta P_f + \Delta P_n$$

(34)

(34)

5. Result and discussion

Process industries mostly use shell and tube heat exchangers. These heat exchangers consist of a shell and number of tubes. Hot fluid flows through tube and cold fluid put into shell side. As per industries requirement, shell size and tube size are taken as constant. Following parameters are used for heat exchanger analysis. Heat transfer takes place between the tubes and the fluid through the surface of tube walls. Heat exchangers of shell and tube type can be manufactured easily in different sizes and flow configurations [24-36]. Several parameters which decide the heat transfer rate are temperature and pressure of feed water, geometry of tube, spacing between the baffles, and diameter of shell. In this paper heat exchanger has been designed by using different parameters for studying the effect of mass flow rate of water and diameter of tube on convective heat transfer coefficient and pressure drop inside the tube. The above formulae are used for analysis. EES simulation program has been used for analysis. The input and output temperature and pressure of water and steam, shell diameter, nozzle diameter, tube length are provided by industry as working conditions. These are displayed in table 5.

Table 5. Input parameters.

Parameters	Values
Inlet Temp of pure water (Ta)	30° C
Outlet temp of pure water (Tb)	140 ° C
Inlet Pressure of water (P_{iw})	3 bar
Mass flow rate of Plant Steam (m_w)	130 kg/hr
Inlet Temp of Plant Steam (T_1)	164 ° C
Outlet temperature of plant steam (T_2)	160°C
Inlet pressure of steam (P_{is})	6 bar
Length of the tube (L)	0.400m

IOP Conf. Series: Materials Science and Engineering 1104 (2021) 012009 doi:10.1088/1757-899X/1104/1/012009

Thickness of tube (Δt)	1.5mm
Diameter of Shell (Ds)	0.300 m

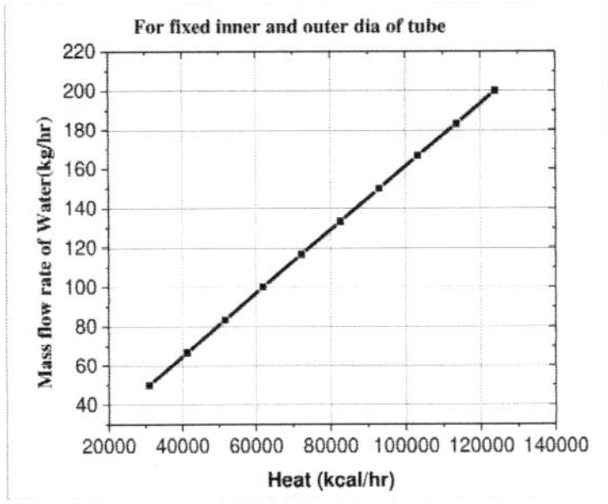

Fig. 2. Variation of heat capacity with mass flow rate of water

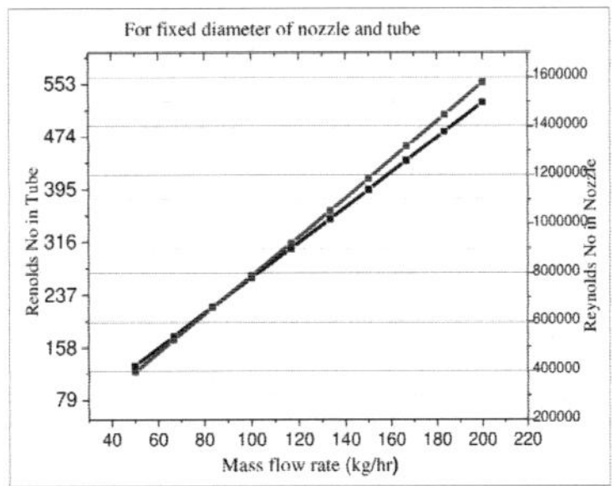

Fig. 3. Variation of Reynolds no in Tube and Nozzle with mass flow rate of water

Fig 2 shows variation of heat capacity for different mass flow rate of water. Curve shows that heat capacity increases with positive change in mass flow rate. Fig 3 shows how Reynolds no in tubes and in nozzle is varies positively with increase in mass flow rate. It shows that increasing value of Reynolds no's in Nozzle increase abruptly after mass flow rate crosses 100kg/hr. That's mean more turbulence is generated inside the nozzle. So in design mass flow rate kept 100kg/hr.

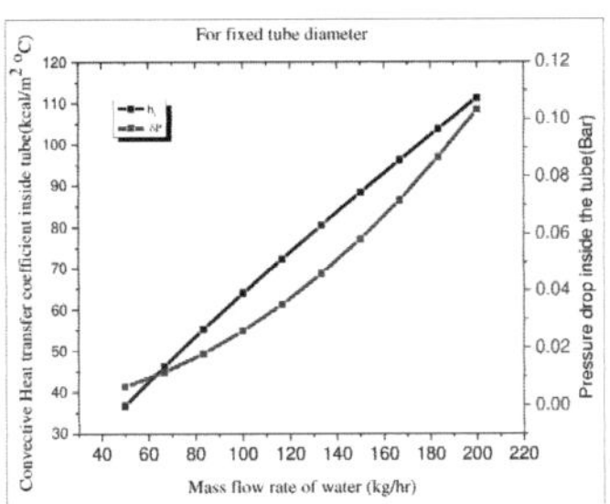

Fig. 4. Variation of pressure drop and convective heat transfer coefficient inside the tube with mass flow rate

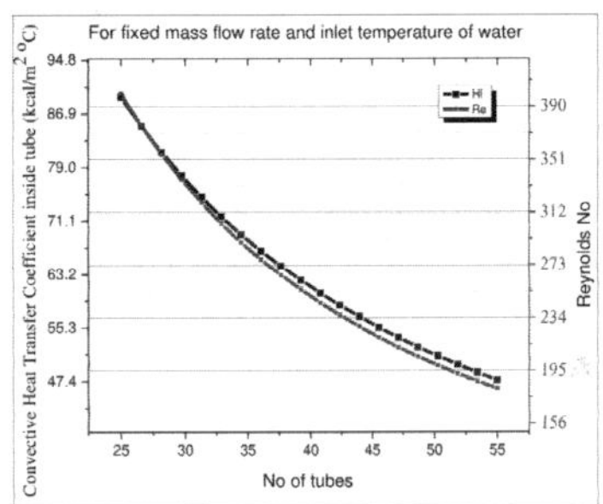

Fig. 5. Variation of convective heat transfer coefficient and Reynold's no inside the tube with variation tube numbers.

Fig 2 shows variation of heat capacity for different mass flow rate of water. Curve shows that heat capacity increases with positive change in mass flow rate. Fig 3 shows how Reynolds no in tubes and in nozzle is varies positively with increase in mass flow rate. It shows that increasing value of Reynolds no's in Nozzle increase abruptly after mass flow rate crosses 100kg/hr. That's mean more turbulence is generated inside the nozzle. So in design mass flow rate kept 100kg/hr.

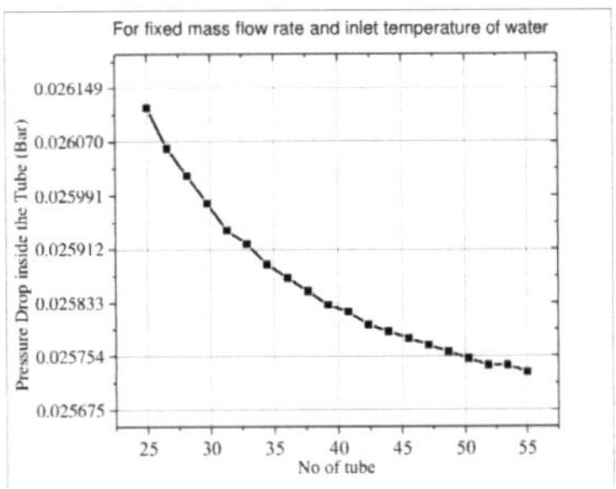

Fig. 6. Variation of pressure drop with tube numbers.

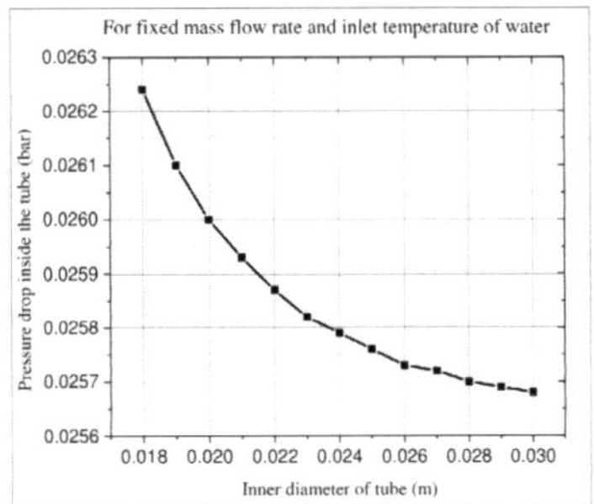

Fig. 7. Pressure drop inside the tube with variation of tube inner diameter

It's observed in Fig 5 and 6 that there is decrease in convective heat transfer coefficient, Reynolds number and pressure drop inside the tube with increase in number of tubes. Fig 7 and 8 shows variation of convective heat transfer coefficient, pressure drop and Reynolds no inside the tube with variation of diameter of tube. It shows increase of inner diameter of tube decreases convective heat transfer coefficient, Reynolds no and pressure drop. Same process we have done for shell side analysis.

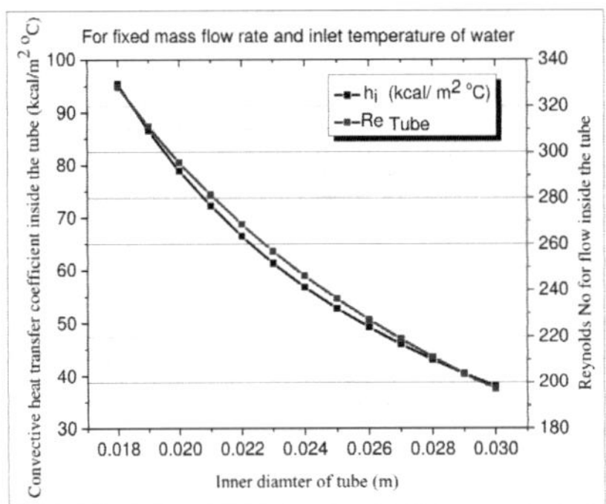

Fig. 8. Variation of convective heat transfer coefficient and Reynold's no inside the tube and with variation of diameter of tube.

6. Conclusion

Shell and tube heat exchanger for Pure Steam Generator of Thermo-siphon Type of capacity 100 kg / hr has been designed here. A Piping and Instrumentation Diagram has been made with all necessary required components of the entire setup. Two phase heat exchanger has been design by manual calculations based on parameters provided by industries and validated using HTRI suite. The mean % error between HTRI results and manual results is 12.05%.Mechanical design of heat exchanger has been done manually in accordance with TEMA guidelines, in which 15 % design error is allowable. So our design is in compliance with the industries standards. The scope of our project included preparation of process and instrumentation diagram, thermal and mechanical design of heat exchanger and its validation, mechanical design of heat exchanger and its validation, part modeling of heat exchanger components and their 2-D drawings. Everything mentioned in scope of our project has been completed on time and is submitted to the industry. The company has appreciated our effort and hard work for completing our project well in time. The Designs and Drawings submitted to the company has been reviewed by their technical team for any discrepancies in design, and further decision on fabrication and testing of the product have been taken.

References

[1] Applied Process Design for Chemical and Petrochemical Volume 3 Third Edition by Ernest Ludwig.
[2] ASME BPE 2009 Guide.
[3] ASME Section II Part D Properties of Material.
[4] ASME Section VIII Division I Rules for Construction of Pressure Vessel.
[5] 'Clean steam in the pharmaceutical industry', Tim Latham, PDH course.
[6] 'Clean steam system', Tim Latham, In Pharmaceutical Engineering Volume. 15, No.2, March/April 1995.
[7] EN 285 Sterilization - steam sterilizers - large Sterilizers (1996).
[8] Health Technical Memorandum (HTM).
[9] HTM 2010 Sterilization (1994) Part 3 validation and verification.
[10] HTRI Design Manual Volume A / B / C (June 2013).
[11] ISPE Baseline Pharmaceutical Engineering Guide: Volume 4 'Water & Steam Systems' (2001).
[12] Perry Chemical Engineering Handbook by Robert Perry and Don W. Green.
[13] Pharmaceutical Production, An Engineering Guide Edited by Bill Bennet and Graham Cole.
[14] Pharmaceutical Water: System, Design, Operation and Validation Second Edition by William V. Collentro.
[15] Pressure Vessel Handbook Eleventh Edition by Eugene F. Megyesy.
[16] Steam Engineering Product Handbook 2 nd Edition by Thermax (January 2013).
[17] Process Heat Transfer by D.Q. Kern.
[18] Process Heat Transfer Principle and Applications by Robert Serth.
[19] Heat Transfer Sixth Edition by J.P. Holman
[20] TEMA Ninth Edition.
[21] The application of steam quality test limits', Keith Shuttle Worth.
[22] Title 21 of the Code of Federal Regulations; Electronic Records; Electronic Signatures (21 CFR Part 11).
[23] Process Heat Transfer DQ Kern
[24] Heat exchanger design data handbook T Kuppan
[25] Mirzaei M, Hajabdollahi H and Fadakar H 2017 *Multi-objective optimization of shell-and-tube heat exchanger by constructal theory*, Appl. Therm. Eng. **125** p 9–19.
[26] Liu L, Ding N, Shi J, Xu N, Guo W and Wu C 2016 *Failure analysis of tube-to-tube sheet welded joints in a shell-tube heat exchanger, Case Study* Eng. Fail. Anal. **7** p 32–40.
[27] Abd A A and Naji S Z 2017 *Analysis study of shell and tube heat exchanger for clough company with reselect different parameters to improve the design, Case Stud*. Therm. Eng. **10** p 455–467.
[28] Shinde S and Chavan 2018 *Numerical and experimental analysis on shell side thermo-hydraulic performance of shell and tube heat exchanger with continuous helical FRP baffles*, Therm. Sci. Eng. Prog. **5**. p 158-171 https://doi.org/10.1016/j.tsep.2017.11.006.
[29] Yu C, Ren Z and Zeng M 2018 *Numerical investigation of shell-side performance for shell and tube heat exchangers with two different clamping type anti-vibration baffles*, Appl. Therm. Eng. **133** p 125-136.
[30] Dizaji H.S, Jafarmadar S and Asaadi S 2017 *Experimental exergy analysis for shell and tube heat exchanger made of corrugated shell and corrugated tube*, Exp. Therm. Fluid Sci. **18** p 475–481.
[31] Alimoradi A and Veysi F *2017 Optimal and critical values of geometrical parameters of shell and helically coiled tube heat exchangers*, Case Stud. Therm. Eng. **10** p 73–78.
[32] Gao B, Bi Q, Nie ZJ and Wu 2015 *Experimental study of effects of baffle helix angle on shell-side performance of shell-and-tube heat exchangers with discontinuous helical baffles*, Exp. Therm. Fluid Sci. **68** p 48–57.
[33] Lemos J C, Costa L H, and Bgajewicz J 2017 *Linear method for the design of shell and tube heat exchangers including fouling modeling*, Appl. Therm. Eng. **125** p 1345–1353.
[34] Yan Li, Xiumin Jiang, Xiangyong Huang, Jigang Jia and Jianhui Tong 2010 *Optimization of High-Pressure Shelland- Tube Heat Exchanger for Syngas Cooling in an IGC*, International Journal of Heat and Mass Transfer, **53,** p 4543-4551.
[35] Sur A 2017 *Experimental investigation on waste heat driven activated carbon-methanol adsorption cooling system.* . Journal of the Brazilian Society of Mechanical Sciences and Engineering ***39****(7)* p 2735–46.
[36] Sur A, Pandya S, Sah R P, Kotecha K and Narkhede S 2020 *Influence of bed temperature on performance of silica gel/methanol adsorption refrigeration system at adsorption equilibrium, Particulate Science and Technology,* DOI: 10.1080/02726351.2020.1778145.

IOP Conf. Series: Materials Science and Engineering 1104 (2021) 012010 doi:10.1088/1757-899X/1104/1/012010

Influence of low-frequency vibration in Die Sinking EDM: a review

Laxmi Devi[1], Kamlesh Paswan[1], Somnath Chattopadhyaya[1] , Alokesh Pramanik[2]

[1]Indian Institute of Technology (Indian School of Mines), Dhanbad, Jharkhand 834001, India.

[2]School of civil and mechanical engineering, Curtin, University, Perth, Western Australia

kamlesh.gcr@rediffmail.com

Abstract. Electric discharge machining (EDM) is a thermal-based machining process where material removal in the form of melting and vaporization occurs due to repeated electrical sparks between electrode and workpiece. EDM process finds wide applications in the manufacturing, automobile, aerospace, and biomedical industries due to its remarkable feature of producing complex geometrical shapes like molds, tools, and dies. Irrespective of materials hardness. In this review, the paper author has given details about a new advancement made in the EDM process, which is low-frequency vibration-assisted Die Sinking EDM. There is also a brief introduction of vibration setups that can be used to generate low-frequency vibration. There is also some discussion about various optimization techniques (e.g., TOPSIS, MOORA, GRA) used to optimize the process parameters that are current, frequency, pulse on time, and pulse off time to get better response parameters (MRR, TWR, and SR). Besides, there is extensive discussion about the effect of current, pulse on time, pulse off time and frequency on MRR, SR, and TWR. In the end, some discussion on the effect of low-frequency vibration on surface topography is also discussed.

1. Introduction

Die-sinking electric discharge machining (EDM) is a process that utilizes the spark energy to shape the surface of molds and tools. It is highly effective in shaping materials that are complex in design irrespective of material hardness. Despite many advantages, EDM faces many technological challenges with the growing complexity of products like in aerospace, automobile, and biomedical industries [1–3]. One major disadvantage is that the machining speed is relatively slow compared to other machining methods. Also, by increasing the discharge energy to machine faster, the machined surface's quality gets deteriorated. Thus to improve the machining performance, it is essential to uniformly disperse the discharge energy and avoid debris accumulation between electrode and workpiece. To teach these disadvantages, researchers have given many new technical solutions such as PMEDM, vibration-assisted EDM and optimization of existing parameters like current, pulse on time, pulse off time, voltage, dielectric fluid pressure, electrode polarity.

In this review, the author has focused on vibration assisted EDM as this is an emerging field these days. Vibration can be integrated into EDM in two ways either in the electrode or in the workpiece. Vibration integration into electrode is quite tricky compared to that of a workpiece in the EDM process as cited in Pandey and Singh [4]. Assisted low-frequency vibration leads to a significant increase in material removal rate (MRR), reduced surface roughness (SR), overcut, and taper angle deviation. Unune and Mali [5] used low-frequency vibration attached to Inconel 718 workpiece, and research showed that MRR increased by 27.6% when the frequency of vibration increased from (F=0-40Hz), electrode wear rate decreased by 6.16%, overcut and taper angle deviation decreased by 31.84% and 18.58% respectively. Assisted ultrasonic vibration increases the process's inefficiency by increasing the MRR and fatigue life and reducing surface roughness and recast layer [6]. A low-frequency vibration on EDM has been assigned to process SS304 stainless steel. A high MRR with reduced Total wear rate (TWR) and SR was achieved [7]. Besides, low-frequency vibration also decreases in short circuit pulses by 80% and discharge pulses by 40% [8]. This leads to an increase in the machinability stability of the process. Vibration contributes to the improvement of the flow velocity of dielectric fluid, thereby increasing the discharge frequency. This improves machining productivity and reduces electrode wear rate. Compared with the conventional EDM, the MRR increases by 187.5%, and EWR decreased by 925% when vibration was assigned to electrode [9]. Several statistical techniques have been introduced like Taguchi method combined with multiple objective optimization techniques (TOPSIS, GRA, Multiple objective optimization techniques based on ratio analysis- MOORA) has been widely used in EDM. Analytical hierarchical process (AHP) is used to determine quality indicators' weight [9,10]. The result of MOORA is better than that of TOPSIS and GRA [11]. Therefore, Taguchi, MOORA and AHP may be suitable for multiple objective optimizations of vibration-assisted EDM. The review paper is to examine the process parameters of vibration-assisted EDM of the SKD61 steel workpiece using a copper electrode. The process parameters include current (I), pulse on time (Ton), pulse off time (Toff), and frequency vibration (F). MRR, TWR, and SR are considered as quality indicators. The Taguchi-MOORA-AHP method is considered for the study. A technological solution for the EDM process shown in figure 1(a).

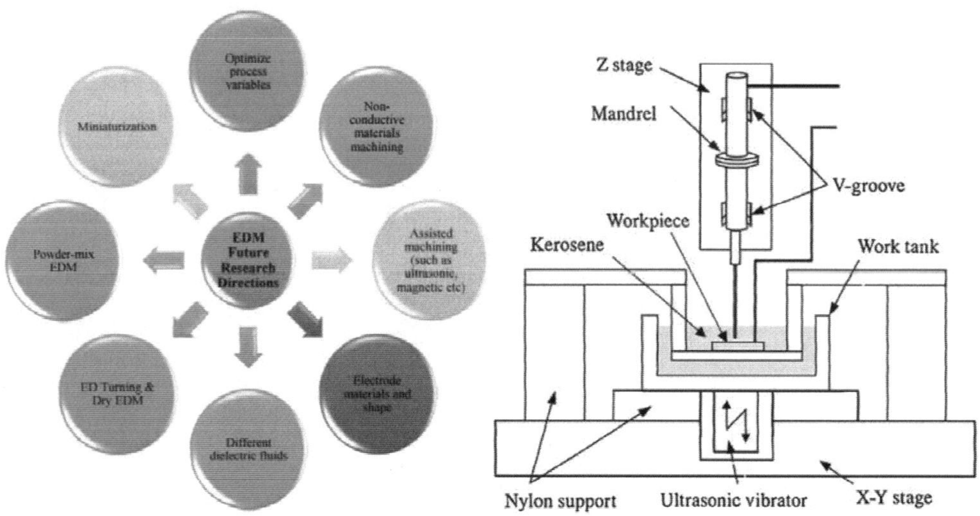

Figure 1. (a) New technological solution for the EDM process[12] and (b) Schematic diagram of vibration-assisted EDM[13]

The basic principle of EDM with vibration integration is shown in the schematic diagram. The whole set up contains a tool, tool holder, workpiece, vibration platform, dielectric, work tank, positioning table and other essential auxiliaries shown in figure 1(b).

Further, there is wide variation in necessary EDM process, e.g. EDM, Wire EDM; powder mixed EDM, dry EDM and micro EDM, which can be selected based on its applications. Every process has its merit and demerit, based on the research trends in the manufacturing industry [12]. Besides, with the growing demand for miniature products into aviation and biomedical industries, many new micro-manufacturing methods came into the picture. Of all the methods present, micro EDM is a complicated process to perform. Despite difficulties, many researchers have mixed micro EDM with other available options, e.g. wire EDM, magnetic field-assisted EDM, vibration-assisted EDM, milling and drilling [14].

Many research papers have been reviewed for deeper insights into vibration assisted EDM process. Some of these papers are listed in table 1 to provide some basic idea about the mechanism, workpiece, electrode, parameters (electrical and non-electrical), and the research outcomes.

Table 1. Evaluation of vibration-assisted Die Sinking EDM

Name of researcher	Workpiece and tool material	Parameters	Finding and remarks
Chern and Chuang [15]	W/P: SUS 304 stainless steel and brass Tool: Tungsten steel	Voltage, capacitance, resistance, frequency, amplitude, rotational speed	a) Punching of 200μm diameter hole on a brass workpiece with an optimum clearance value with improved geometry and surface finish using a vibration-assisted micro EDM. b) SUS304 steel punching is brutal and lacking in quality compared to brass due to higher strength and low machinability.
Xu et al.[16]	W/P: YT15 cemented carbide Tool: Cu	Frequency, amplitude, current, voltage, pulse on time, rotational speed, flashing a pressure	a) Material removal methods of ceramic carbides are discussed. b) MRR increased because the material gets easily separated due to integrated vibrations.
Yu et al.[17]	W/P: Stainless steel Tool: W	Voltage, capacitance resistance, frequency, amplitude	a) The aspect ratio of micro-hole drilled by micro EDM integrated with ultrasonic vibration and planetary of the tool is 29. b) Aspect ratio with only planetary movement is 16.9 c) In standard micro EDM drilling, the aspect ratio is 14.7

Jahan, Wong, and Rahma[18]	W/P: WC Tool: W	Voltage, current, frequency, amplitude	a) Discharge pulse rate can be increased by approximately 40% by using ultrasonic vibrations.
Iwai, Ninomiya, and Suzuki[19]	W/P: PCD Tool: Cu	Current, voltage, frequency, pulse on time	a) Integrated axial ultrasonic vibration EDM leads to 86% decrease in TWR than standard EDM. b) SR height in the range of 2.72 ~3.55 micrometre. c) MRR was almost three times more than standard EDM.
Ichikawa and Natsu[20]	W/P: SUS 304 stainless steel Tool: W	Current, frequency, amplitude, rotational speed	a) With a change in amplitude from 4 to 6 μm, the average feed rate increases by 5%, resulting in an improvement of MRR.
Kurniawan et al.[21]	W/P: CFRP composite Tool: WC- Co, Cu, Al, Brass	Voltage, frequency, amplitude, pulse on time, resistance, capacitance, rotational speed, feed rate	a) In CFRP material, Cu is best suited as electrode material. b) BRR and deburred hole quality increase considerably.
Hsue, Hab, and Lin [22]	W/P: SKD61 high-speed steel Tool: Cu-Cr	Frequency, amplitude, current, pulse on time, pulse off time, voltage, rotational speed	a) Rotary ultrasonic-assisted EDM leads to decrement in MRR by 51% at 1500 rpm, whereas Ultrasonic assisted EDM leads to better performance.
Li, Tang, and Bai[23]	W/P: Ti-6Al-4V Tool: W	Voltage, capacitance, resistance, rotational speed	a) Machining with circular ultrasonic vibration-assisted on tool electrode leads to greater consistency of hole dimensions b) Outlet and inlet diameters dimensions accuracy increased by 2.8% and 22% respectively.

IOP Conf. Series: Materials Science and Engineering 1104 (2021) 012010 doi:10.1088/1757-899X/1104/1/012010

2. Experimental setup and equipment

There are various types of EDM machine that we can select based on the operations we will perform on a device like deep hole drilling, curved hole drilling, turning. In this review paper, we are concerned with modifying essential EDM into the ultrasonic EDM process. Therefore, the experiments can be performed on any primary EDM machine equipped with a vibration unit. A basic experimental setup is shown in Figure 2(a). Some of the vibration setups are discussed in this paper. High-speed cameras can also be attached to a machine to record minute details of the whole machining process [24]. There is also a proper fixture mechanism to hold the workpiece rigidly during vibration, which enhances the machine's stability and accuracy. For verification and validation of results, many trials have been performed. The vibration unit is mainly based on the ultrasonic vibration principle in this machine, but the operating frequency range is narrow (100-5000Hz). The maximum limit of vibration amplitude is fixed in the range of 2-3 mm. The MRR and TWR are measured by using a digital weighing device [25]. Surface roughness is measured by surface profilometer (MITOTOYU, Japan), and surface morphology is presented by using an electronic microscope.

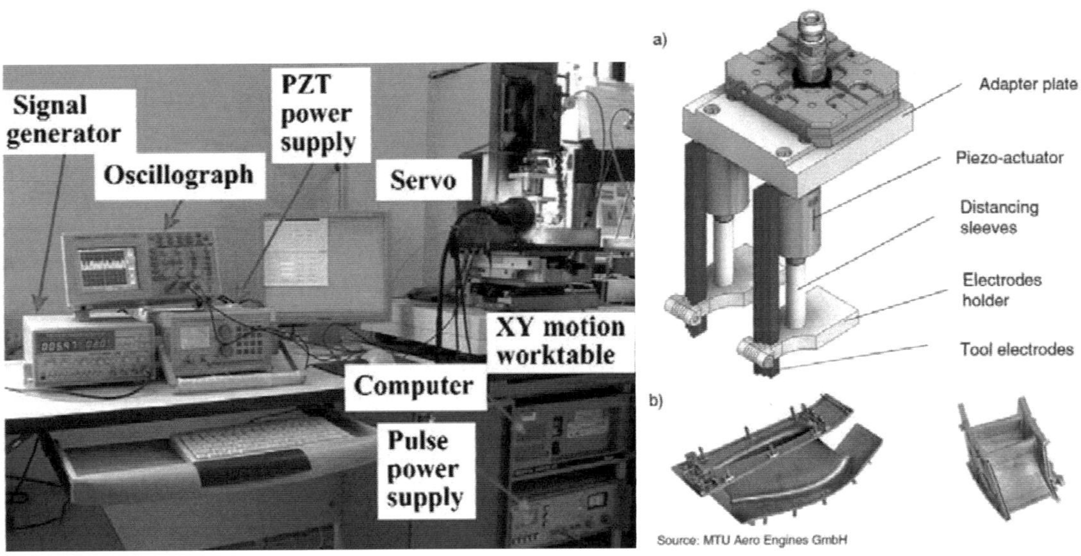

Figure 2. (a) Schematic diagram of vibration Assisted EDM [26] and (b) Three-dimensional models of a Piezo unit [27].

2.1 Vibration Set up

There are many types of equipment that can generate vibration to a workpiece. Below are listed four types of vibration generating methods and equipment.

2.1.1 Piezo Unit

The Piezo Unit can be specifically designed for particular applications, as depicted in Figure 2(b). The unit comprises of two identical Piezo actuator, function generator, and charge amplifier. These Piezo actuators are driven synchronously with the help of a function generator. The function of the charge amplifier is to determine the dynamics and movement characteristics of the Piezo Unit. The designed unit must be compact to restrict the machine's movement within the order and also fully insulated. It does not hinder the EDM machine working principle and avoid short circuits is presented in Figure 3. A Piezo Unit is specified by the frequency range from (0-1000Hz), the maximum deflection should be an order of 30 micrometres, and amplitude must not exceed 15 micrometres. Besides, to realize the actual load condition, the extra weight of 0.5 kg of the electrode and electrode holder should be considered per actuator. This designed unit can be verified by doing various experiments in the frequency range of (0-1000Hz) with a variation of amplitude from (2-16um), and the results can be validated by using Laser Doppler Vibrometer Sensor [27]. However, the two actuators behaved similarly at low frequency and amplitude. Whereas, there is a considerable change in the amplitude at higher frequencies (Figure 4). The best result, material removal rate also increased by 11.1%, is obtained at the frequency of (500-800Hz), and amplitude of 2um compared to EDM without vibration. The influence of vibration generated by the piezo unit has an advantage in making deep seal slots with improved MRR, decreased TWR, and better surface finish can be observed in Figure 5.

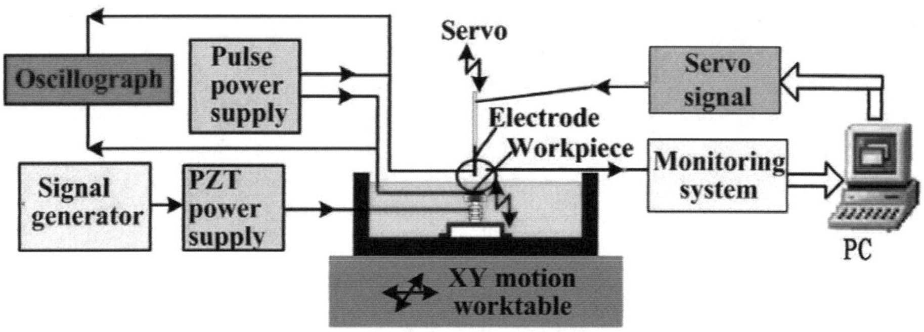

Figure 3. Piezo unit assisted EDM set up [26]

2.1.2 Ultrasonic vibration platform based on voice coil motor

Ultrasonic vibration-assisted EDM is the most preferred solution for improving machining efficiency. Despite that, ultrasonic vibration's primary challenge is the loss of vibration energy and heating effect due to fluctuation. Also, the vibration exciter operates in a very narrow range of frequency. Therefore, to avoid these drawbacks, a new method has been introduced based on voice coil motor associated with a vibration platform. The frequency range utilized in these platforms is very less than the ultrasonic range. Voice coil motor comprises a permanent magnet, coil, and coil frame, and these all assembled into a shell. This voice coil motor used in making the vibration platform, which consists of a metal flat diaphragm, Teflon diaphragm, seal ring, bottom board, and hull, is depicted in Figure 6. The metal diaphragm is an essential component and used for providing vibration. So it must have good strength, elastic modulus, hardness.

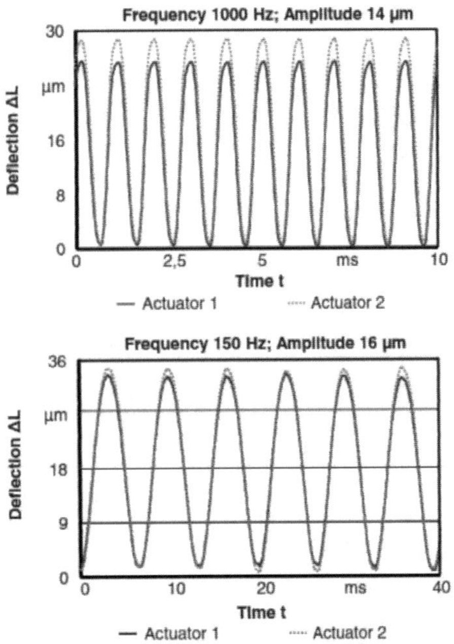

Figure 4. Vibration behaviour of two identical actuators at a different frequency and amplitude[27]

Electrical and control parameters at machine tool:

I_e	= 29 A	$\hat{u}_i$	= 270 V
t_i	= 32 µs	t_0	= 14 µs
Uburn._L	= 60 V	UOL_L	= 160 V
WUB_Lp	= 200	WUB_Ln	= 300

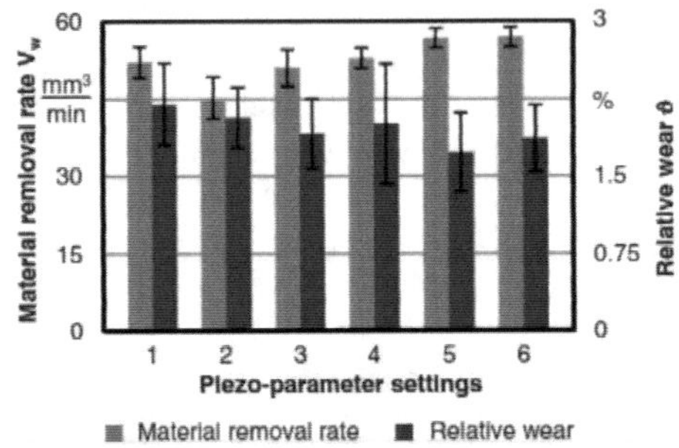

Piezo-Parameters	1	2	3	4	5	6
Frequency [Hz]	0	100	100	150	800	700
Amplitude [µm]	0	16	6	6	2	2

Figure 5. Influence of vibration on machining efficiency of high depth seal slots [27]

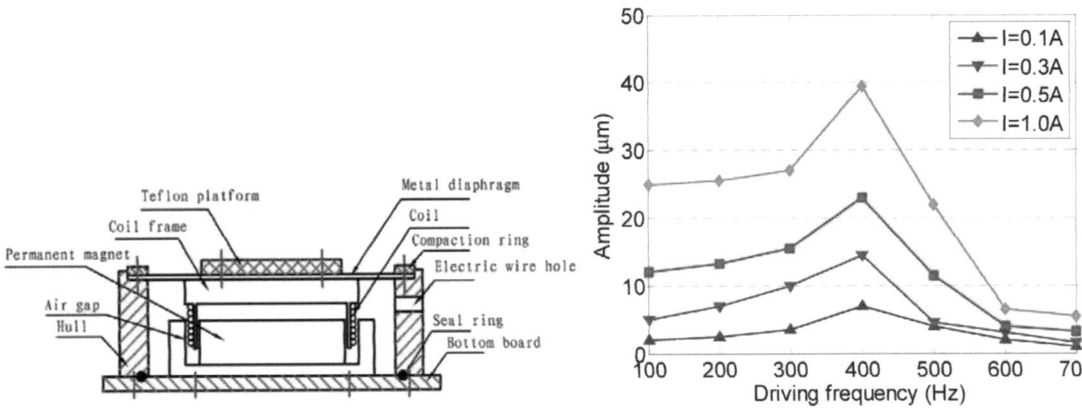

Figure 6. Design of Vibratory Platform and (b) Relation of frequency and amplitude of the vibration platform [28].

Brass is the best-suited material used in the making of flat diaphragm because it has good plasticity and higher strength. Teflon platform is provided for electrical isolation and placed above the metal diaphragm (Figure 6). So that short circuit does not happen. Various experiments are done to check the dynamic performance with a frequency range of (100-700Hz) and excitation current from (0.1-1A) [28]. The relation between vibration amplitude and frequency is shown in Figure 12. Also, there is a remarkable improvement in the material removal rate, almost five times without vibration assisted EDM process [29].

2.1.3 Turbine vibrator

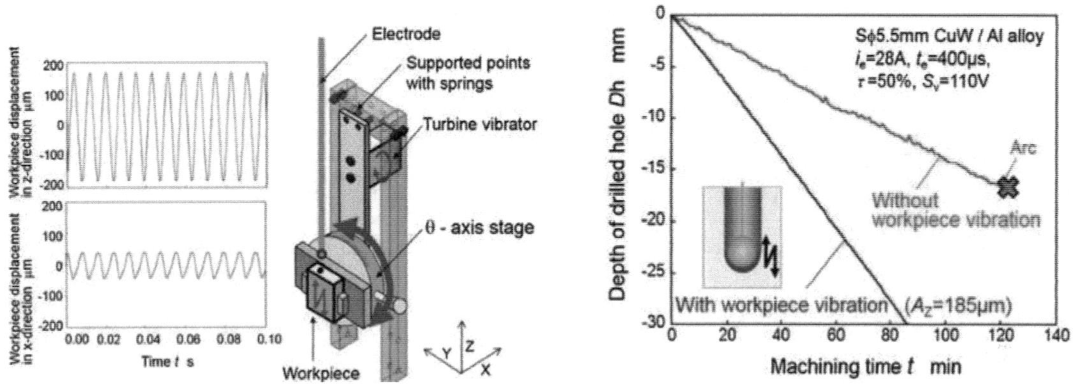

Figure 14. Workpiece vibration with turbine vibrator and vibration behaviour [29] and (b) Time variation with drilled hole depth with and without vibration with ball-shaped electrode [29]

In the turbine generator, generated vibrations are transmitted to the workpiece through an L shaped support structure combined with springs are depicted in Figure 14. It is effectively used for drilling operations, whether curved or straight. The turbine generator is used for drilling curved holes, and results are remarkable. There is no hindrance in drilling operation due to arc stop or short circuit, drilling speed is approximately twice than standard EDM drilling [29]. The time required to drill a hole gets halves with vibration than conventional EDM drilling are plotted in Figure 15. The significant advantage is that the vibration-assisted process makes it possible to dig a deep curved hole.

2.1.4 Langevin type ultrasonic vibrator

In this ultrasonic vibrator, the sine wave generated from waveform generator gets amplified with amplifier passes through a current transformer and then as input into vibrator which starts vibration with sine wave frequency, I.e. natural frequency of ultrasonic vibration shown in Figure 16. The natural frequency determined by measuring the current through the current transformer and natural frequency can be changed by changing the electrode's length. This ultrasonic vibrator is best suited for electrode vibration. Machining results get improved considerably with frequency magnitude of up to 66.5 kHz. However, the MRR did not improve in the frequency range of (27.6-57.3)kHz, also due to high vibration frequency surface roughness increased by 1 micrometre, as shown in Figure 17.

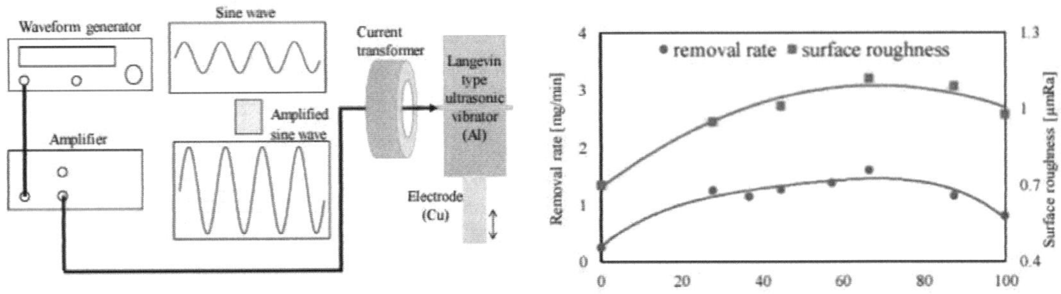

Figure 16. Schematic diagram of Langevin type ultrasonic vibrator Natural frequency (kHz)[30] , (b)The relation between response parameters and frequency[30]

3. Design of experiments and optimization

Current (I), frequency(F), pulse on time(Ton), and pulse off time(Toff) are taken as process parameters in the range of 3-8A, 128-512Hz, 12-50us, and 5.5-12.5 us respectively. MRR, TWR, and SR are taken as quality indicators (response parameters). Usnig0 can obtain optimized results. The Taguchi method is cost-efficient and easy to implement as the Taguchi method can not optimize the parameters solely. A combination of other optimization techniques like TOPSIS, MOORA, PSI, GRA, and ANOVA is required [31]. MOORA method gives the best optimization results among available options. The hierarchical analytical process (AHP) can calculate the weightage of process parameters on quality indicators. So a combination of MOORA-AHP-Taguchi can be used for better results.

4. Results and discussion

MOORA, along with AHP and Taguchi method, used to simultaneously optimize the quality indicators that are MRR, SR, TWR. Calculation by AHP method showed that currently has the highest priority than followed by frequency (F), pulse off time (Toff), and pulse on time (Ton), respectively [9]. Various experiments have been done, and the results are plotted graphically.

4.1 Effect of current on quality indicators.

The current has the highest contribution in the given parameter and is a measure of power delivered to the workpiece. As current increases, spark energy increases, which leads to an increase in the melting and evaporation process; hence, MRR increases drastically. The current increases from 3A - 8A, the MRR increases with almost 357.81%, TWR and SR also increase by 65.5% and 55.57%, respectively, shown in Figure 18(a). However, an increase in TWR and SR harms machining efficiency, but all together increase in MRR compensate for the decrement and enhances the machining efficiency. Simultaneously, the machining tool steel by Gr electrode current increased from 5A-50A results in an increase of tool wear rate by approximately 36% [9].

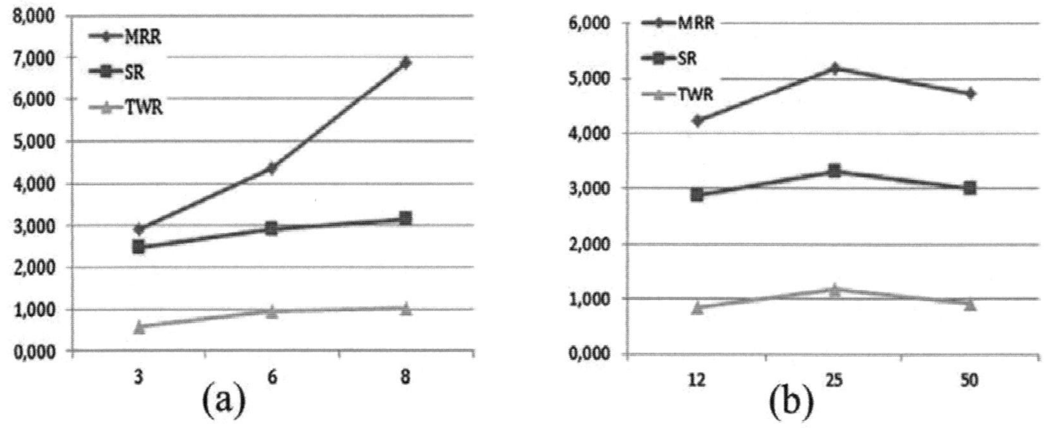

Figure 18. The relation between (a) I and results and (b)Ton and results[9].

4.2 Effect of Pulse on time (Ton) on quality indicators

As pulse on-time increase MRR, TWR and SR increase, however, the increment is small in comparison to that of current because with an increase in pulse on time spark timing increases; hence spark energy also increased. This increment of quality indicators occurs to optimal Ton after that decrement starts (Figure 18(b)). As Ton increases, Toff decreases, which leads to a reduction in time available to push the debris out of the discharge gap; hence, the Stop of arc or short circuit takes place. It is contribution ranked fourth out of given parameters. The increment of quality indicators (MRR, TWR, SR) is 22.9%, 36.25%, and 14.76%, respectively. While machining AISI D3 Die steel using copper electrode when Ton increased from 50us-100us, MRR decreased considerably [32].

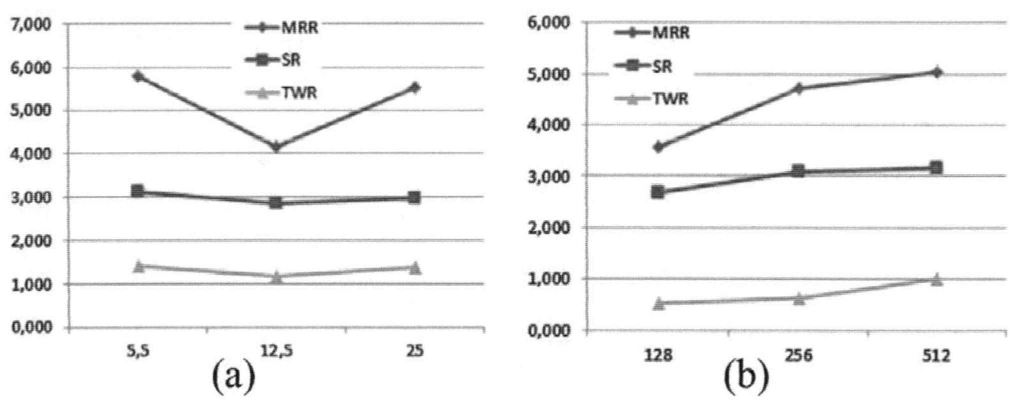

Figure 19. The relation between (a)Toff and outputs and (b) F and outputs [9]

4.3 Effect of Pulse off time on quality indicators.

It is ranked third out of the given parameters. As pulse off time (Toff) increases from (5.5-12.5us), MRR decreases drastically up to a certain point, i.e., 12.5 us because it results in a decrease in spark timing and hence spark energy also lower. Melting and evaporation of material decreases by 28.08%. With an increase in pulse off time. Debris flows from the discharge gap increase, and hence surface roughness increased by 8.85%, and TWR decreased by 16.96%. After this, there is an increment in quality indicators shown in Figure 19 (a).

4.4 Effect of Frequency on quality indicators

Frequency comes into the picture because it is a vibration-assisted EDM process. The vibration rate plays a vital role in the flow of debris and the recirculation of dielectric fluid. This increases the stability of the process and a short circuit. Due to vibration, the workpiece moves upward and downward continuously. When the workpiece moves upwards, it creates a pumping effect, making debris flow out of the discharge gap under pressure. Also, the space between workpiece and tool reduced that means tool and workpiece came into close contact, which results in an increase in spark frequency with an increase in spark energy. When the workpiece moves downward, fresh dielectric fluid came into play. These all factors cumulatively increase MRR, TWR, and SR by 64.49%, 20.30%, and 18.47%, respectively, at an optimal frequency of 512 Hz shown in Figure 19(b). The experiments were done in a frequency range of (200-600Hz). At F=200Hz, 400Hz and 600Hz MRR increased by 10.22%, 13.81%, and 34.9% respectively on the same workpiece material (i.e., SKD61 steel) with the same EDM process [8]. At the F=200Hz tool wear rate decreased by 16% compared to regular EDM. In the case of machining of seal slots in temperature resistant MAR-M247 workpiece because of vibration machining time decreases from 30.2 min to 17.4min, which means an increase in material removal rate by 42.38%, TWR decreased 21%, and surface roughness is less than 6.4um. Vibration generated by voice coil motor in the range of (100-700Hz) results in a significant decrease in perforation timing that means an increase in MRR [28]. Dimensional accuracy and machining efficiency increased by 10.5um and 18 times, respectively, while machining using a piezo actuator with vibration accuracy of 6000Hz and amplitude of 3um by using a tungsten electrode[26]. In vibration-assisted micro EDM of a brass workpiece with a frequency of 1000Hz and amplitude of 1.5 um using piezo actuator gives shortest machining time means the highest material removal rate [33].

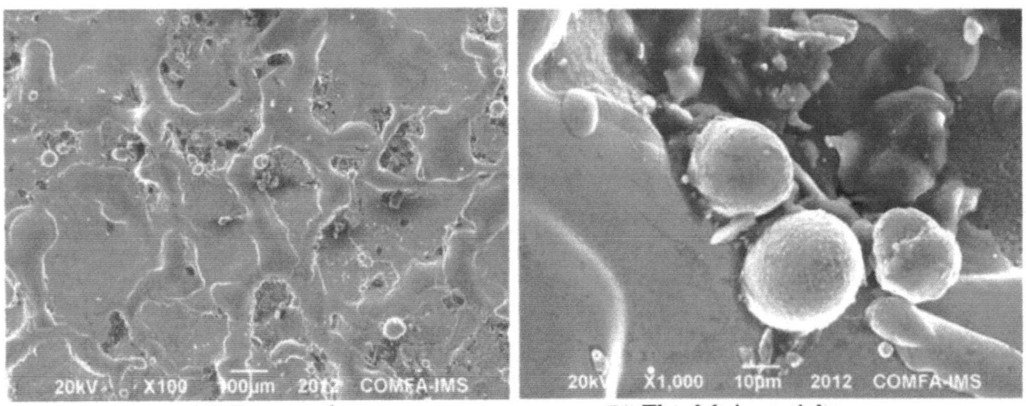

(a) Topography **(b)** The debris particles

Figure 20. Machined surface after standard EDM (a) Topography and (b) Debris particle [9]

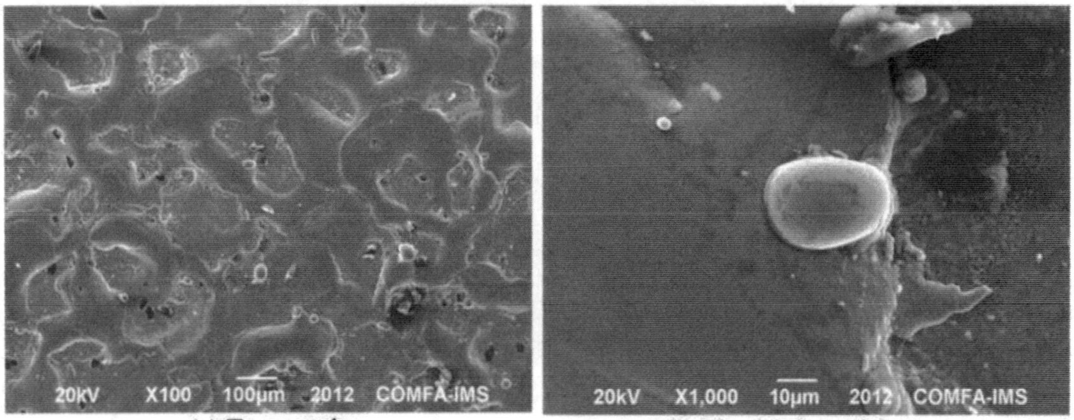

(a) Topography **(b)** The debris particles

Figure 21. Machined surface after vibration assisted EDM (a) Topography and (b) Debris particle [9]

4.5 Surface Topography and Integrity

EDM is a thermal-based machining process in which spark generates at a very high temperature of approximately 10000 degree Celsius. Which results in high thermal stress due to the higher thermal gradient when this thermal stress increases beyond critical point micro crack formation start in material[16]. Whereas in vibration-assisted EDM there is proper circulation of dielectric fluid, resulting in low thermal gradient; hence, crack formation is reduced considerably. Also in vibration-assisted EDM process, surface topography is relatively uniform that that of normal EDM process. In standard EDM, debris particles stick to the workpiece and electrode and create non-uniformity on the surface shown in Figure 20. While in vibration-assisted EDM there is a proper flow of debris particles and recirculation of motion because of to and fro movement of workpiece or electrode. This creates stability of the process and reduces short circuit due to which there is no hindrance in tool motion. Material removal takes place continuously with approximately equal volume removed (in crater shape with the same depth) in each spark. Also, very less debris remained on the surface shown in Figure 21. The surface's topography can be seen by using a finite element scanning electron microscope [34].

5. Conclusion

There is a detailed analysis of low-frequency vibration assisted EDM on SKD61 steel, various vibration setups, process, and response parameters in the current work. Following conclusions can be drawn;

- Low-frequency vibration-assisted EDM leads to a significant increase in machining efficiency.
- Discharge current, frequency, pulse off time, and pulse on-time decrease the order of influence on response parameters.
- Vibration assisted EDM results in improved machining stability conditions.
- The number of sparks increased which results in a higher amount of available energy per unit time

- In SKD61 steel, optimum results are obtained in the frequency range of 400 Hz to 512 Hz.
- Low-frequency vibration leads to better surface finish in comparison to normal EDM process.
- It is easier to assist low-frequency vibration to workpiece than to electrode.

So, assisting vibration into workpiece or electrode is a new technology with great potential and enormous future experimentation scope.

References

[1] Paswan K, Ranjan R, Hembram S, Bajaj R, Rai Dixit A, Mandol A and Kumar Das A 2016 Experimental Investigation on Machining Performance using Deionized Water as Dielectric in Die-Sinking EDM *Indian J. Sci. Technol.* **9** 2–6

[2] Paswan K, Pramanik A and Chattopadhyaya S 2020 Machining performance of Inconel 718 using graphene nanofluid in EDM *Mater. Manuf. Process.* **35** 33–42

[3] Paswan K, Chattopadhyaya S and Pramanik A 2018 Effect of composition and grain structure on machining performance in EDM-A review *IOP Conf. Ser. Mater. Sci. Eng.* **377**

[4] Pandey A and Singh S 2007 Current research trends in variants of Electrical Discharge Machining: A review *Int. J. Mach. Tools Manuf.* **47** 1214–28

[5] Unune D R and Mali H S 2016 Experimental Investigations on Low Frequency Workpiece Vibration in Micro Electro Discharge Drilling of Inconel 718 *6th Int. 27th All India Manuf. Technol. Des. Res. Conf.* 1413–7

[6] Kremer D, Lhiaubet C and Moisan A 1991 A Study of the Effect of Synchronizing Ultrasonic Vibrations with Pulses in EDM *CIRP Ann. - Manuf. Technol.* **40** 211–4

[7] Raju L and Hiremath S S 2016 A State-of-the-art Review on Micro Electro-discharge Machining *Procedia Technol.* **25** 1281–8

[8] Nguyen H P and Ngo N V 2019 Study on effects of low frequency vibration on efficiency of die sinking electrical discharge machining *Univers. J. Mech. Eng.* **7** 330–5

[9] Huu P N, Tien L B, Duc Q T, Van D P, Xuan C N, Van T N, Duc L N, Jamil M and Khan A M 2019 Multi-objective optimization of process parameter in EDM using low-frequency vibration of workpiece assigned for SKD61 *Sadhana - Acad. Proc. Eng. Sci.* **44**

[10] Paswan K, Pramanik A, Chattopadhyaya S and Basak A K 2019 A novel approach towards sustainable electrical discharge machining of metal matrix composites (MMCs) *Int. J. Adv. Manuf. Technol.* **106** 1–10

[11] Jaiswal A, Peshwani B, Shivakoti I and Bhattacharya A 2018 Multi response Optimization of Wire EDM Process Parameters *IOP Conf. Ser. Mater. Sci. Eng.* **377**

[12] Abu Qudeiri J E, Zaiout A, I. Mourad A-H, Abidi M H and Elkaseer A 2020 Principles and Characteristics of Di ff erent EDM Processes in Machining Tool and Die Steels *Appl. Sci.* **10** 1–46

[13] Chavoshi S Z and Luo X 2015 Hybrid micro-machining processes: A review *Precis. Eng.* **41** 1–23

[14] Mukhopadhyay P and Sarkar B R 2019 Advancement in ultrasonic vibration and magnetic field assisted micro-EDM process: An overview *Int. J. Adv. Res. Eng. Technol.* **10** 362–73

[15] Chern G L and Chuang Y 2006 Study on vibration-EDM and mass punching of micro-holes *J. Mater. Process. Technol.* **180** 151–60

[16] Xu M G, Zhang J H, Li Y, Zhang Q H and Ren S F 2009 Material removal mechanisms of cemented carbides machined by ultrasonic vibration assisted EDM in gas medium *J. Mater. Process. Technol.* **209** 1742–6

[17] Yu Z Y, Zhang Y, Li J, Luan J, Zhao F and Guo D 2009 High aspect ratio micro-hole drilling aided with ultrasonic vibration and planetary movement of electrode by micro-EDM *CIRP Ann. -*

Manuf. Technol. **58** 213–6
[18] Jahan M P, Wong Y S and Rahman M 2012 Evaluation of the effectiveness of low frequency workpiece vibration in deep-hole micro-EDM drilling of tungsten carbide *J. Manuf. Process.* **14** 343–59
[19] Iwai M, Ninomiya S and Suzuki K 2013 EDM properties of newly developed PCD made up of electrically conductive diamond particles *Procedia CIRP* **6** 140–5
[20] Ichikawa T and Natsu W 2013 Realization of micro-EDM under ultra-small discharge energy by applying ultrasonic vibration to machining fluid *Procedia CIRP* **6** 326–31
[21] Kurniawan R, Kumaran S T, Prabu V A, Zhen Y, Moon K, In Y, Islam M and Jo T 2017 Measurement of burr removal rate and analysis of machining parameters in ultrasonic assisted dry EDM (US-EDM) for deburring drilled holes in CFRP composite *Measurement* **110** 98–115
[22] Hsue A W J, Hab T J and Lin T M 2018 Pulse Efficiency and Gap Status of Rotary Ultrasonic Assisted Electrical Discharge Machining and EDM Milling *Procedia CIRP* **68** 783–8
[23] Li Z, Tang J and Bai J 2020 A novel micro-EDM method to improve microhole machining performances using ultrasonic circular vibration (UCV) electrode *Int. J. Mech. Sci.* **175** 105574
[24] Okada A, Uno Y, Nakazawa M and Yamauchi T 2010 Evaluations of spark distribution and wire vibration in wire EDM by high-speed observation *CIRP Ann. - Manuf. Technol.* **59** 231–4
[25] Paswan K, Pratap Singh S, Chattopadhyaya S and Pramanik A 2020 Experimental investigation on the effects of aqueous solution in electric discharge machining *Mater. Today Proc.*
[26] Tong H, Li Y and Wang Y 2008 Experimental research on vibration assisted EDM of micro-structures with non-circular cross-section *J. Mater. Process. Technol.* **208** 289–98
[27] Uhlmann E and Domingos D C 2013 Investigations on vibration-assisted EDM-machining of seal slots in high-temperature resistant materials for turbine components *Procedia CIRP* **6** 71–6
[28] Jiang Y, Zhao W, Xi X, Kang X and Gu L 2012 Vibration Assisted EDM of Small-hole Using Voice Coil Motor *Procedia CIRP* **1** 645–50
[29] Okada A, Yamaguchi A and Ota K 2017 Improvement of curved hole EDM drilling performance using suspended ball electrode by workpiece vibration *CIRP Ann. - Manuf. Technol.* **66** 189–92
[30] Hirao A, Gotoh H and Tani T 2018 Some Effects on EDM Characteristics by Assisted Ultrasonic Vibration of the Tool Electrode *Procedia CIRP* **68** 76–80
[31] Paswan K, Chattopadhyaya S and Pramanik A 2018 A Review Paper on Machining of Metal Matrix Composite and Optimizing Methods used in Electrical Discharge Machining *Mater. Today Proc.* **5** 24428–38
[32] Singh H and Singh A 2012 Effect of Pulse On/Pulse Off Time On Machining Of AISI D3 Die Steel Using Copper And Brass Electrode In EDM *Int. J. Eng. Sci.* **1** 19–22
[33] Endo T, Tsujimoto T and Mitsui K 2008 Study of vibration-assisted micro-EDM-The effect of vibration on machining time and stability of discharge *Precis. Eng.* **32** 269–77
[34] Salecha M, Paswan K and Chattopadhyaya S 2020 Effect of SiC micro and nanopowder in the electrical discharge machining process *Mater. Today Proc.*

IOP Conf. Series: Materials Science and Engineering 1104 (2021) 012011 doi:10.1088/1757-899X/1104/1/012011

Decision making on the machining parameters of Electrical Discharge Machined AISI D2 tool steel by AHP and PROMETHEE method

Rakesh Kumar Patel* and Mohan Kumar Pradhan
Department of Mechanical Engineering, Maulana Azad National Institute of Technology, Bhopal, India.
E-mail: rakesh1993patel@gmail.com

Abstract. This paper determines the optimum process parameters of the non-conventional electrical discharge machining (EDM). The performance of the EDM machine depends upon the process parameter used. In this analysis weight percentage, pulse current (Ip), discharge voltage (V) and pulse duration (Ton) used as process parameters. The optimized output parameters are MRR, TWR, surface roughness and flatness. By using face centered composite design nine trials were conducted on the work piece which is made up of AISI D2 tool steel. The trial results obtained were used in AHP and PROMETHEE multi attribute decision making method. The weights of the responses were determined by the AHP method and select the best alternative by AHP and PROMETHEE method. machining parameters and accuracy of the component produce by EDM can be control by using the obtained results

1. Introduction

Electrical discharge machine (EDM) is the well know advanced machining process for making component which has complex geometry and difficult to machine. Now a day's interest on the advanced materials is more because they are used in aerospace industry, complex mould and die making, and medical appliances. The component use in these industries requires properties such as light weight, high strength, better surface properties, high damping corrosion resistance and low thermal expansion. The biggest challenge for these industries is to produce high dimensional accuracy and better surface property which leads to high cost of machining of these components [1]. AISI D2 tool steel have such properties and it should be explored to survive in steel industries [2]. From the last few years EDM extensively used to make desired shape, size and high dimensional accuracy of the advanced material which is difficult to machine. EDM remarkably used in aerospace, automobile, medical and consumer electronics industries [3-4]. EDM process accuracy, productivity and versatility increases due to the technological advancement. In this research the key interest is to set the optimum input parameter which gives increased MRR and accuracy as well as reduction in the radial over cut, TWR, surface roughness and flatness.

To create several holes in AISI D2 tool steel, Prathipati et al. [5] used EDM. Peak current, pulse off time and pulse on time considered as input parameters. Cu used as a tool electrode. Taghuchi method L18 orthogonal array used to design the experiment and the experimental results are analyzed by statical approach. EDm performance MRR and TWR calculated.

Dhar et al. [6] used aluminum alloy Al–4Cu–6Si and mixed with 10% by weight silicon carbide (SiCp) particulate to make composites which is difficult to machine. And for machining of such hard material EDM is used. To analyse the process output such as Material removal rate (MRR), Tool wear rate (TWR) and Radial overcut (ROC) it uses three level full factorial design with Ip, V, and Ton as process input parameters. They found that MRR, TWR and ROC increases by increasing Ip, V and Ton. Pradhan et al. [7] used AISI D2 tool steel as workpiece material and copper as electrode. Two neuro-fuzzy and one neural network models are used for study of the MRR, TWR and radial overcut performance parameters. For planning the experiment uses full factorial design and Ip, V, duty cycle and Ton of various level as input process parameters. The most influential parameters for MRR and ROC were found to be Ip. And V is for TWR Chiang and Chang [8] used wire EDM and Ton, Toff, cutting radius, wire feed, and V as input process parameter and as output parameters MRR and surface roughness (Ra). They were used Grey relational analysis (GRA) for multi objective analysis to optimize the wire EDM process. Salonitis et al. [9] use Ip, V, and Ton as input parameters and MRR, surface roughness (Ra) as output parameters. They found that by increasing the Ip, V, and Ton results in increase of MRR and surface roughness (Ra) and by reducing the idling time MRR increased.

Moreover, the process recognized as good when the proposed model gives similar behaviour at vital parameters. For constructing the model, it is important to correlate the different aspects of the process. It will be better to predict the process by the model instead of doing trial directly with the process because it is expensive and time consuming. By the literature and working characteristics input parameter of the machine chosen. Pulse current (Ip), discharge voltage (V), pulse of time (Ton), and pulse of time are the input process parameters in this analysis (Toff). These were used as input parameters by some of the EDM researchers. Multi criteria decision-making (MCDM) approach used to pick the best process parameters Analytic Hierarchy Process (AHP) and Preference Rating Organization Method for Enrichment of Assessment (PROMETHEE) of AISI D2 tool steel.

The experimenters have made several attempts to enhance the performance of EDM, but the hybridization of the AHP and PROMETHEE process on AISI D2 tool steel EDM was never used for optimization. Thus, the aforementioned combination attempted in this study.

2. Description of the Experiment

Workpiece material as AISI D2 tool steel

The chemical elemental composition of the tool steel material follows

Table 1. Chemical composition of workpiece material

Elements	**C**	**Si**	**Mn**	**Mo**	**Cr**	**Ni**	**V**	**Co**	**Fe**
Wt. %	1.5	0.3	0.3	1.0	12.0	0.3	0.8	1.0	Balance

Cu as tool electrode and EDM oil used as dielectric. The machining time for each experiment set at 15 minutes.

Material removal rate (MRR)

The MRR is obtained as the loss of weight of workpiece per unit time and then modified to volumetric erosion per unit time

$MRR = \Delta V_w/T = \Delta W_w/ (\rho_w\ g\ T)$

Where, ΔV_w is the loss of volume from the work piece, ΔW_w is the loss of weight from the workpiece (it is calculated by taking the difference of mass of workpiece before and after machining), T is processing time and $\rho_w = 7{,}700$ kg/m^3 is the mass density of the workpiece.

Tool wear rate (TWR)

TWR is presented as the tool volumetric loss per unit time, presented as

$TWR = \Delta V_t /T = \Delta W_t / (\rho_t\ g\ T)$

Where ΔV_t the loss of volume from the electrode, ΔW_t is the loss of weight from the electrode (it is calculated by taking the difference of mass of tool before and after machining), T is the time of machining and ρ_t= 8,960 kg/m3 is the mass density of the copper electrode.

Figure 1. Copper tool and machined AISI D2 work piece

To see the effect of all the selected input parameter in the smaller number of experiment proper planning of experiment requires. Here uses Taguchi L9 Orthogonal array design of experiment [10] to plan the experiment.

Table 2. Input parameters and its level

Input process parameters	**Low Level**	**Medium**	**High**
Ip (A)	4	8	12
Ton (μs)	0.5	1	1.5
Toff (μs)	5	10	15
Voltage (V)	45	50	55

Table 3. Experimental results

Alternative. No.	**Ip(A)**	**Ton (μs)**	**Toff (μs)**	**Voltage (V)**	**MRR** (mm^3/min)	**TWR** (mm^3/min)	**Ra (μs)**	**Flatness (μm)**
1	4	0.5	5	45	1.1102	0.0273	4.7400	0.0370
2	4	1	10	50	2.4297	0.0663	5.2800	0.0220
3	4	1.5	15	55	2.8174	0.1047	4.4600	0.0400
4	8	0.5	10	55	1.4466	0.0220	4.1800	0.0440
5	8	1	15	45	2.0232	0.0756	3.6000	0.0230
6	8	1.5	5	50	3.0052	0.1563	2.6600	0.0470
7	12	0.5	15	50	1.4015	0.0368	2.4600	0.0110
8	12	1	5	55	2.3034	0.0571	2.2600	0.0480
9	12	1.5	10	45	2.8895	0.1763	3.0200	0.0260

3. Analysis Method

3.1 Analytic Hierarchy Process (AHP) Method: [11]
Step 1: First, take the decision table as we produce the experimental results in Table 3.

Step 2: Form a standardized table to create a useful and non-beneficial attribute on a common scale without disturbing the differences in the range of values. Making the values between 0 to 1.

Step 3: Make a comparison matrix using a scale of relative importance to find the contribution (i.e. weight) of each attribute. The attribute compared with itself the value will be assigned as 1. The numbers 2, 3, 4, 5, 6, 7, 8 and 9 correspond to the 'moderate importance', 'strong importance', 'very strong importance' and 'absolute importance' of verbal judgments.

Attributes	B_1	B_2	B_3	-	-	B_M
B_1	1	b_{12}	b_{13}	-	-	b_{1M}
B_2	b_{21}	1	b_{23}	-	-	b_{2M}
B_3	b_{31}	b_{32}	1	-	-	b_{3M}
-	-	-	-	-	-	-
B_M	b_{M1}	b_{M2}	b_{M3}	-	-	1

This square matrix denoted as A_1.

The geometric mean of each row is determined and all geometric means are calculated by;

$$GM_J = \sqrt[M]{[\prod_{J=1}^{M} b_{ij}]}$$

And $w_j = GM_J \div \sum_{J=1}^{M} GM_J$

To test the reliability of the judgement, perform these steps.

Make A_2 with the help of w_j values.

$$A_2 = \begin{matrix} w_1 \\ w_2 \\ \_ \\ w_j \end{matrix}$$ and find A_3.

$A_3 = A_1 * A_2$ and $A_4 = A_3 \div A_2$

Determine the maximum Eigen value λ_{max} that is the average of matrix A_4.

Calculate the Index of consistency CI= (λ_{max}-M)/ (M-1)

M is the order of matrix A_1.

Calculate the consistency ratio (CR) by using the table of Random index (RI) values given below

Attributes	3	4	5	6	7	8
RI	0.52	0.89	1.11	1.25	1.35	1.4

CR = CI / RI

If the CR value is less than one, the measured weights are accurate and significant, with a maximum of 10% error.

Step 4: The scores of each alternative are determined by using the measured weights.

The score of an alternative can be calculated by

$$P_{i=} \sum_{j=1}^{M} w_j m_{ij}$$

Where,

$$w_j = weight\ of\ each\ attributes$$

$$m_{ij} normalized\ value\ of\ alternative\ regarding\ to\ each$$

$$attribute$$

$$P_i\ overall\ or\ composite\ score\ of\ the\ alternative$$

The higher value of P_i is considered as the best alternative.

Step 5: give the ranking to the alternative in the decreasing order of score calculated.

3.2 *PROMETHEE Method:*

In this method alternatives are compared with alternatives during making the matrix for a particular attribute [11].

Step 1: The dominant alternative with other respect to other assign the value 1 and 0 for non-dominating alternatives with respect to other. The no. of matrix is equal to the no. of attributes

Step 2: make an overall matrix by using the weights assigned to each attribute.

Step 3: Calculate the sum of each row and each column in the overall matrix.

Step 4: The difference between the corresponding row and column gives the score for the alternatives.

Step 5: Arrange the score in the descending order and accordingly gives the ranking to the alternatives.

By using these two-method selecting best alternatives

4. Result and Discussion

Taguchi based designed experiments perform on the electrical discharge machine. During the workpiece and instrument measurement weight trial, before and after machining using the weight measuring system, and also time of machining of each experiment measured by using stop watch. Responses such as MRR, TWR, Ra and flatness are determine using above mentioned formula. From the 9 experiment it is difficult to find which value of input parameter giving optimum value of responses because responses are different natures i. e. MRR required maximum value and other three TWR, SR, and flatness required minimum value. So, to find the optimum value of responses at selected range of parameter used decision making method MCDM, AHP and PROMETHEE. By following the above-mentioned steps of AHP and PROMETHEE find the score of each alternative.

Score Calculated by AHP method ranking of the alternatives can be
6-9-3-2-8-5-4-1-7

Score Calculated by PROMETHEE method ranking of the alternatives can be
6-9-8-3-2-5-7-4-1

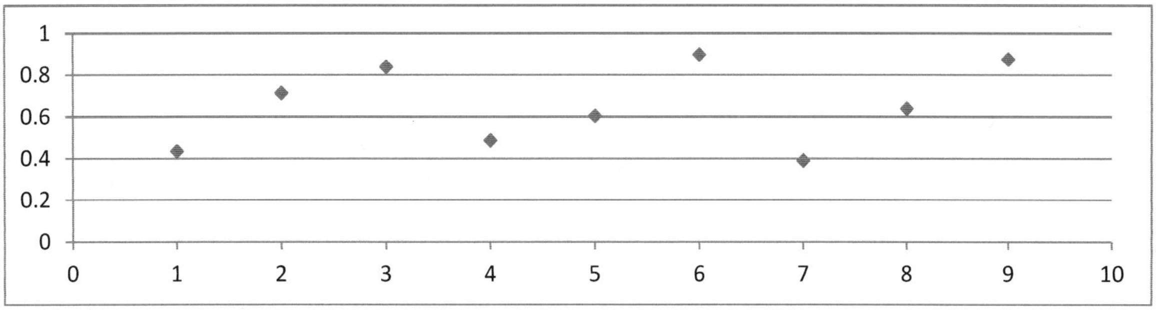

Figure 2. Score of AHP method

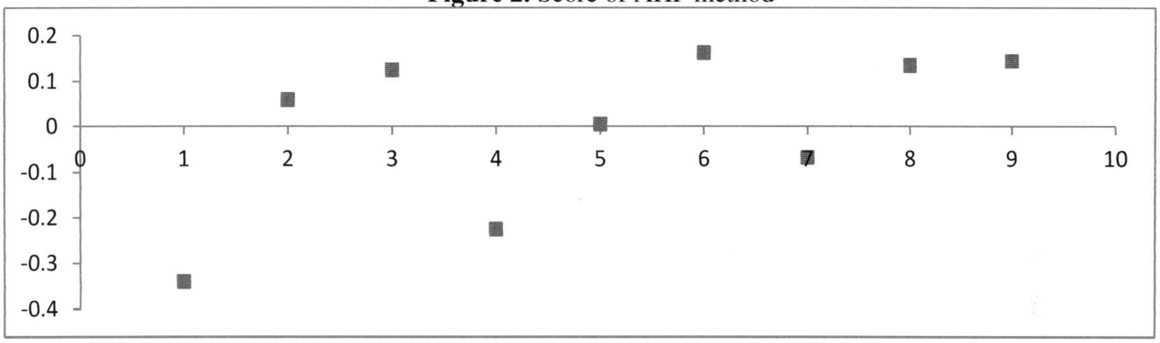

Figure 3. Score of PROMETHEE method

The process provides the same results for both AHP and PROMETHEE analysis i. e. In the selected set of input parameters, alternative 6 has the best EDM input parameter for AISI D2 tool steel. Alternative 6 has input parameter current (Ip) 8 A, Ton 1.5 µs, Toff 5µs and discharge voltage (V) 50 V and gives maximum MRR 3.0052 mm3/min in the selected range of parameter. In this input parameter output TWR, Ra and flatness got optimum value 0.1563 mm3/min, 2.66 µm and 0.0470 µm respectively.

5. Conclusion

This study determines the values of input process parameters on the EDM of AISI D2 tool steel to get the optimum value of MRR, TWR Ra and Flatness. The method used AHP and PROMETHEE give the same result. i. e. alternative number 6 as the best alternative. The optimal value of input parameter determines by the AHP and PROMETHEE methods are given as: current (Ip) 8 A, Ton 1.5 µs, Toff 5µs and discharge voltage (V) 50 V. These outcomes would help to reduce the cost of machining, error, and time needed for the process of machining and also increase the machine's surface characteristics and productivity. In other manufacturing industry to enhance the performance these methods can be used to find the optimum responses in multi response system.

References

[1] Kozak J. and Rajurkar, K.P. 2000 September Hybrid machining process evaluation and development. In *Proceedings of 2nd international conference on machining and measurements of sculptured surfaces Keynote Paper Krakow* 501-536

[2] Bin Abdul Rahim, M.A.S., bin Minhat, M., Hussein, N.I.S.B. and bin Salleh, M.S. 2018 A comprehensive review on cold work of AISI D2 tool steel. Metallurgical Research & Technology **115(1)** 104

[3] Ho, K.H. and Newman, S.T., 2003 State of the art electrical discharge machining (EDM). *International Journal of Machine Tools and Manufacture*, **43(13)** 1287-1300

[4] Kung, K.Y., Horng, J.T. and Chiang, K.T. 2009 Material removal rate and electrode wear ratio study on the powder mixed electrical discharge machining of cobalt-bonded tungsten carbide. *The International Journal of Advanced Manufacturing Technology* ***40*****(1-2)** 95-104

[5] Prathipati, R.P., Devuri, V., Cheepu, M., Gudimetla, K. and Kiran, R.U. 2018 Machining of AISI D2 tool steel with multiple hole electrodes by EDM process. In *IOP Conf. Ser. Mater. Sci. Eng* **330** 012067

[6] Dhar, S., Purohit, R., Saini, N., Sharma, A. and Kumar, G.H., 2007 Mathematical modeling of electric discharge machining of cast Al–4Cu–6Si alloy–10 wt.% SiCP composites. *Journal of materials processing technology* **194(1-3)** 24-29

[7] Pradhan, M.K. and Biswas, C.K., 2010 Neuro-fuzzy and neural network-based prediction of various responses in electrical discharge machining of AISI D2 steel. *The International Journal of Advanced Manufacturing Technology* **50(5-8)** 591-610

[8] Chiang, K.T. and Chang, F.P., 2006. Optimization of the WEDM process of particle-reinforced material with multiple performance characteristics using grey relational analysis. *Journal of Materials Processing Technology*, **180(1-3)** 96-101

[9] Salonitis, K., Stournaras, A., Stavropoulos, P. and Chryssolouris, G., 2009 Thermal modeling of the material removal rate and surface roughness for die-sinking EDM. *The International Journal of Advanced Manufacturing Technology* **40(3-4)** 316-323.

[10] C Montgomery, D., 1997 Montgomery Design and Analysis of Experiments.

[11] Patel, R.K. and Dwivedi, R.K., 2020 Determination of Critical Component Failure in Thermal Power Station by Using Multi-criteria Decision-Making Methods *Journal of Failure Analysis and Prevention* 1-5

POTENTIAL FOR ENERGY SAVING IN MUD WALL CONSTRUCTION

Dr. Subhash Mishra[1], Dr. H.K. Rai[2] and Mr. Pradeep Barnwal[3]

[1] **Associate Professor, Department of Mechanical Engineering, IPEC, Ghaziabad, (UP)**
[2] **Professor and Head, Department of Mechanical Engineering, IPEC, Ghaziabad**
[3]**Assistant Professor, Department of Mechanical Engineering, IPEC, Ghaziabad**

Abstract. Energy saving refers to method used to reduce energy consumption. By using proper insulation material and efficient utilisation of energy resources, energy saving can be achieved. A considerable amount of energy may be obtained by minimizing the loss of heat across the outer surface of building walls. The finding of this study is energy saving is maximum at optimal insulation thickness. The heat loss is mainly depending upon the thickness of walls and thermal conductivity of insulation for building construction. In building wall, the thickness of insulation material has important role in saving of energy. Due to lower thermal conductivity and higher thermal mass of mud wall, it is better to constructed mud house as compared to brick construction according to energy saving point of view. During the analysis of energy saving, dimension of demonstrative room is considered as 6 m x 5 m x 3 m. The result shows that 576 kWh of energy in year 2019 has been saved when slurry of mud dung insulation is considered for mud wall construction.

Key words- Energy saving, Critical insulation thickness, payback period, life cycle Method

1. INTRODUCTION

We all know that the development of any countries is basically depends on Energy consumption and its effective utilization. During this modern time, the energy consumption rate is gradually increased. But energy resources have limited value. So, minimizing the heat loss, effective utilization and energy saving is very important for any country development. Energy saving strategy is primary steps for maintaining the energy demands. Energy saving is also needed due to high cost of non-renewable energy sources, emission of greenhouse gases by burning of energy fuel and abruptly climatic changes. A number of methods are used to minimize the heat loss across the wall. Thermal insulation is one of the best methods among them. This study tells us that for decreasing the heat loss across the outer surface of wall, energy saving is necessary. In general, heat loss will occur through surface of walls i.e floor, roof, surrounding wall surface and ventilation. In this paper, we have considered the heat losses across surrounding surface.

According to the aim of paper, a literature review is written as follows. Binici et al.[1] had investigated mechanical properties of fibre reinforcement mud brick as wall materials. Tarik and Bolatturk[2] had determined the energy saving amount by using permeable tuff stone in outside walls of buildings in the zone of cold climate. Dombayci, O.A.[3] had investigated that when optimum thickness of insulation is used, then emission of greenhouse gases decreases by 41.53%. Kaynakli[4] investigated the energy requirement of building in Bursa for various architectural design. Ekici and Aksoy[5] had used artificial neural networks for determining the consumption of energy for building in Elazig region. Ozel. M.[6] calculated the heating load and cooling load yearly after the application of different thickness of insulation. Mahlia and Izbal[7] had determined the cost saving and emission reduction by the application of insulation at optimum thickness. Kaynakli[8] had investigated the effect of different parameter on energy saving, payback period and optimal insulation thickness. Kayfeci et al.[9] determined the optimum thickness of insulation for building construction. By using the optimization model, optimum insulation thicknesses, and energy saving for exterior walls of buildings are calculated for electricity tariff. Dhaka et al.[10] defining the condition of human comfort for naturally ventilated buildings material.

On the basis of literature review, the objective of this paper is to calculate the amount of energy saved annually when slurry of mud dung insulation is considered for mud wall construction.

2. STRUCTURE OF BUILDING AND ITS MATERIALS

Mud is suitable material for building construction due to some specific properties like high thermal mass, low thermal conductivity, thermal buffer and high thermal capacity. For economic point of view, Mud house is better than brick wall construction according to thermal comfort condition. Generally, mud house was constructer manually or less usage of mechanical mean. During the analysis of energy saving for mud house in rural area (India), dimension of demonstrative room is considered as 6 m x 5 m x 3 m. For construction of mud house, agriculture waste material like straw, special variety of grass, mud, cow dung, clay, timber and bamboo are used. Mud wall has some specific properties like compressive strength has high value but low tensile strength so useful life of mud house is less as compared to brick wall construction. Useful life of mud wall can be improved by using fibrous constitutes like straw with mud. A suitable thickness of mud plaster on inside side of mud brick is done to maintain the thermal comfort. Generally, in rural area of India, mud, straw and clay is selected as a building material for wall. Some special type of grass and khapra is used for roof construction. During the analysis of energy saving, thickness of wall is considered as 27 cm and 2.25 cm thickness of mud dung salary insulation on both side of walls. In this analysis, we have to consider two room sets (approx. 70 m^2 Area) for mud house construction. Fig. 1 shows the composition of different material used in mud house.

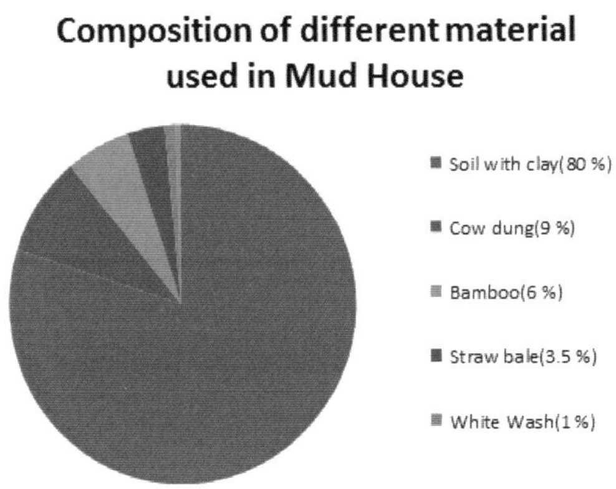

Fig.1. Composition of different material used in Mud house

2.1 METHOD OF OPTIMIZATION FOR INSULATION THICKNESS AND DETERMINATION OF ENERGY SAVING OVER THE USE FULL LIFE

Life cycle cost method is applied to determine the different cost like fuel cost, insulation cost and total cost. During analysis of total cost estimation, 8 years is taken as total useful life of Mud wall. The express of annually fuel consumption is written as;

$$M_{fc} = \frac{DDV \times 86400}{LCV \times \eta_{fs}\left(\frac{x}{K} + R_w\right)} \quad \text{.................. (1)}$$

Where DDV is the degree-day value of atmosphere, LCV is lower calorific value of fuel, η_{fs} is the efficiency of heating system. R_w, k and x are the thermal resistance, thermal conductivity and thickness of insulation of wall material respectively.

A number of parameters like thickness of wall(x), insulation cost (C_i), useful life of wall(N) and present worth factor (PWF) is required to calculate the total heating cost.
The expression for cost of heat in the case of mud house is written as,

$$C_t = C_i \times x + PWF \times C_A \quad \text{...................} (2)$$

For achieving the expression of optimum insulation thickness($X`_{opt}$), the first derivative of total cost(C_t) must will zero. Therefore, the expression of optimum thickness can be written as

$$X_{opt} = 291(K \times DDV \times PWF \times C_t / \eta_s \times C_i \times LCV)^{0.5} - (R_w \times K) \quad \text{...} (3)$$

The expression of Energy saving (E_s) can be written as

$$E_s = C_{tcf} - C_i \quad \text{............................} (4)$$

Where C_{tcf} is the total cost of fuel.

3. RESULTS AND DISCUSSION

In Ghaziabad (India) climatic region, during eight months cooling is required for comfort condition and remaining four-month cooling is required. When outer surface of building is covered with optimum insulation thickness of suitable insulation material, then some amount of energy (Electricity) has been saved. That amount of saving is closely related with energy saving. On outcome basis, it is concluded that the lowest and highest energy is required in the month of February and June. Mud wall construction has higher energy saving value per square meter as compared to brick wall construction.

After analysis of energy saving for mud house, energy saving in the terms of cooling load and heating load is shown in Table 1. The difference of heating load with and without insulation gives the heating energy saving. In this analysis, slurry of mud dung insulation is used over building envelope of mud house. For mud wall construction, amount of cooling load and heating saving are 416 Kwh and 160 Kwh respectively.

The heating will be required for four months (January, February, November and December) and cooling will be required for eight months (March to October) in the year 2019. The energy saving required per annum without insulation for mud wall is 1650 Kwh. The energy saving required per annum in presence of slurry of mud dung insulation for mud house is 1074Kwh. The energy saving has been analyzed which is 35 % by mud wall construction with slurry of mud dung insulation. Energy saving required in different month of year 2019 for mud house is represented in Table 2.

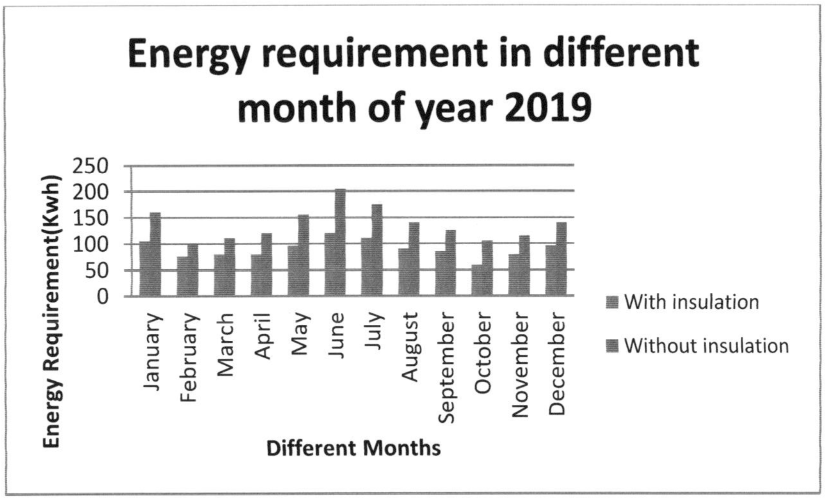

Fig. 2 Energy requirement in different month of year 2019 for mud house

Energy requirement in mud wall house for every month of year 2019 with and without insulation are shown in figure 2. Energy saving has been determined by the difference of energy cost with optimum insulation and energy cost without optimum insulation. It is visualized that in the month of June and February, the requirement of energy saving has highest and lowest value respectively. During the calculation of energy savings in the terms of Rupees, we have to considered as 5.5 Rs for per kW-hr. From result, the cost of minimum and maximum energy requirement is 550 Rs and 1128 Rs respectively without optimum insulation for February and June month. The cost of minimum and maximum energy consumptions is 413 Rs and 660 Rs respectively with optimum insulation for February and June month. From result, the cost of minimum and maximum energy requirement is 137 Rs and 468 Rs respectively for February and June month.

Table (1): Yearly heating/cooling Energy conservation potential of Mud house in Ghaziabad

S.No.	Parametres	kW-hr	Cost (Rs)
1.	Heating Load without insulation (HL)	515	2833
2.	Cooling Load without insulation (CL)	1135	6243
3.	Heating Load with insulation (HLW)	355	1953
4.	Cooling Load with insulation (CLW)	719	3955
5.	Heating Energy Saving = (HL-HLW)	160	680
6.	Cooling Energy Saving = (CL-CLW)	416	2288

Table (2): Energy saving in different month of year 2019 for mud house construction

S.No	Month	Energy Potential With insulation (kW-hr)	Cost of Energy With insulation (Rs)	Energy potential Without insulation (kW-hr)	Cost of Energy Without insulation (Rs)	Energy Saving(E_s) (Rs)	Energy Saving(E_s) (%)
01	January	105	578	160	880	302	10
02	February	75	413	100	550	137	4
03	March	80	440	110	605	165	5
04	April	80	440	120	660	220	7
05	May	95	523	155	853	330	10
06	June	120	660	205	1128	468	15
07	July	110	605	175	963	358	11

08	August	90	495	140	770	275	9
09	September	85	468	125	688	220	7
10	October	59	325	105	578	253	8
11	November	80	440	115	633	193	6
12	December	95	523	140	770	247	8

5. CONCLUSIONS

From results and discussion, some conclusions are drawn, which are following;

1. Mud act as an insulation due to higher thermal mass and low thermal conductivity.
2. After the application of insulation thickness, approximately 35 % energy has been saved for mud house construction.
3. Based on the analysis of energy saving, it is concluded that 576 Kwh/year energy has been saved In year 2019 by the application of insulation to the mud house.
4. At optimum insulation thickness with proper insulation material, the energy saving has maximum value.
5. The heat transfer rate through wall surface is depending upon insulation material so for selection of insulation material, the cost and Thermal conductivity must be considered.

References

1. Hanifi Binici, Orhan Aksogan and Tahir Shah , " Investigation of fibre reinforced mud brick as a building material", Journal of Construction and Building Materials, Vol.19, pp. 313–318, 2005
2. O.H. Tarik and A. Bolatturk, "The use of tuff stone cladding in buildings for energy conservation", Journal of Construction and Building Materials, Vol.20, PP.435-440, 2006.
3. O.A. Dombayci, "The Environmental impact of optimum insulation thickness for external walls of buildings", Journal of Building and Environment, Vol. 42, PP. 3855-3859, 2007.
4. O. Kaynakli, "A study on residential heating energy requirement and optimum insulation thickness", Journal of Renewable Energy, vol. 33, pp 1164-1172, 2008.
5. B.B.Ekici and U.T. Aksoy, "Prediction of building energy consumption by using artificial neural networks". Journal of Advances in Engineering, Vol. 40,PP. 356-362,2009.
6. M. Ozel," Thermal performance and optimum insulation thickness of building walls with different structure materials", Journal of Applied Thermal Engineering, Vol. 31 (2010),PP. 3854-3863, 2010.
7. T.M.I. Mahlia and A. Izbal, "Cost benefits analysis and emission reductions of optimum thickness and air gaps for selected insulation materials for building walls in Maldives", Journal of Energy, Vol. 35, PP. 2242-2250, 2010.
8. O. Kaynakli, "Parametric investigation of optimum thermal insulation thickness for external walls", Journal of Energies, Vol. 4, PP. 913-927, 2011.
9. M. Kayfeci, A. Kecebas and E.Gedik, "Determination of optimum insulation thickness of external walls with two different methods in cooling applications". Journal of Applied Thermal Energy, Vol.50,PP. 217-224,2013
10. S. Dhaka, J. Mathur , G. Brager, and A. Honnekeri, "Assessment of thermal environmental conditions and quantification of thermal adaptation in naturally ventilated buildings in composite climate of India", Journal of Building and Environment ,Vol. 86,pp. 17-28 , 2015.

A Review on 3D printed force sensors

H Choudhary[1*], D. Vaithiyanathan and H Kumar
[1]National Institute of Technology Delhi, Delhi - 110 040, India
[*]E-mail: harshvardhanchoudhary@nitdelhi.ac.in

Abstract:
A force sensor plays major role in robotics, wearable electronics and meteorological applications. Three dimensional (3D) printing is an exceptional technology and provides easy and economical fabrication process for the force sensors. Some salient features of 3D printing and development of force sensors in recent times have been represented in this paper. All possible forms of force sensors, for instance flexible, transparent, stretchable and highly sensitive can be developed using 3D printing technologies in an economical, quick and environmental-friendly approach. Over the years, fused deposition modeling (FDM) is gaining attention for the fabrication of force sensors. Also, due to advancements in materials and fabrication process, these sensors are being developed with enormous benefits. Some recent developments include flexible capacitive force sensor, wearable sensor and tactile sensor that are popular with 3D printing.
Keywords: 3D printing, force sensors, fused deposition modeling (FDM)

1. Introduction

Force sensor is an important measuring tool that converts the applied force into a corresponding output (electrical) signal. Depending on the required applications, there is a vast range of force sensors, ranging from few Newton's to mega Newton's. They have been used in various applications, including measurement of cutting forces in machining processes, thrust measurement, electronic weighing machine and transportation[1-2]. They include different measuring devices or indicators, including dial gauged, digital dial gauged, strained gauged, tuning fork etc for the realization of applied force. Among these, strain gauges have been prominently utilized [2-3].

Three dimensional (3D) printing (additive manufacturing) is utilized to fabricate geometrically and functionally complex objects and have various advantages, including time saving, cost effective and economical manufacturing process. Using this technology, force sensor has been utilized in various areas, like robotics, biomedical environment, and meteorological applications [3]. Also, it can be utilized to develop new, cost effective and custom-made force sensor. In recent times, it is possible to manufacture force sensor with small size and in an inexpensive manner in the field of robotics [4-6].

This paper represents an overview of various 3D printed force sensors; developed using fused deposition modelling (FDM) in recent times. The paper has been organised as per following particulars. In section 2, followed by section 1, a brief introduction of 3D printing has been described. A general discussion on 3D printed force sensors has been represented in section 3. Next, discussions and conclusions have been narrated in section 4 and 5 respectively.

2. 3D printing : a brief introduction

3D printing (also termed as additive manufacturing) is a resourceful procedure that performs manufacturing of physical items from 3D computer-aided design (CAD) model in an additive (usually, layer) approach. Numerous features can be addressed, including better utilization of materials, customized products on demand, development of multipart structure with easy fabrication and quick prototyping [7]. In every manufacturing method, CAD files can be exchanged among users from one end to other. Furthermore, significant developments in the 3D printing lead to develop customized electrical passive components with easy approach [8-9].

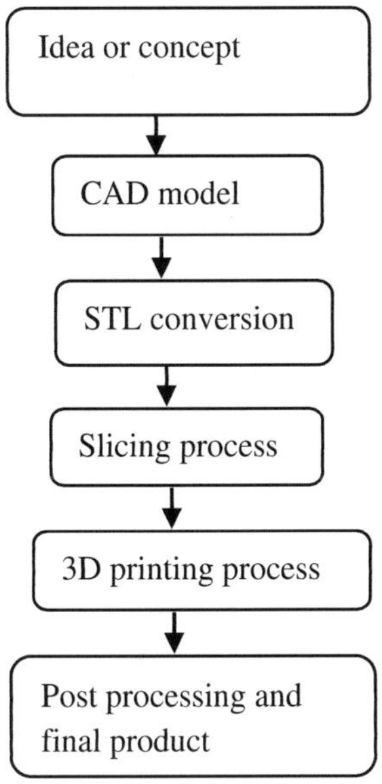

Figure 1: Various steps in a 3D printing technique [10-11]

Fused deposition modelling (FDM) and inkjet printing method pave the way to manufacture a number of sensors in different forms in many applications. Moreover, FDM is popularly used due to its features, like simple, high speed and cost-effective manufacturing process. Also, other techniques play essential role, including selective laser sintering (SLS), stereolithography (SLA), laminated object manufacturing (LOM) and direct energy deposition (DED). Figure 1 shows various steps in a 3D printing technique.

In the first place, the product concept has been transformed into a 3D model using any popular CAD software, like CATIA or AUTO CAD. A standard tessellation language (STL) file is commonly termed as standard in each 3D printing method. It converts the continuous geometry in the CAD file into small triangles [10-11]. It is sliced into thin cross-section layers using precise software. Then, the part is fabricated using one of the 3D printing technologies. With post-processing, the final product is prepared [12]. Table 1 lists numerous advantages and challenges of 3D printing.

Table 1: 3D printing at a glance [12-13]

S.No.	Advantages	Disadvantages
1.	Manufacturing of customized and complex products	Limited availability of materials for production.
2.	Material wastage is minimum; reduction in weight and size of final product.	Generally post processing is required to get final product.
3.	3D CAD model data can be shared.	Some intellectual property rights issues associated with products.
4.	Some cost effective and simple printing techniques for force sensors, for example FDM.	Material for support structure is wasted.
5.	Development of capacitive force sensor, flexible force sensor or tactile sensor, sensors for electronics industry.	Conductivity issue in electronic sensors
6.	It allows functional design and a great time saving in product development.	The multifunctional products are in development stage

7.	A good approach towards zero waste manufacturing.	Some material wastage cannot be recycled during post processing.
8.	Some applications: aerospace, robotics, architecture, chemical and biomedical	Dimensional inaccuracy, Limited selection of 3D-printable ceramics, Weak mechanical properties for polymers and composites.

3. Application of 3D printing: force sensors

Due to significant developments in 3D printing, it is possible to incorporate force sensing elements into an array of parts and objects. In a study [5], low cost force and tactile sensors have been designed and developed that can be incorporated into 3D-printed robot parts. Moreover, in the next study [6], a low-cost fingertip 3D-printed force sensor has been presented for robotic manipulation. In recent times, force sensors have been manufactured in a number of ways with attractive properties including sensitivity, flexiblity and stretchability. Table 2 lists development of different force sensors.

4. Discussions

3D printing technology pave the way to manufacture cost effective, reliable and durable force sensors. Moreover, 3D printing technologies, including FDM and microextrusion method have been utilized to manufacture flexible force sensors with an inclusion of sensitivity and stretchability. Also, they have been commonly utilized in various areas, such as stretchable electronics, robotics, biomedical, health care sector etc.

Table 2: Development of 3D printed force sensors [14-21]

S. No.	Force sensors	3D printing methods	Working principle and materials	Research findings
1.	Capacitive force sensor	1)FDM filament extrusion 2) Fibre encapsulation additive manufacturing (FEAM), thermoplastic elastomer additive manufacturing (TEAM). 3)Multifunctional and multi material printing	1) Parallel plate capacitor; 2) Sense the change between plates with the application of force 3) BendLay (ABS-based material), thermoplastic elastomer, copper wire	1) Formation of composite capacitor using FEAM and TEAM. 2) FEAM enables the production of integrated electrical circuit path such as fibre, metallic wire inside 3D printed parts. TEAM is used to produce soft thermoplastic elastomers for dielectric material. 3) Linear response, between change in capacitance and applied force. 4) Post processing has been done to avoid preloading of the sensor.
2.	Flexible force sensor	3D printing	1) Parallel plate capacitor filled with sensitive silicon rubber as dielectric layer 2) Poly lactic acid(PLA), silver conductive ink, silicone rubber Ecoflex 00-30	1) Silicone rubber is found to be main material for the fabrication. 2) It has attractive features, including good elasticity and performances, easy commercial availability and fabrication , versatile, large deformation, and integration into different sensors structures. 3) The sensor can measure force up to 4N 4) Linear behaviour is found between capacitance and the applied force
3.	Fully FDM flexible capacitive and resistive transducer	FDM	1) Capacitive sensing principle 2) Resistive temperature detector (RTD) principle 3) Thermo-Plastic polyurethane (TPU) and semi-rigid carbon based polylactic acid (PLAcb)	1) FDM allows fabrication of all components of transducer, including substrates, electrodes and active elements and it is called as full printable device. 2)FDM process affects the thermo-electro-mechanical properties, such as resistivity and dielectric constant due to process parameters, like printing temperature, speed and flow and the thickness of the layer 3)A non- linear behaviour between resistance and temperature was observed for resistive transducers.

4.	Flexible tactile sensor	3D printing mold and electro spinning	1) Piezoelectric System 2) Polydimethylsiloxane (PDMS),piezoelectric nanofiber mat and gold sputtering electrode.	1) The sensor system consists of PDMS substrates and a piezoelectric nanofiber mat, developed by casting process. 2) Finger tip and wrist form sensor with high sensitivity to micro scale deformation.
5.	Whisker Inspired Flexible tactile sensor	1) FDM 2)3d multi-material printing	1) Piezoelectric System 2) Conductive TPU 3) Ninjaflex	1) The sensor is used in contact imaging and 3D position measurement. 2) It is used to determine force acting on whisker and its direction.
6.	Cost effective Flexible capacitive force sensor	FDM	1) Parallel plate capacitor 2) Flexion X60 Ultra-Flexible Filament and Conductive thermoplastic poly-urethane (PI-ETPU)	1) FDM is used to print flexible capacitive force sensor using conductive TPU. 2) The change in capacitance due to applied force in measured using an LCR meter and microcontroller based read out circuit. 3) Read out circuit is cost effective and the sensor is used in soft robotic and prosthetic devices.
7.	Transparent, flexible, highly sensitive and biocompatible force sensor	Microextrusion 3D printing method	1) Piezoresistive Principle 2) Resistance change of hydrogel during a force is applied to the sensor. 3) Slow-gelling alginate hydrogel and PDMS	1) Highly sensitive and stable due to its sandwiched structure. 2) The zig-zag patterned hydrogel and PDMS layer provide stable water content and recording. 3) The sensor can be fabricated as wearable sensor for improvement in skin sensory capability of patients. 4) It can be used as wearable sensor for skin sensory recovery
8.	3D Printed Stretchable Tactile Sensor	A multimaterial, multiscale, and multifunctional 3D printing	1) Silver nanoparticle and modified silicone ink, Dragon Skin 10	1) The sensor has been designed and fabricated using the combination of nanocomposite ink optimization, 3D imaging, and multimaterial 3D printing. 2) It is capable of detecting and differentiating human movements, including radial pulse, finger pressing and bending. 3) Further work: optimization of the inks, including incorporation of semiconducting materials and devices 4) Development of other crucial devices such as, temperature sensors to monitor tissue

5. Conclusions

In the development of force sensors, 3D printing provides design freedom, simple and cost effective printing process, better utilization of materials etc. Furthermore, efforts have been made to develop a family of economical force sensors that are useful for 3D-printed robot objects and other industrial applications. Under stressful circumstances, these sensors perform well with a fair amount of reliability and robustness. In table 2, development of force sensors in various types using FDM and other techniques has been represented along with working concept, different materials or inks and research findings. It can also be shown that FDM has a great potential in the cost effective fabrication of stretchable force/tactile sensors, capacitive force sensors etc. 3D printed force sensors include not only traditional mechanical properties, but also embedded optical and electrical properties. Some of the challenges include conductivity issue in wearable sensors, material types and processing issues.

References

[1] Liu M, Zhao Y, Shao Y, Zhang Q and Liu C 2018 *Proc. IEEE Sensors* **2018** pp 1–4.
[2] Kumar R and Maji S 2016 *Eng. Solid Mech.* **4** pp 81–90.
[3] Xu Y *et al* **2017** *Sensors* **17** 1166.
[4] Kesner S B and Howe R D 2011 *IEEE/ASME Trans. Mechatronics* **16** pp 866–70.

[5] Wasserfall F, Hendrich N, Fiedler F and Zhang J 2017 *20th Int. Conf. on Climbing and Walking Robots and the Support Technologies for Mobile Machines, Porto, Portugal* 485-92.
[6] Xu Z Kolev S and Todorov E 2014 *Proc. IEEE Int. Conf. Robot. Autom.* pp 2749–56.
[7] Gao W *et al.* 2015 *CAD Comput. Aided Des.* **69** pp 65–89.
[8] Ford S and Despeisse M 2016 *J. Clean. Prod.* **137** pp 1573–87.
[9] Tan H W Tran T and Chua C K 2016 *Virtual Phys. Prototyp.* **11** pp 271–88.
[10] Wong K V and Hernandez A 2012 *ISRN Mech. Eng.* **2012** pp 1–10.
[11] Ngo T D Kashani A Imbalzano G, Nguyen K T Q and Hui D 2018 *Compos. Part B Eng.* **143** pp 172–196.
[12] Thompson M K *et al* 2016 *CIRP Ann. - Manuf. Technol.* **65** pp. 737–760.
[13] Tofail S A M Koumoulos E P Bandyopadhyay A, Bose S O'Donoghue L and Charitidis C 2018 *Mater. Today* **21** pp. 22–37.
[14] Saari M Xia B Cox B Krueger P S Cohen A L and Richer E 2016 *3D Print. Addit. Manuf.* **3** pp 137–141.
[15] Kisic M, Blaz N, Zivanov L and Damnjanovic M 2019 42nd international spring seminar on electronics technology (ISSE) Wroclaw, Poland 1–5.
[16] Aeby X, Van Dommelen R and Briand D *2019 20th Int. Conf. Solid-State Sensors, Actuators Microsystems Eurosensors XXXIII (Transducers Eurosensors XXXIII)* pp 2440–2443.
[17] Lee H B, Kim Y W Yoon J Lee N K and Park S H 2017 *Smart Mater. Struct.* **26** 045032.
[18] Eijking B, Sanders R and Krijnen G 2017 *Proc. IEEE Sensors* **2017** pp 1–3.
[19] Schouten M, Sanders R, and Krijnen G 2017 *Proc. IEEE Sensors* **2017** pp 1–3.
[20] Yang R *et al* 2018 *Micro Nano Lett.* **13** pp 1460–64.
[21] Guo S Z Qiu K Meng F Park S H and McAlpine M C 2017 *Adv. Mater.* **29** pp 1–8.

IOP Conf. Series: Materials Science and Engineering 1104 (2021) 012013 doi:10.1088/1757-899X/1104/1/012013

An investigation into microstructure and mechanical properties of maraging steel weldment

Deepak Kumar Gope[1], Priyanshu Kumar[1], Somnath Chattopadhyaya[1], Gowrishankar Wuriti[1,2], Tessy Thomas[2]

[1]Indian Institute of Technology (ISM), Dhanbad, Jharkhand 834001, India.

[2]Defence Research and Development Organization, Hyderabad, 500058, India

dkrg05@gmail.com

Abstract. Maraging steel is an alloy of Ni-Fe having very low carbon content. These steel is having superior properties like toughness and strength without losing its malleability. In the present work 5mm thick Maraging steel plates were welded using GTAW process with maraging steel filler wire. This paper attempts to analyse the microstructure and mechanical properties of the weldment. Post weld heat treatment (PWHT) was carried out at 480^{0}C. Strength of welded joint of as-weld and PWHT was tested using ultimate tensile machine (UTM). Both the tensile specimen (As-weld and PWHT) breaks at base metal zone. The quality of joint at micro-level was analysed using optical microscopy. PWHT sample have finer grain compare to weld sample. Different phases transformation was observed in microstructures. Microhardness was also observed at different location in and around the joint. It was also observed that the fusion zone and heat affect zone (HAZ) of PWHT has higher value of Microhardness compare to weld sample.

1. Introduction

Maraging steel is 18% Ni-Fe alloy. It is a low carbon martensitic steel with addition of cobalt, molybdenum, titanium and aluminium. It has superior properties such as toughness and strength. The strength of any steel is depend upon the carbon % ,but in maraging steel the strength is depend upon precipitation of intermetallic compound. These alloys are extensively used in aerospace and military application.
GTAW is one of the most widely used welding process for aerospace materials. It provide flexibility in welding design with higher weld quality. During GTAW process these steel get heated to very high temperature which changes the crystal structure and morphology at different zone of the welded plate. In order to retain mechanical and metallurgical properties PWHT have been carried out.
Post weld heat treatment of direct ageing (480^{0}C for 3 hours). Heating between 400 and 480^0 C, Ni_3Ti and Ni_3Mo precipitate is formed which is responsible for the hardness gained during initial stage. Aging at higher temperature e.g. 650^{0}C for 3 hours, promotes the dissolution of both the phases and formed Fe_2Mo stable precipitates and Ni in enrichments of matrix, which causes the formation of austenite phase. The fully austenite phase is obtained by subjecting the steel to the solution treatment in the temperature range 815^{0}C to 900^{0}C followed by quenching will lead to transformed into martensitic matrix. At 500^{0}C to 700^{0}C the reverted austenite phase formed which is located on the lath boundary of martensitic, it is mainly a micro-segregation and decomposition products which promote the fracture propagation and lead to the crack formation. Some of the researchers have shown that many of the welding processes are less receptive towards SCC. However, gas tungsten arc welding is extensively employed for this welding because of its high weld quality at comparatively low cost. In most cases, the tensile efficiency is more than 90%. However, the toughness of the fusion zone is an important concern. This is mainly due to micro segregation

of alloying elements. Formation of reverted austenite is particularly enhanced when alloying elements segregate to the cell and dendrite boundaries [1]. Kenyon [2] used a filler materials having reduced amount of titanium and molybdenum. It has been observed that there is significant enhancement of toughness and the associated formation of austenite pools were reduced. In GTA-welds, there is a corrosion susceptible region at the interface of austenite pool and surrounding martensite matrix [3]. Segregation of Ti and Mo, which otherwise results in the formation of compounds like Ti (CN) and Mo (FeC) resulting deterioration of SCC resistance in GTA-welds [3]. Coarse grain and grain size are also important factor which is more responsive to produce crack. Refinement of grain is very much necessary to reduce crack propagation. The process of PWHT has eliminated the segregation of alloying elements and avoided the formation of reverted austenite. The homogenized and aged welds exhibited higher corrosion resistance compared to the directly aged, solutionized and aged conditions attributes to absence of austenite/martensite interface. The Stress corrosion cracking resistance was found to be the highest for maraging steel welds after the post weld heat treatment condition of homogenization [4]. Sakai et.al compared the microstructure with different welding process and shows that the width of LBW has significant reduction in the width of HAZ and FZ as compared to TIG and PAW welding processes. This difference does not have significant affect in the mechanical properties of the maraging steel [5]. During the fracture, the rupture was generally initiated at the interface between the FZ and HAZ, spreading into the region of lower hardness of the steel. Anodic path dissolution and hydrogen embrittlement are the proposed mechanism for SCC in maraging steel [6]. Cracking mainly occurs in a plane and inclined towards pre-cracking [7]. It has been observed that the crack growth velocity is much faster in water than in oil [8]. Properties of maraging steel has been studied in different environmental conditions which shows the vulnerable effects of those towards SCC [9-12]. So the SCC of maraging steel may highly responsible for environment also, but it has not critically studied. V.Venkateswara et.al studied the influence of PWHT on microstructure and mechanical properties of GTAW of maraging steel weldments [13]. P.Venkata Ramana et.al studied the microstructure and residual stress distribution of similar and dissimilar electron beam welds-maraging steel to medium alloy medium carbon steel [14]. H. Hou studied the effect of austenitizing temperature on the mechanical properties of high-strength maraging steel [15]'.VIJAY et.al studied the precipitation Reaction and Strengthening Behavior in 18 Wt Pct Nickel Maraging Steels [16]. Shuo Yin et.al investigated the influence of aging temperature and ageing time on the mechanical and tribological properties of selective laser melted maraging 18Ni-300 steel [17].G.Madusudhan Reddy et.al studied the microstructure and corrosion behaviour of gas tungsten arc welds of maraging steel [18]. M.Vinoth kumar et.al found that the stress corrosion cracking behaviour of gas tungsten arc welded super austenitic stainless steel joints [19]. Deepak Kumar Gope et.al studied the Experimental investigation of plug cutter embedded TIG welding of Ti-6A1-4V titanium alloy [20].

From the literature review it can be observed that very few literature is available on maraging steel. This material has many exceptional properties but it is rarely investigated. Welding of maraging steel is a complex process and it require lot of investigation on the effect of process parameters and it PWHT.

This paper made an attempt to analyse the welding of maraging steel using GTAW process. It investigate the microstructure and mechanical properties of the welded joint. It also investigate the effect of post weld heat treatment on microstructure and mechanical properties of the welded joint.

2. Materials and Methods

Maraging steel is an alloy of Fe-Ni and having very low carbon content. It has superior properties of toughness and strength without losing its malleability. In the present study 18% Ni maraging steel has been welded using GTAW process which is maraging steel grade 250. The filler material used in this process is of maraging steel. The chemical composition of as received maraging steel plate and filler wire is shown in table 1. The chemical composition of both filler wire and plates are very similar.

Table 1: Chemical composition of base metal and filler wire

Element	Weight % (Plate)	Weight % (Filler Wire)
C	0.013	0.01
Si	0.033	0.042
Mn	0.004	0.005
P	0.003	0.004
S	0.001	0.001
Co	7.66	7.74
Ni	17.95	17.83
Mo	4.74	4.67
Ti	0.435	0.40
Al	0.085	0.092
Fe	Bal	Bal

The as received plates were cut in dimension of 100mm X 100mm X 5mm using abrasive cutter which is shown in fig 1.

Fig 1: Maraging Steel Plate samples.

Fixture Design

Fixture is an essential component of welding design. Fixture has direct effect on the quality of welding. If the fixture is not properly design the welding may have misalignment defect or have distortion in geometry due to heating of the materials. In GTA welding an inert gas is used to protect the weld pool from environmental contamination and oxidation due to high heat input. This inter gas is supplied through GTAW torch on the top surface of the weld pool. But the bottom side of the weld pool gets affected due to high heat input and environmental contamination. In order protect the back side of the welding plates, a back purge fixture have been design to provide an inert environment from the back side of the welding. The back purge fixture is designed in accordance with AWS standards. In this fixture a pipe of dia 2.5 mm is perforated with series of small pores of dia 1 mm is inserted in the base of the fixture. At the top of the fixture and at the gap position similar pores have been created to get the inert gas in proper direction. The back purge fixture design along with the fixture is shown in fig 2.
After cutting the plates in dimension, a groove was provided at one edge of each plate with bevel angle of 35^0. The root gap is taken as 1.5 mm and root face is 2 mm. The configuration in which the welding of the plates to be carried out is shown in fig.3.

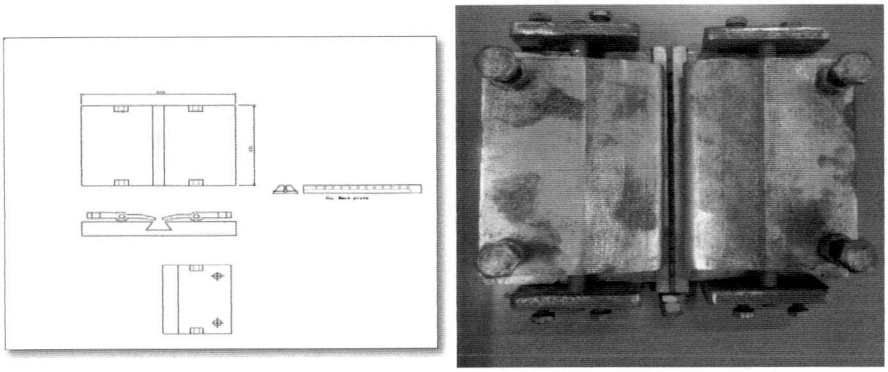

Fig 2: (a) Back-Purge Fixture Design; (b) Image of back purge fixture

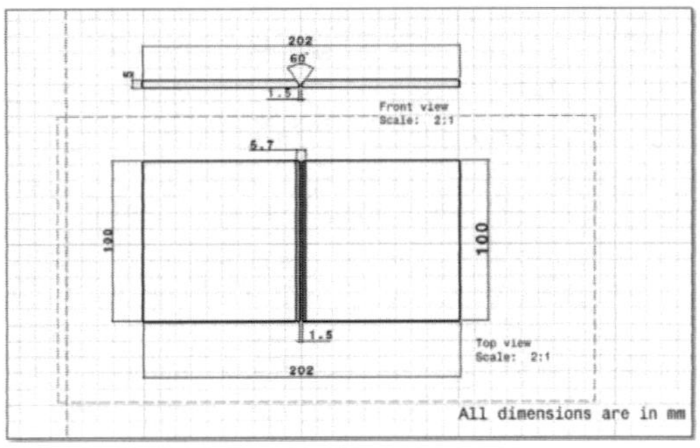

Fig 3: Drawing of welding specimen

Welding Process

Based on several trial on experiments the process parameters to carry out the welding is shown in table 2.

Table 2: Process parameters for Experiment

Current (A)	Speed (mm/s)	Voltage (V)	Gas Flow Rate (L/min)
100	1	12	15

The joining of the two maraging steel plates was carried out with the above shown parameters using 18Ni250 filler wire on Lincoln GTAW machine. The welded plate is shown in fig. 4.

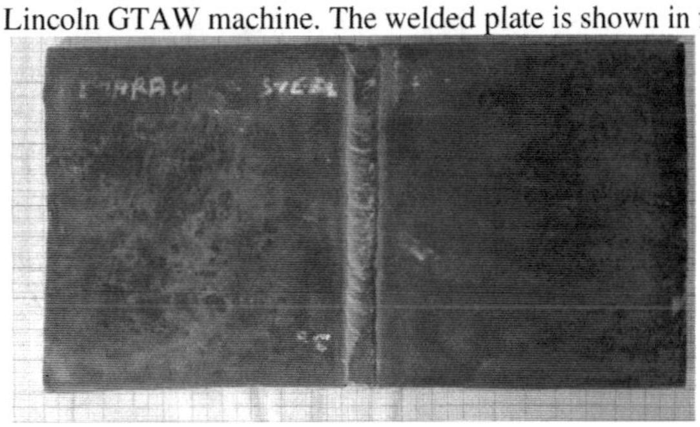

Fig. 4: The welded plates

Two set of welding was performed with same parameters. One sample is taken for PWHT and other sample is kept in as weld condition. PWHT was carried out in electric furnace with heating capacity of 1000^{0}C. PWHT of the sample was performed at 480^{0}C for 3 hours and then it is furnace cool for next 8 hours.

Testing and analysis

After welding the different mechanical and metallurgical samples have been cut out for their testing and analysis. For tensile test the samples have been cut as per ASTM E8 standard. Two tensile specimen have been cut out. One for PWHT and another is for as weld condition which is shown in fig 5. The tensile strength test was performed in universal testing machine.

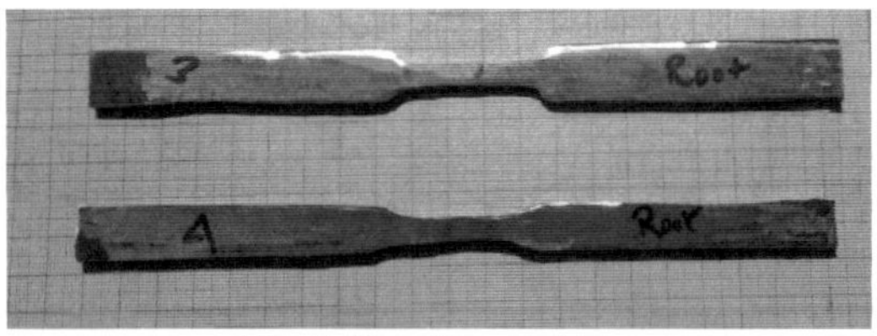

Fig. 5: Tensile specimen

For metallurgical analysis, sample in the dimension of 35 mm X 6mm X 5mm has been cut out from the welded plate. Then samples were polished from different grit size of emery paper. Starting from 200 grit size emery paper, the samples were gradually polished to 1500 grit size emery paper. After polishing from emery paper the samples were polished in alumina powder. Then at final stage diamond paste was used to give final polishing. For Microstructural analysis the polished samples were etched with chemical prepared by mixing 50 ml of HCl, 25 ml of HNO_3, 1gm of $CuCl_2$ and 150ml of H_2O. Microstructure was observed in metallurgical microscope at different magnifications.

For Microhardness testing the sample have been cut in the dimension of 35 mm X 6mm X 5mm. Microhardness analysis was performed in Vicker's Microhardness testing machine. The indentation was made with 1kgf load and dwell time of 10 sec. Micro-indentation was done at three zone which are as follow:

- Fusion zone (FZ),
- HAZ (heat affected zone),
- Base material

3. Result and Discussion

Microstructural Analysis

Microstructural analysis was carried out on two sample. One is for as weld condition and another is for PWHT condition. On each sample microstructure was observed on three different location i.e. base metal, fusion zone and HAZ. Microstructure of as weld samples in as weld condition is shown in fig. 6. It was observed that multidirectional columnar grain was distributed over the fusion zone. The grains are found to be coarse in this zone as compared to other zones. Fig. 6 (b) shows the HAZ region as well as the weld interface boundary. A clear boundary between weld and base metal can be observed. It also shows complete penetration and complete fusibility of the materials. The HAZ region also have coarse grain but it is comparatively less coarse than FZ. Microstructure of the base metal is shown in fig.6 (c).

IOP Conf. Series: Materials Science and Engineering 1104 (2021) 012014 doi:10.1088/1757-899X/1104/1/012014

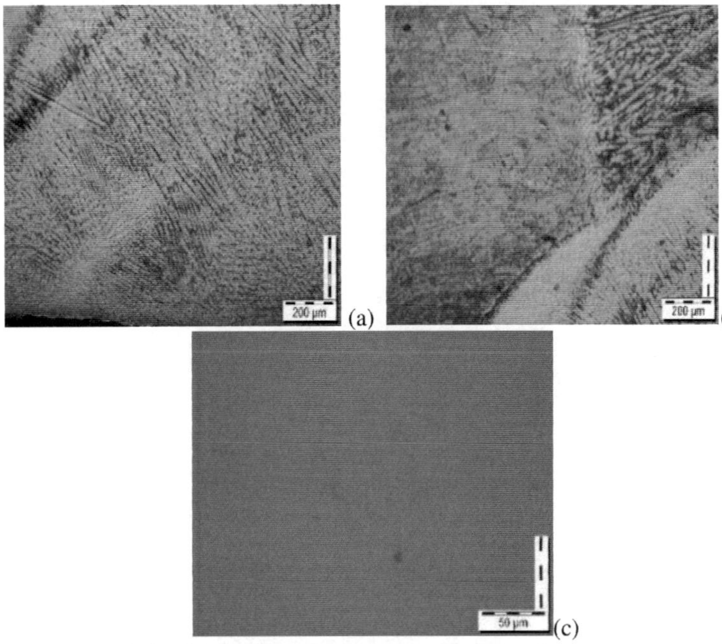

Fig. 6: Microstructure of As welded sample

Microstructure of PWHT sample is shown in fig.7. PWHT has significantly affect the microstructure in all regions. From fig. 7(a) it can be observed that the grains become finer and unidirectional. The PWHT condition gives martensitic phase with cellular structure. The multidirectional grain which was observed in fig.6 (a) give rise to dislocation defect which was overcome by PHWT as observed in fig.7 (a). The interface between weld metal and base metal is shown in fig.7 (b). It also shows HAZ. It also shows significant improvement in grain structure. Base metal region shows evenly distribution of austenite phase in fig. 7 (c).

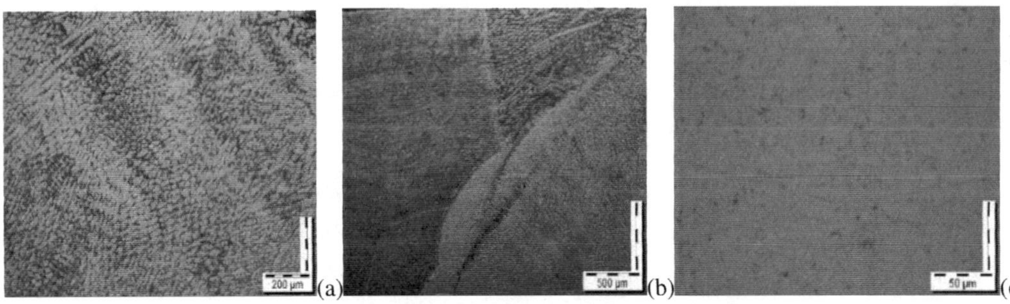

Fig. 7: Microstructure of PWHT samples

Microhardness Analysis

Microhardness values of the work piece has been taken on three different regions namely base material, fusion zone, HAZ of both the sample which is graphically shown in fig. 8. The hardness value of weld material varies in the range from 272 -458 HV. Hardness value of HAZ is varies from 295-492. Hardness value of parent metal is in range of 301-473. Grain size also affect the hardness of the material in the following way - fine grain size with heat treatment enhances the hardness value while coarse grain has lower hardness. The hardness value of HAZ is found to be greater than fusion zone in both as weld condition and PWHT condition. In HAZ the grain size decreases because of its distance from the fusion zone. Whereas at fusion zone the microstructure is coarser which were transform from austenite phase during cooling. In certain cases the formation of small amount of reverted austenite result in lower hardness value of HAZ compare to the base metal hardness. It can be observed that PWHT sample has significantly higher value compared to as weld sample. This is mainly due to precipitate formation during PWHT.

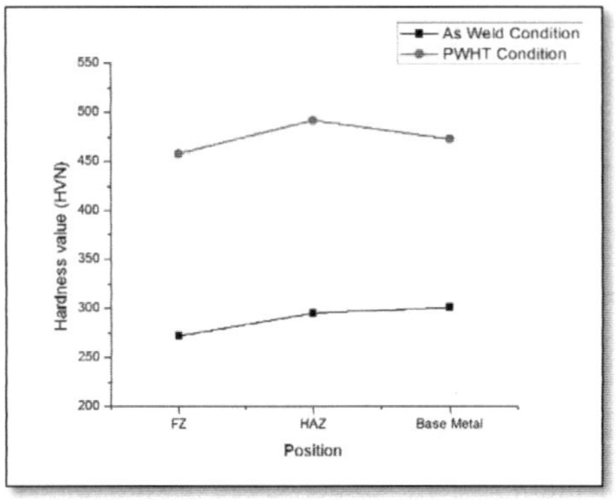

Fig.8: Graphical representation of hardness value

Tensile Strength

The ultimate tensile strength of any material is depend upon the grain structure inside that material. The reduction in the ultimate tensile strength was mainly due to coarser grain structure in the heat treated samples. The coarser grain structure is formed due to slower cooling rate which was mainly due to higher heat input. The tensile strength value obtained from universal testing machine for as weld condition and PWHT condition is shown in table 3 and broken samples are shown in fig. 9. Here CSA stands for Cross section area.

Table 3: Details of Tensile test

Sample	As Weld	PWHT
CSA (mm^2)	30.10	30.10
Elongation (%)	17	18
Location of fracture	HAZ	Base metal
Tensile strength (MPa)	1080.75	1245.25
Breaking Strength (MPa)	825.56	1130.85

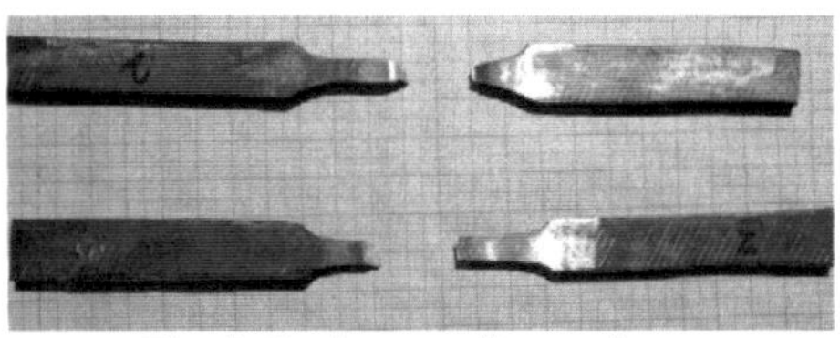

Fig. 9: Broken tensile specimen

The post-weld heat treated sample is having high tensile strength comparing to as weld condition. The improve in strength is mainly due to precipitation hardening effect and it was observe that at age-treated weld sample from 360^0C to 480^0C a precipitate is form which is Ti and Mo [1].

4. Conclusion

The experiment conducted in this paper presents the welding of maraging steel grade 250 with same filler wire. Simultaneously a critical comparison was made for PWHT condition and as weld condition. Following conclusion con be drawn from this study:

- During welding the FZ as well as HAZ undergoes complex changes due to high heat input which adversely effects its grain structure arrangement.

- PWHT at 480^{0}C for 3 hours shows significant improvement in its microstructure at FZ and HAZ. The grains become unidirectional.
- Hardness value at FZ (272 HVN) and HAZ (295 HVN) decrease as compared to base metal (301 HVN) in both condition. This is mainly due to coarsening of grain structure.
- The hardness value of PWHT samples in FZ (458 HVN), HAZ (492 HVN) and base metal (473 HVN) is significantly higher which is mainly due to precipitate formation during PWHT and finer grain structure.
- Tensile test shows that the strength value of PWHT sample have higher strength. This is mainly due to precipitate formation.

Acknowledgement
The author would like to acknowledge the financial support provided by RCI, DRDO, Hyderabad, (India) under grant no. DRDO (11)/2018-2019/604/MECH.ENGG.

References

1. Venkateswara Rao, V., Madhusudhan Reddy, G., & Sitarama Raju, A. V. (2010). Influence of post-weld heat treatments on microstructure and mechanical properties of gas tungsten arc maraging steel weldments. *Materials science and technology*, *26*(12), 1459-1468.
2. Kannan, M. B., Dietzel, W., Blawert, C., Riekehr, S., & Kocak, M. (2007). Stress corrosion cracking behavior of Nd: YAG laser butt welded AZ31 Mg sheet. *Materials science and engineering: A*, *444*(1-2), 220-226.
3. Fu, Z. H., Gou, G. Q., Zhu, Z. Y., Ma, (C. P. Yang, W. Q. Zhang, L. L, & Gao, W. 2018). Stress corrosion cracking behavior of SUS301L-MT
4. G. S. Wuriti, T. Thomas, and S. Chattopadhyaya, "Prediction of tensile failure load for maraging steel weldment by acoustic emission technique," *Lect. Notes Mech. Eng.*, vol. 1, pp. 431–442, 2019.
5. Sakai, P. R., Lima, M. S. F., Fanton, L., Gomes, C. V., Lombardo, S., Silva, D. F., & Abdalla, A. J. (2015). Comparison of Mechanical and Microstructural Characteristics in Maraging 300 Steel Welded by three different processes: LASER, PLASMA and TIG. *Procedia Engineering*, *114*, 291-297. .
6. Reddy, G. M., & Rao, K. S. (2015). Microstructure and corrosion behaviour of gas tungsten arc welds of maraging steel. *Defence Technology*, *11*(1), 48-55.
7. Makabe, C., & Socie, D. F. (2001). Crack growth mechanism in precracked torsional fatigue specimens. *Fatigue & Fracture of Engineering Materials & Structures*, *24*(9), 607-615.
8. Chaudhury, M. K., & Whitesides, G. M. (1992). How to make water run uphill. *Science*, *256*(5063), 1539-1541.
9. Meshram, S. D., Reddy, G. M., & Pandey, S. (2013). Friction stir welding of maraging steel (Grade-250). *Materials & Design*, *49*, 58-64
10. Ramana, P. V., Reddy, G. M., Mohandas, T., & Gupta, A. V. S. S. K. S. (2010). Microstructure and residual stress distribution of similar and dissimilar electron beam welds–Maraging steel to medium alloy medium carbon steel. *Materials & Design*, *31*(2), 749-760.
11. Li, K., Shan, J., Wang, C., & Tian, Z. (2017). The role of copper in microstructures and mechanical properties of laser-welded Fe-19Ni-3Mo-1.5 Ti maraging steel joint. *Materials Science and Engineering: A*, *681*, 41-49.
12. Meshram, S. D., Paradkar, A. G., Reddy, G. M., & Pandey, S. (2017). Friction stir welding: An alternative to fusion welding for better stress corrosion cracking resistance of maraging steel. *Journal of manufacturing processes*, *25*, 94-103.
13. Venkateswara Rao, V., Madhusudhan Reddy, G., & Sitarama Raju, A. V. (2010). Influence of post-weld heat treatments on microstructure and mechanical properties of gas tungsten arc maraging steel weldments. *Materials science and technology*, *26*(12), 1459-1468.
14. Ramana, P. V., Reddy, G. M., Mohandas, T., & Gupta, A. V. S. S. K. S. (2010). Microstructure and residual stress distribution of similar and dissimilar electron beam welds–Maraging steel to medium alloy medium carbon steel. *Materials & Design*, *31*(2), 749-760.
15. Hou, H., Qi, L., & Zhao, Y. H. (2013). Effect of austenitizing temperature on the mechanical properties of high-strength maraging steel. *Materials Science* and Engineering: A, 587, 209-212.

16. Vasudevan, V. K., Kim, S. J., & Wayman, C. M. (1990). Precipitation reactions and strengthening behavior in 18 Wt Pct nickel maraging steels. *Metallurgical Transactions A*, *21*(10), 2655-2668.
17. Yin, S., Chen, C., Yan, X., Feng, X., Jenkins, R., O'Reilly, P., ... & Lupoi, R. (2018). The influence of aging temperature and aging time on the mechanical and tribological properties of selective laser melted maraging 18Ni-300 steel. *Additive Manufacturing*, *22*, 592-600.
18. Reddy, G. M., & Rao, K. S. (2015). Microstructure and corrosion behaviour of gas tungsten arc welds of maraging steel. *Defence Technology*, *11*(1), 48-55.
19. Kumar, M. V., Balasubramanian, V., Rajakumar, S., & Albert, S. K. (2015). Stress corrosion cracking behaviour of gas tungsten arc welded super austenitic stainless steel joints. *Defence Technology*, *11*(3), 282-291.
20. Gope, D. K., Kumar, U., Chattopadhyaya, S., & Mandal, S. (2018). Experimental investigation of pug cutter embedded TIG welding of Ti-6Al-4V titanium alloy. *Journal of Mechanical Science and Technology*, *32*(6), 2715-2721.
21. Srivastava, M., Hloch, S., Tripathi, R., Kozak, D., Chattopadhyaya, S., Dixit, A. R.& Adamcik, P. (2018). Ultrasonically generated pulsed water jet peening of austenitic stainless-steel surfaces. *Journal of Manufacturing Processes*, *32*, 455-468.

A review on laser beam machining process modeling and optimization

V B Magdum [1] and J K Kittur [2]

[1] Research scholar, Department of Mechanical Engineering, KLS Gogte Institute of Technology, Belagavi, Karnataka, India, 590008, vbmagdum@rediffmail.com
[2] Professor, Department of Mechanical Engineering, KLS Gogte Institute of Technoly, Belagavi, Karnataka, India, 590008, jkk@git.edu

Abstract. Machining of advanced materials by conventional methods consequences in increased high cutting temperature, high cutting force magnitudes, tool wear, lesser tool life and meagre machined surface. Machining of these materials by conventional methods are originate to be uneconomic. Researches are attracted towards laser beam machining (LBM) between the various nonconventional machining methods. This paper focuses the probable usage of LBM for various materials machining, current progress, benefits and challenges in machining, process parameters and performance characteristics, modeling and optimization. There is need to agreeably designed process parameters which is appropriate to LBM. It is resolute that investigational based modeling and optimization methods are essential to generate model that gives very good fitting with experiments, while determining the effects of several process parameters.

1. Introduction

Advanced engineering materials having the intrinsic physical, mechanical properties. However, machining is related with converting these materials into products. The machinability characteristics pretense various challenges to researcher such as greater cutting force, severe design requirements, complex shape, infrequent size, lesser tool life, upper cutting temperature and meager surface integrity etc. and hence measured as problematic to machining and inefficient. LBM is used for machining nearly entire variety of materials [1-4]. Lasers are extensively used in manufacturing sector. Laser grounded processes are discriminated into various types, including machining, forming, sintering, coating and welding etc. [3]. The figure of LBM is shown in Fig. 1 [5]. In 1960, Theodore H. Maiman built first laser. In LBM, concentrate the laser beam on the workpiece results in melting of material. The melted material is removed by pressurized gas jet. The material removed depends upon the fundamental conditions [4]. Various LBM process parameters affects the performance characteristics are laser power, pulse frequency, pulse width, cutting speed, current, stand off distance, operation mode, wavelength, workpiece material, thickness etc. Thus, selecting the optimal process parameters that would produce the best machining output is the main challenge [2]. The important performance characteristics are surface roughness (R_a), kerf taper (K_t), kerf width (K_w), kerf deviation (K_d), material removal rate (MRR), heat affected zone (HAZ) etc. Kerf taper, kerf width and kerf deviation are three main physiognomies of laser machined kerf and used for decide the kerf geometry [5-8]. The figure of machined kerf is shown in Fig. 2 [9].

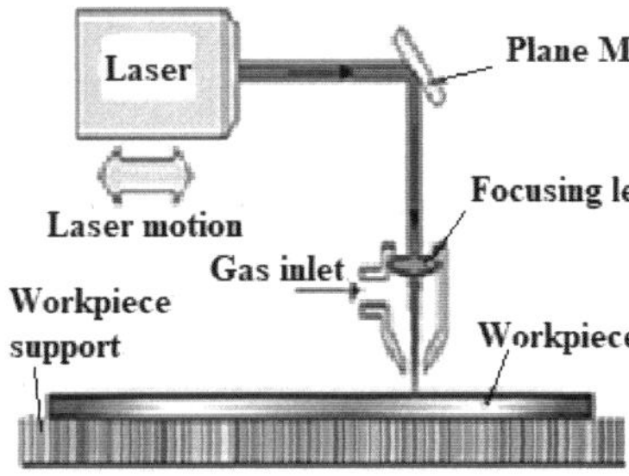

Fig. 1. Laser beam machining

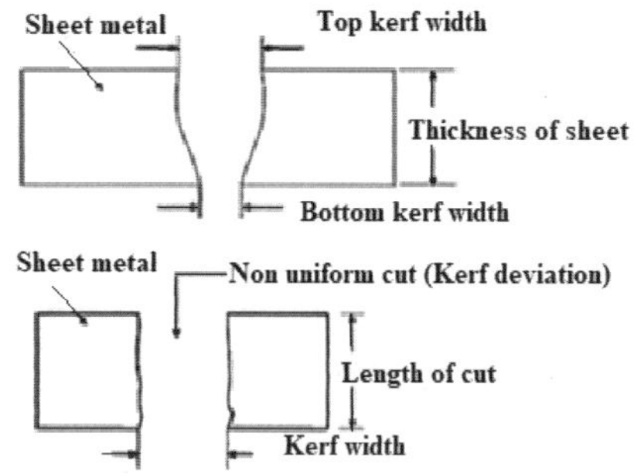

Fig. 2. Machined kerf

2. Material

This segment presents notable studies that have been directed in the field of LBM on material. The important materials which are frequently used for LBM are discussed below:

Ferrous alloys materials are iron based material classified under stainless steel, low carbon ductile steels and hardened steel. These alloys found its application in defense, automotive, chemical, construction, aircraft and medical industry [1,10-12]. Aluminum and its alloys are possessing important properties as high strength to weight ratio, light weight and better electrical, thermal properties. The Nd:YAG laser useful for the machining of these material due to their property of short wavelength, well focusing characteristics and power [10,13]. The outstanding properties like high strength to weight ratio, nontoxicity, strong erosion and corrosion resistance and capability to sustain high strength at high temperature makes titanium and its alloys widely held in automotive, aerospace, nuclear, biomedical and gas turbines industries [1,10,14]. Inconel, waspaloys, hastelloys and udimet are the nickel based alloys. These alloys are another eye catching material over titanium alloys. The abrasives TiC, CrC, MoC make the machining of nickel alloys tough. This effects in shorter tool life, low cutting speed, meager surface quality and results in increase in machining cost [1,15-18]. Mullite, zirconia, alumina and silicon nitride are frequently used ceramic materials in manufacturing industry. Due to their excess brittleness and hardness, low thermal conductivity and diffusivity; machining of ceramics is become difficult. The expeditious improvements in LBM makes it most convenient processing tools [1,10]. Composites are generally formed by mixing or bonding of particles, whiskers in a matrix and fiber. Mixture of fiber or reinforce particle improves the properties like stiffness, thermal properties, abrasive, adhesive, diffusion wear resistance, hardness. High tool wear and successive destruction in the work piece occurred during machining [1,10,19]. Laser beam striking on workpiece results in thermal interface. Very high temperature of laser beam altered physical, chemical and mechanical properties of workpiece. HAZ is form at the surface of cutting edge [10,20-21]. Modeling of HAZ is very complex because of extremely local heat energy to the workpiece and various machining parameter [22-23]. Mathematical, numerical, analytical and soft computing methodologies used for predictions of HAZ [24-27].

3. Modeling and optimization

Manufacturers are interested in LBM due to optimization of the process parameters. Contribution of several process parameters makes LBM highly nonlinear and complex process. Well designed model developed for different performance characteristics [28-31]. The main parameters affecting LBM are beam characteristics, laser characteristics, machining characteristics, quality characteristics, geometry characteristics and metallurgical characteristics [32-35]. Various Modeling and optimization tools are classified as: a. Hard computing methods b. Soft computing methods

3.1. Hard computing methods

Traditional numerical models are considered as hard computing methods. Various hard computing models used for LBM are given below:

Dr. Genichi Taguchi developed Taguchi method (TM). An orthogonal array (OA) was employed by Taguchi's design of experiment (DOE). It is the shortest possible experimental matrix of process parameters combinations. Significance of process parameters on performance characteristics studied through main effect plots by analysis of variance (ANOVA) method [36-41]. The methodology used is select process parameters, recognize the performance characteristics, select an OA, design the experimentation matrix, conduct experiments and record the responses and study the results using S/N ratio and ANOVA [42-45]. Deng in 1982 introduced a Grey relational analysis (GRA) for meagre, inadequate and uncertain system. The experimental data first of all normalized between gray relational grade 0 and 1. Later data preprocessing gray relational coefficient (GRC) are calculated. Grey relation grade (GRG) is the summation of weighted relational coefficients [45-48]. RSM is a collecting of arithmetical and mathematical models applied for reviewing effect of multiple process parameters on performance characteristics and optimization [49-55]. A regression model develops to create relationship among process parameter and performance characteristics. Implication of the developed model studied by predicting a coefficient of determination (R^2). R^2 value ranges between 0 to 1. For higher order interaction 2nd order polynomial is appropriate and it gives better result [26,30,56]. Principal component analysis (PCA) is used orthogonal transformations for bring out patterns in experimental data to mention their similarities and differences. Quality losses are calculated from principal components [57]. Desirability function analysis (DFA) used for converting the multiple response characteristics into single response characteristic named ‘composite desirability’ [58]. The desirability objective function diverges in between ‘0’ and ‘1’ [59].

Hybrid optimization methods used for selecting weighting factor owing to possible relationships among the multi performance characteristics optimum [57]. In TM and PCA, TM used for single objective optimization and S/N ratios got from TM applied in PCA designed for multi objective optimization (MOO). The outcomes of MOO comprise the forecast of optimized process parameter level as well as their proportional inference on multi performance characteristics [57]. TM coupled with DFA is used for optimization of process parameter. The optimum levels of process parameters are recognized on basis of parameters composite desirability value and effect of each process parameter is determined by ANOVA. This hybrid method of TM and DFA is an effective means for optimizing multi response characteristics [60-62]. In RSM based DFA, RSM used for employing to develop experimental models and for optimization of process. When various performance characteristics evaluated, the optimum parameters grasped separately for individual process parameter did not match in entirely cases. In such cases Integrated RSM based DFA used for optimization of process parameter [63-64]. In TM based RSM (TRSM), TM using S/N ratio and orthogonal array applied for finding the effect of process parameters and ANOVA prepared to find the effect of process parameter. The models generated using joint method precisely predicting the responses in machining. Application of TRSM was initiate to be time saving and cost effective [65-66]. A collective method of grey based RSM was discovered for predicting the optimum combination of process parameters. The shortcoming of regression analysis compensates by the grey theory analysis. A quality characteristic of all the responses is calculated as grey relational grade on basis of the grey theory. This was further modelled using RSM for optimization of performance characteristics [67].

3.2. *Soft computing methods*

A huge volume of experimental data required for modelling and optimization using the traditional statistical tools which is time consuming and nonefficient. These problems are overcome by soft computing methods. These methods provide modeling and optimization with the restricted data [68-71].

Genetic algorithm (GA) improves and optimize performance characteristics of LBM by setting process parameters at optimum level [72-75]. Five phases of GA are initial population, fitness function, selection, crossover and mutation. A fittest individual is select by natural selection [73]. Artificial neural network (ANN) was enthused by the comportment of biological neural networks. Neurons are weighted and interconnected in allotted layers. A multifaceted relationship among process parameters and performance characteristics shown in network can be estimated. The ANN modeling can give good results from incomplete and noisy data, and produce nonlinear and complex relationships [76-78].

A hybrid approach of ANN and GA can accelerate the learning. ANN require huge data and GA requires less data. Hyper parameter are the values required to ANN for perform properly in given problem. In such cases GA use to learn best hyper parameters to ANN. The predicted model is found well in agreement with the experimental data [79-80]. Fuzzy logic (FL) model is developed on the basis of fuzzy set theory. Process parameter denoted as linguistic terms with the grade in between 0 to 1. FL modeling have distinct advantages like mathematical concepts simple, easy to construct and understand, solution to complex problems. FL may not give accurate results but it gives acceptable results [81]. In FL modeling and DFA, Multi response predictive model are established using FL model and MOO of various process parameters are carried out with DFA [81]. Grey Fuzzy method is the combination of GRA and fuzzy logic model. Grey relational coefficients are assigned for individual performance characteristics are deliberate from the experimentally detected data. These GRC are used as input process parameter in the FL modeling to determine the grey relation grades based on fuzzy. Also, GRC are used to determine mutual performance index for various performance characteristics [82].
Many researches attracted due to advantages of LBM over conventional machining for enhancement of viability and machinability benefits difficult to cut material. Modeling and optimization methods are developed to study the significance of process parameters on performance characteristics and their shared interactions effect to forecast the optimal LBM parameter setting.

4. Another section of your paper

The major process parameter and their effect on performance characteristics is discussed below:

4.1. Surface Roughness

It is considered as significant performance characteristic for LBM specimens. It is defined by using of surface finish, crack propagation and striation formation. Surface roughness was measures by using surface measuring devices like surface tester, Talysurf etc. It is measured in 'microns' [53]. Influence of laser power, gas pressure, pulse frequency, pulse width, cutting speed are significant while other parameters like nozzle distance and diameter are less effected. Surface roughness is more influenced by laser power, gas pressure and cutting speed. The percentage influence of cutting speed and gas pressure on surface roughness is 56.62% and 29.38% respectively [36]. Gas pressure and speed have influence of 60.8% and 33.5% respectively on roughness value [38]. Gas pressure is the utmost important factor for surface roughness during LBM [44]. It is found that laser power is more significant as compared to gas pressure and cutting speed on surface roughness [64].

4.2. Kerf width

It specifies the degree of accuracy [72]. It is measured by using microscope [53]. Kerf width is of two types: a) Bottom kerf width b) Top kerf width. It is increases by increasing laser power, gas pressure, pulse frequency and current intensity. It is decreases with decrease in cutting speed. Laser power, gas pressure and cutting speed having percentage influence 88.74%, 7.3% and 3.15% respectively on kerf width [36]. Higher kerf widths are generated due to increase in laser energy. With low pulse energy and pulse frequency the machining process is more reliable and results in small kerf width [72].

4.3. Kerf taper

It is calculated by using eq. (1)

$$\text{Kerf taper} = ((\text{Top kerf width} - \text{Bottom kerf width}) \times 180)) / (2\pi \times \text{Workpiece thickness}) \quad (1)$$

The kerf taper decreases by increase in laser power and increases with increase in the piercing time [13]. The result show that small values of pulse width and gas pressure combined with middling to higher values of cutting speed and pulse frequency gives optimum outcomes for kerf taper [73]. It is significantly reducing with increase of pulse width having higher workpiece thickness [76].

4.4. Kerf deviation

It is measured with the help of microscope [53]. It is calculated by using eq. (2)

Kerf deviation= Maximum top kerf width - Minimum top kerf width (2)

As cutting speed rises from 1000 to 3000 mm/min, interaction time among workpiece and laser beam increases results in decreased kerf deviation. But as laser power increases, high energy penetrated into cutting zone and SiC particles presents distributed energy unevenly results in increment of the kerf deviation [53].

4.5. Heat affected zone

Increase in laser power lead to decrease in HAZ due to fast temperature falling off caused by higher thermal conductivity of aluminum sheets [7]. The width and depth of the HAZ enlarged by increase in the laser power and reduced by increase in the laser scan speed and the laser spot size [24]. Cutting speed and laser power are important for HAZ [26]. Amount of HAZ increases by increase in pulse frequency and pulse width but it decreases with workpiece thickness and power [76].

4.6. Material removal rate

MRR decides the rate of production and financial side of machining [72]. Pulse frequency has substantial effects on MRR afterward peak power and pulse width. MRR also increases by increase in workpiece thickness [47]. MRR increases by increase in gas pressure and decreases by increase in modulation frequency and wait time [55]. Low pulse energy and pulse frequency produces less thermal energy focused on workpiece is of lesser extent in low MRR. Also, lower values of pulse width and cutting speed utilized laser beam heat entirely to molt the material initiating high MRR. Low range of pulse width and high pulse energy, incident laser beam effects in penetrating melting and vaporizing along the whole workpiece thickness results in rapid increase in MRR. Intensity has noticeable effect on MRR while frequency and pulse overlaps impact are not causative much for the variation in MRR values [75]. Process parameters and performance characteristics used in literature are shown in Fig.3

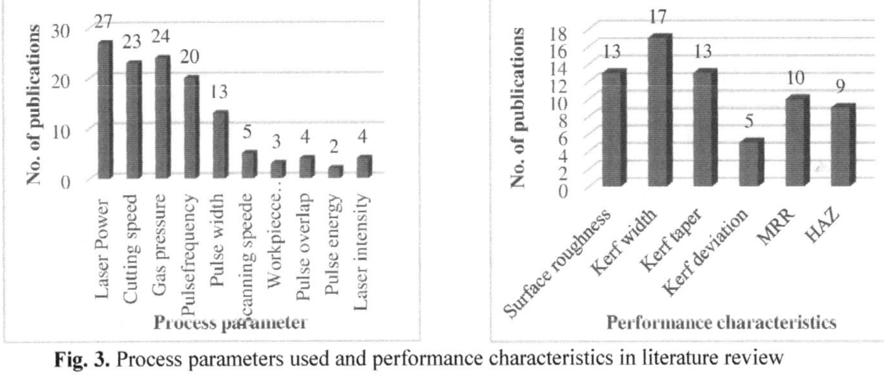

Fig. 3. Process parameters used and performance characteristics in literature review

5. Conclusion

In this study, an effort has been made to review the research carried out in LBM of advanced materials. The outcomes of review are concise as follows:

- LBM can be used to increase the process effectiveness of machining of advanced materials compared to conventional methods.
- Study the effect of process parameters on performance characteristics and optimization of process by modeling and optimization tools.
- Some studies are described multiple process parametric effect on machining of advanced materials. However, no method has been acceptable in a complete way.
- More study is essential to have a good understanding of effect of process parameters and to select the process parameter in the proper way.

References

[1] Venkatesan, Ramanujam and Kuppan, 2014 Elsevier, *Global Congress On Manufacturing and*

Management, PP.1626–1636.
[2] A D Shinde and Kubade, 2016 *International Journal of Scientific Engineering and Research*, PP.101-104.
[3] Lee, Woo, Baek and Kim, July 2016 *International Journal of Precision Engineering and Manufacturing,* PP.973-985.
[4] Arun Kumar Pandey and Avanish Kumar Dubey, 2012 *Elsevier, Optics and Lasers in Engineering* **50**, PP.328–335.
[5] R Venkata Rao and V D Kalyankar, 2014 *International Journal of Advance Manufacturing Technology,* Springer-Verlag London **73**, PP.1159–1188.
[6] Mahmoud Moradi, Omid Mehrabi, Taher Azdast and Khaled Y Benyounis, 2017 *Elsevier, Optics and Lasers Technology* **96**, PP. 208–218.
[7] V V S Nikhil Bharadwaj, V N Saikiran Aka, V Gopichand and Parthasarathy Garre, 2014 *International Journal Of Research In Aeronautical And Mechanical Engineering,* Vol. 2, Issue 3, PP. 270-274.
[8] Chithirai Selvan M, Nethri Rammohan and Sachidananda HK, 2015 *European Journal of Advances in Engineering and Technology*, PP.70-76.
[9] Avanish Kumar Dubey and Vinod Yadava, 2008 *Elsevier, Optics and Lasers in Engineering,* PP.124–132.
[10] Girish Dutt Gautam and Arun Kumar Pandey, 2018 *Elsevier, Optics and Laser Technology* **100**, PP.183–215.
[11] D F Pessoa, P Herwig, A Wetzig and M Zimmermann, *Elsevier, Engineering Fracture Mechanics,* PP.227–240.
[12] W Moćko, Radziejewska, A Sarzyński, M Strzelec and J Marczak, 2017 *Elsevier, Optics and Laser Technology* **90**, PP.165–173.
[13] A Bharatish, H N Narasimha Murthy, B Anand, K N Subramanya, M Krishna and P V Srihari, 2017 *Elsevier, Measurement* **100**, PP.164–175.
[14] Sagar Telrandhe, Ashish Saxena, Sutar and Mishra, 2016 *Elsevier, 44th Proceedings of the North American Manufacturing Research Institution of SME,* Vol. 5, PP.282–294.
[15] Basim A Khidhir and Bashir Mohamed, 2010 *Springer, Journal of Mechanical Science and Technology,* **24**, PP.1053-1059.
[16] K Parthiban, Muthukannan Duraiselvam and R Manivannan, 2018 *Optics and Laser Technology* **102**, PP.32–39.
[17] K Venkatesan, R Ramanujam and P Kuppan, *Springer, The International Journal of Advanced Manufacturing Technology* **91**, PP.3807–3821.
[18] Naveed Ahmed, Darwish, A Alahmari and M Shar, 2016 *Springer, The International Journal of Advanced Manufacturing Technology,* PP.1955–1968.
[19] Meltem Altin Karatas and Hasan Gokkaya, 2018 *Elsevier, Defence Technology,* PP.1-9.
[20] Zhipeng Pan, Yixuan Feng, Tsung-Pin Hung, Yun-Chen Jiang, Hsu, Wu, Lin, Ying Lu and Liang, 2017 Elsevier, *Journal of Manufacturing Processes* **30**, PP.141–147.
[21] Georg Wälder and Jacques Richard, 2016 *Elsevier, 18th CIRP Conference on Electro Physical and Chemical Machining, Procedia CIRP* **42**, PP.475–480.
[22] Mostafa Kashani, M Movahhedy, M Ahmadian and R Razavi, 2016 *Elsevier, HPC 2016 – CIRP Conference on High Performance Cutting, Procedia CIRP* **46**, PP.575–578.
[23] Marcin Kubiak, Wiesława Piekarska, Zbigniew Saternus and Tomasz Domański, 2016 Elsevier, *The 20th International Conference: Machine Modeling and Simulations, Procedia Engineering,* PP.88–94.
[24] Obrad Anicic, Jović, Skrijelj and Nedić, *Elsevier, Optics and Laser in Engineering,* PP.1-4.
[25] Lijun Yang, Ye Ding, Cheng, Abattouy Mohammed and Wang, 2017 *Springer, International Journal Advance Manufacturing Technology* **91**, PP.1171–1180.
[26] Pathik Patel, Sheth and Patel, *Elsevier, International Conference on Innovations in Automation and Mechatronics Engineering, Procedia Technology,* PP.406 – 413.

[27] Yecheng Cai, Lijun Yang, Hongzhi Zhang and Yang Wang, 2016 *Elsevier, Optics and Lasers in Engineering* **82**, PP.173–185.
[28] Goyal and Dubey, *Journal of Mechanical Science and Technology,* PP.1281-1293.
[29] K Ahmmed, Ling, Servio and Kietzig, 2015 *Elsevier, Optics and Lasers in Engineering* **66**, PP.258–268.
[30] Yuchao Bai, Yongqiang Yang, Zefeng Xiao, Mingkang Zhang and Di Wang, 2018 *Elsevier, Materials and Design* **140**, PP.257–266.
[31] Amit Sharma and Vinod Yadava, *Elsevier, Optics and Laser Technology,* PP. 264–280.
[32] Bessmeltsev and Bulushev, 2014 *Optoelectronics, Instrumentation and Data Processing,* Vol.50, PP. 533–548.
[33] K Venkatesan, 2017 *Journal of Advanced Research* **8**, PP. 407–423.
[34] G Casalino, Campanelli, N Contuzzi and A D Ludovico, 2015 *Elsevier, Optics and Laser Technology* **65**, PP.151–158.
[35] Bekir Sami Yilbas, Mian Mobeen Shaukat and Farhan Ashraf, 2017 *Elsevier, Optics and Laser Technology* **93**, PP.67–73.
[36] P V Argade, Dr. Rachayya and R Arakerimath, July 2016 *International Journal of Engineering Trends and Technology*, Vol. 37, Number 6, PP.311-316.
[37] Flaviana Tagliaferri, Giacomo Leopardi, Ulrich Semmler, Michael Kuhl and Biagio Palumbo, 2013 *CIRP Conference on Modeling of Machining Operations*, PP.170–175.
[38] Rudramuni Khed and N K Kamble, Sept. 2015 *International Journal of Engineering Research and Technology,* Vol. 4, Issue 09, PP.326-333.
[39] Xianjun Kong, Yang, Zhang, Chi and Wang, 2017 *Springer, The International Journal of Advanced Manufacturing Technology* **89**, PP.529–542.
[40] Zhichao Wang, Jinkai Xu, Huadong Yu, Zhanjiang Yu, Yiquan Li and Qiang Du, 2018 *Springer, Int J Adv Manuf Technol* **94**, PP.3679–3690.
[41] M Boujelbene, 2018 *Elsevier, International Conference on Materials Manufacturing and Design Engineering,* PP.259-264.
[42] Ribeiroa, Césara and Lopes, September 2017 *International Conference on Structural Integrity,* PP.355-362.
[43] Rajeev Ranjan and Ajay Mishra, 2016 *Springer, Arabian Journal of Science and Engineering* 41, PP.4607–4612.
[44] Amitkumar D Shinde and Pravin R Kubade, 2017 *International Journal of Engineering Research and Technology,* Vol. 10, Number 1, PP. 621-625.
[45] Anup Malik and Alakesh Manna, 2018 *Springer, Journal of the Brazilian Society of Mechanical Sciences and Engineering*, 40:148, PP.1-21.
[46] Priyadarshini, Pattnaik, Mishrac, Panda and Dhalmahapatra *International Conference on Materials Processing and Characterization,* PP.1518– 1532.
[47] Mishra and Yadava, 2013 *Optics and Laser Technology,* PP.461–474.
[48] Tamrin, Nukman and Choudhury, *Optics and Lasers in Engineering,* PP.57–65.
[49] Alahmari, Darwish and Ahmed, 2016 *Springer, The International Journal of Advanced Manufacturing Technology,* PP.2411–2431.
[50] Murat Sarıkaya and Volkan Yılmaz, *Springer, Neural Computing and Applications.*
[51] Ahmed, Alahmari, Darwish and Naveed, *Applied Physics A Material science and processing.*
[52] K Venkatesan, R Ramanujam and P Kuppan, 2016 *Elsevier, Optics and Laser Technology* **78**, PP.10–18.
[53] Vikas Sharma and Vinod Kumar, 2016 *Journal of the Brazilian Society of Mechanical Engineering* **38**, PP.1221–1238.
[54] Sachin C Borse and M S Kadam, 2018 *Elsevier, 2nd International conference on materials manufacturing and design Engineering, Procedia manufacturing* **20**, PP.213-218.
[55] Arindam Ghosal and Manna, 2013 *Elsevier, Optics and Laser Technology* **46**, PP.67–76.
[56] K Venkatesan and Ramanujam, *Measurement,* PP.97–108.

[57] M Lakshmi Chaitanya, Santhosh and Srikanth, December 2014 *International Journal of Emerging Technology and Advanced Engineering,* Vol. 4, Issue 12, PP.320-326.
[58] S Indumathi, V Diwakarreddy and G Krishnaiah, Feb 2014 *International Journal of Industrial Engineering and Technology,* PP.19-26.
[59] Venkatesan and Ramanujam, *International Journal of Machining and Machinability of Materials,* PP.252-272.
[60] M P Jenarthanan and R Jeyapaul, 2013 *International Journal of Engineering, Science and Technology,* Vol. 5, PP.23-36.
[61] Reddy, Dec. 2014 *All India Manufacturing Technology, Design and Research Conference, IIT Guwahati, India,* PP.338:1-6.
[62] R Ramanujam, R Raju and N Muthukrishnan, 2010 *Journal of Studies on Manufacturing,* Vol. 1, PP.120-125.
[63] A Tamilarasan and D Rajamani, 2017 *Sprimger, Journal of Mechanical Science and Technology* **31**, PP.813-821.
[64] Amit Sharma and Vinod Yadava, 2013 *Elsevier, Optics and Lasers in Engineering* **51**, PP.77–88.
[65] D J Kotadiya and D H Pandya, 2016 *Elsevier, 3rd International Conference on Innovations in Automation and Mechatronics Engineering, Procedia Technology* **23**, PP.376 – 382.
[66] R Adalarasan, M Santhanakumar and S Thileepan, *Springer, The International Journal of Advanced Manufacturing Technology,* PP.305-317.
[67] R Adalarasan, M Santhanakumar and M Rajmohan, 2015 *Elsevier, Measurement* **73**,PP.596-606.
[68] Arun Pandey and Avanish Dubey, June 2011, *International Journal of Modeling and Optimization,* PP.107-112.
[69] Giuseppe Casalino, 2018 *Elsevier, Optics and Laser Technology* **100**, PP.165–175.
[70] Sudipto Chakia, B Shanmugarajanb, Sujit Ghosalc and G Padmanabhamd, 2015 *Elsevier, Applied Soft Computing,* PP.365–374.
[71] Mansour Aminian and Reza Teimouri, 2015 *Springer, Soft Computing*, 19:793-810, PP.793–810.
[72] D Kondayya and A Gopala Krishna, 2013 *Springer, The International Journal of Advanced Manufacturing Technology* **65**, PP.259–274.
[73] Phipon and Pradhan, *International Journal of Computational Engineering Research,* PP.1510-1516.
[74] Muneer Khan Mohammed, Usama Umer and Abdulrahman Al-Ahmari, 2017 *Springer, The International Journal of Advanced Manufacturing Technology,* PP.2017–2029.
[75] Usama Umer, Muneer Khan Mohammeda and Abdulrahman Al-Ahmari, 2017 *Elsevier, Measurement* **95**, PP.181–192.
[76] Casalino, Facchini, Mortello and Mummolo, *Elsevier, IFAC-Papers OnLine,* PP. 378–383.
[77] Sanjay Mishra and Vinod Yadava, 2013 *Elsevier, Optics and Lasers in Engineering*, PP.461–474.
[78] Aixi Sun, Xue Jin and Chang, 2017 *Springer, The International Journal of Advanced Manufacturing Technology,* PP.3485–3498.
[79] Arun Kumar Pandey and Avanish Kumar Dubey, 2013 *Springer, Journal of Mechanical Science and Technology* **27**, PP.2115-2124.
[80] Kuldip Singh Sangwana, Sachin Saxena and Girish Kanta, 2015 *Elsevier, CIRP conference on Life Cycle Engineering,* PP.305–310.
[81] Sharma, Tamang, Devarasiddappa and Chandrasekran, *Elsevier, 3rd International Conference on Materials Processing and Characterisation,* PP.1805–1814.
[82] Arun Kumar Pandey and Avanish Kumar Dubey, 2013 *International Journal of Advance Manufacturing Technology* **65**, PP 421–431.

IOP Conf. Series: Materials Science and Engineering 1104 (2021) 012016 doi:10.1088/1757-899X/1104/1/012016

Simulation Driven Approach For Deriving The PID Parameters For The Control Of An Electric Resistance Furnace (ERF)

Prabhat Dev[1], Sidharth Jain[2] and Harish Kumar[1$]

[1]National Institute of Technology Delhi, Delhi, India

[2]Meerut Institute of Engineering and Technology, Meerut, India

[$]E-mail: harishkumar@nitdelhi.ac.in

Abstract

Modelling plays a major role in predicting the behaviour of any system. It is also one of the crucial parts in the design of the control system and tuning of the controller. Tuning parameters are developed based on the transfer function obtained from the modelling process. The primary goal of this paper is to model and simulate a research grade Electric Resistance Furnace (ERF) under transient conditions and derive the transfer function using simulation data. Using this transfer function the gain parameters (PID, Proportional, Integral and Derivative) of the controller will be derived using conventional tuning approaches. The secondary goal is to provide an optimized model for future development of a Neural Network based controller for the furnace. The controller will be equipped with an auto tuning algorithm based on Model Predictive Control (MPC) which will have a Recurrent Neural Network (RNN) as a system model. The training data set for the RNN will be obtained by using the PID parameters derived in this paper. This would enable precise control of the furnace with lower rise and settling time, reduced or no overshoot, maximum stability and ramp hold ability.

Keywords: Transfer function, Transient thermal analysis, Control system, Zeigler-Nicholas method, Cohen-Coon method, PID tuning

1. Introduction

System modelling involves analysing every aspect of a particular system and developing a model that predicts its behaviour. The element of compromise is quite large in mathematical modelling because of the complexity of the real world systems involved [1]. Various assumptions are to be made to simplify the system. This paper deals with modelling and simulation of an Electric Resistance Furnace (ERF) under transient conditions and obtaining transfer function that will be used in the future for design of a PID controller for the furnace. A precise model is the key element in obtaining effective control over any system [2]. Transient problems can be tricky because of the unsteadiness involved. The temperature varies with respect to both time and spatial coordinates.

1.1 Problem definition

The Electric Resistance Furnace (ERF) used here consists of 4 heating elements each made of high grade silicon carbide and body made of Ceramic. The primary mode of heat transfer will be radiation. A transient analysis will be performed to find out the temperature distribution at various locations in the furnace. Material properties of different components of furnace as follows. From the transient simulation results a transfer function will be derived using system identification which will be used in derivation of the PID parameters.

1.2 PID Tuning

PID controllers are either tuned by the following methods: Analytical methods, Heuristics methods, frequency response methods, optimization methods or by adaptive tuning methods [5]. In this paper we will be using frequency response method and derive the parameters based on step response test by implementing Zeigler Nicholas method and Cohen Coon method and discuss various possibilities of

IOP Conf. Series: Materials Science and Engineering 1104 (2021) 012016 doi:10.1088/1757-899X/1104/1/012016

optimization. Most applications involving thermal systems make use of PID controllers for temperature control. But conventional controllers cannot tackle discontinuities in the system [4]. In [6] authors have used lag and unit reaction rates of the system as quantitative measures and this concept serves as a basis for developing the PID parameters based on the Zeigler-Nicholas method. The PID parameters are obtained by performing an open loop step response test on the plant, from which process gain and delay are obtained. For this purpose a mathematical model of the system is required which may be transfer function model, state space model or differential equations defining the system. In this paper, a transfer function model is used for analysing of the system behaviour and to obtain the tuning parameters.Future work of this paper involves precise modelling of the system using artificial neural network and later implementing Model Predictive Control (MPC) for precise temperature control. The PID parameters derived from this paper will serve as base parameters for obtaining input/output data for system identification.

2. Development of geometric model

The geometry was modelled using Ansys DesignModeler which was found suitable for this problem. The model was developed based on a bottom up approach (Figure 1).

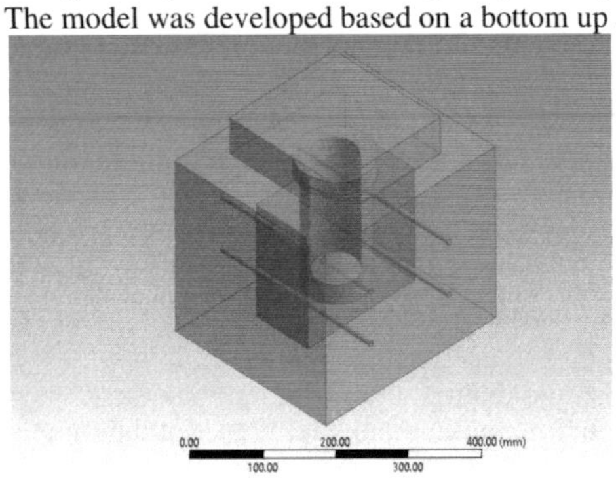

Fig.1.1 Isometric view

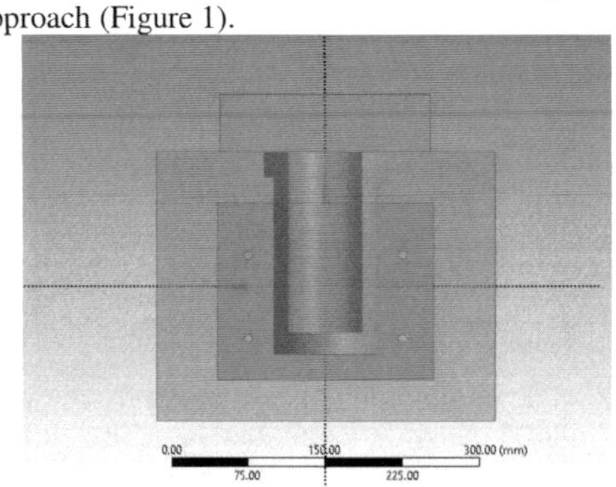

Fig.1.2 Sectional view

After the geometric modelling process the model was imported to Transient Thermal module of Ansys workbench for Meshing and simulation (Figure 2).

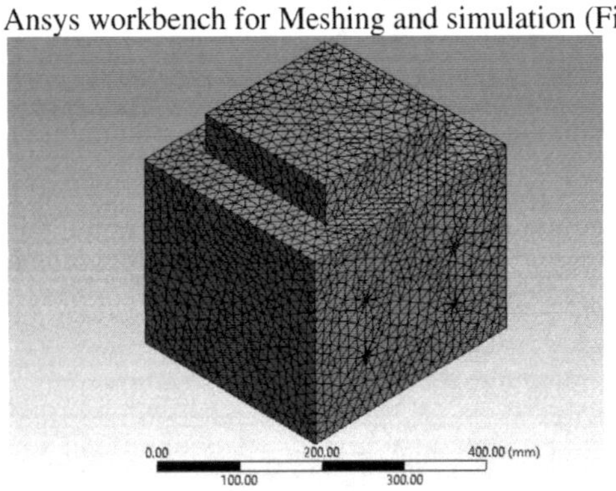

Fig.2.1 External mesh

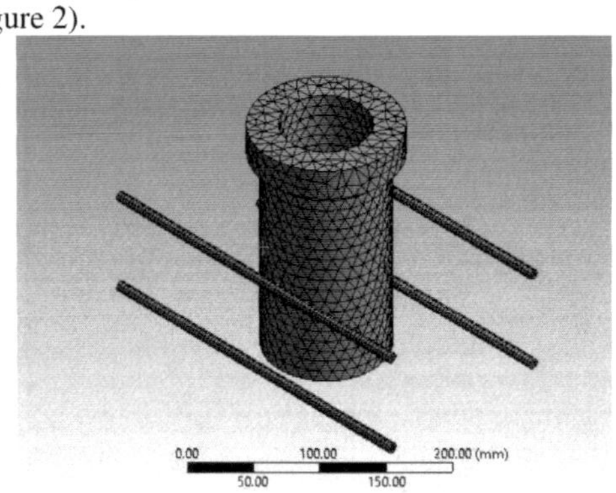

Fig.2.2 Internal Mesh

The mesh size was carefully chosen to obtain proper temperature distribution even in the thinnest regions of the model.

2.1 Boundary conditions and loading

The thermal loading was specified in terms of internal heat generation at the heating filaments.
Calculation of internal heat generation
Equivalent resistance of the filaments, R = 20 Ω (4 Filaments, 5 Ω each connected in series)

$$Voltage = 220\,V$$
$$Power = \frac{V^2}{R}$$
$$P = 2420\,W$$

Combined volume of heating elements

$$Volume = 4 * \pi * (0.004)^2 * 0.290\,m^3$$
$$v = 5.827\,e - 5\,m^3$$
$$Heat\ generation\ rate = Power\ (P)/Volume\ (v)$$
$$HGR = 4.1503\,e + 7\,\frac{W}{m^3}$$

Boundary conditions: Surface to Surface radiation (S2S) between filaments, crucible and furnace walls, convection at Filament-Fluid, Fluid-body and Fluid-Crucible interfaces, convection between furnace outer body and outside environment. Setting used in analysis are summarized in table 1.

Table 1: Analysis settings were used for the simulation

Total number of steps	200
Step end time	1 sec
Step interval	0.5 sec

2.2 Simulation results

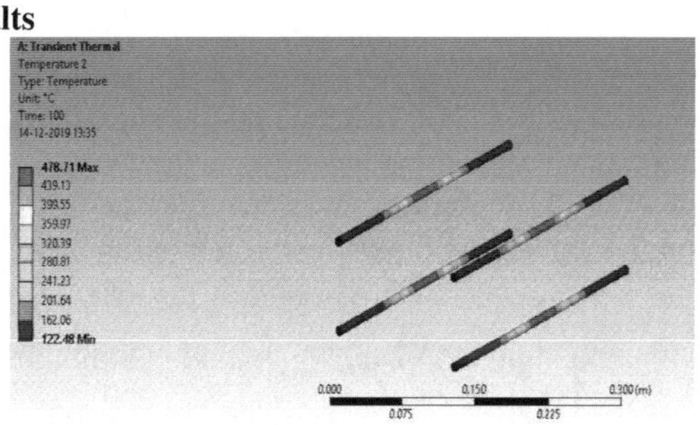

Fig.3.1 Temperature distribution in filaments

Temperature near the filaments was measured using probe feature to simulate a thermocouple and the plotted against time (Figure 3).

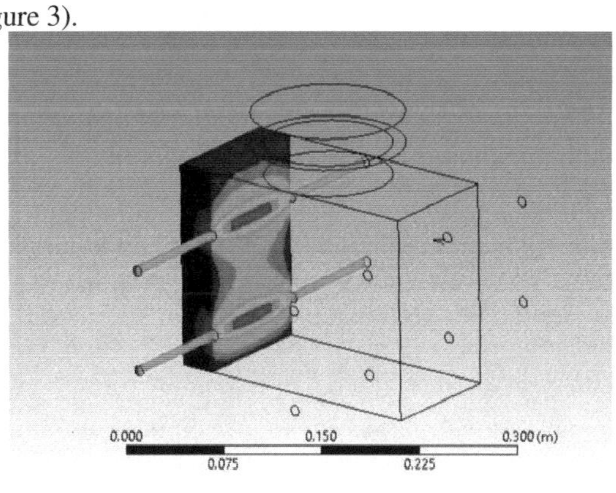

Fig.3.2 Temperature distribution in fluid domain

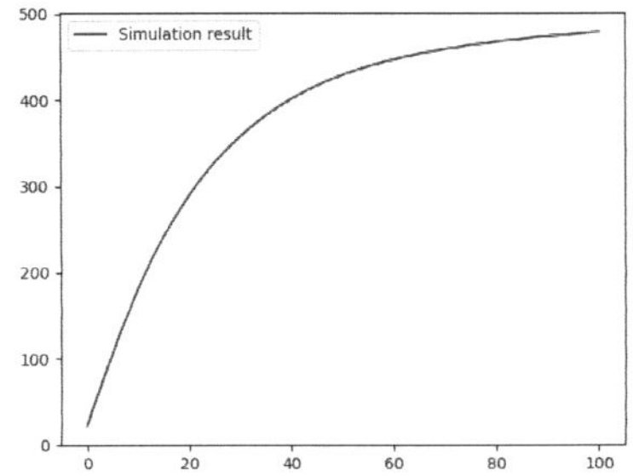

Fig.3.3 Time vs. Core Temp from simulation

2.3 Mathematical Modelling

Governing equations

For unsteady transient problems,

$$\frac{\partial T}{\partial t} = \alpha \nabla^2 T \qquad (1)$$

Where T- Temperature, α- Thermal diffusivity $\alpha = \frac{k}{\rho C_p}$

Suitable approach

Lumped parameter approach is an effective way for solving transient problems, where the system parameters are function of time. Whole body is assumed to be a single lump at a certain temperature. Bi<0.1 is the necessary condition for proceeding with Lumped parameter approach. Bi- Biot number, which compares internal conduction and external convection resistance.

Lumped parameter equations for the heating filament

Input energy is given by

$$I^2 Rt \qquad (2)$$

where I is the current through the circuit and R is the resistance

Change in temperature of the filaments is given by

$$m\, c_p \frac{dT}{dt} \qquad (3)$$

Heat transfer due to convection is given by

$$h\, A\, (T_t - T_i) \qquad (4)$$

Where A is the combined surface area of the filaments and h is the convective heat transfer coefficient

Heat radiated from the surface of the filaments is given by

$$\frac{\sigma\, (T_t^4 - T_i^4)}{\frac{1-\epsilon}{\epsilon\, A}} \qquad (5)$$

Where T_i is the initial temperature of the filament surface and T_t is the temperature after time 't' and ϵ is the emissivity of the filament surface.

Applying energy balance we get,

$$I^2 Rt = m\, c_p \frac{dT}{dt} + h\, A\, (T_t - T_i) + \frac{\sigma\, (T_t^4 - T_i^4)}{\frac{1-\epsilon}{\epsilon\, A}} \qquad (6)$$

Where I is the current flowing through the filaments and R is equivalent resistance of the filaments (Series connection).

The above equation can be used to study the nature of the system by obtaining a transfer function from it. It can be done by taking Laplace Transforms of each term and combining into a single transfer function.
Laplace transform of a function $f(t)$ is given by

$$F(s) = \int_0^\infty f(t)e^{-st}\,dt \quad (7)$$

2.4 Transfer function using system identification

Once the simulation results were obtained, from the Time vs. Temp data, a transfer function was extracted using the MATLAB system identification tool.
MATLAB command used: *ident*
Input and outputs were specified in the workspace for later use in the system identification
From the mathematical model of the thermal system it can be seen that it is a first order system and hence and hence the number of poles and zeros were defined so as to obtain a first order transfer function. After defining the time domain data obtained from Ansys, the model was solved. The transfer function obtained was found to be 99.08% fit to the simulation data. The following transfer function was obtained from MATLAB

$$\frac{9.046\,s + 5.073}{s + 0.04371}$$

2.5 Estimation of PID parameters

Control systems implemented using Zeigler-Nicholas and Cohen-Coon methods are tuned for quarter-wave damping in which overshoot of any given wave is 1/4th of the previous wave. It involves obtaining a compromise between excessive oscillations and settling time [7]. Controller tuning can be done either by closed loop method or process reaction method. In this paper process reaction method is used.

$\frac{PID\ parameters}{Tuning\ methods}$	Controller gain	Integral time	Derivative time
Z-N method	$\frac{1.2\,\tau}{T_d.K}$	$2\,T_d$	$0.5\,T_d$
Cohen-Coon method	$\frac{1}{K}\frac{\tau}{T_d}\left[\frac{16\tau + 30}{12\,\tau}\right]$	$\frac{T_d[32 + 6(T_d/\tau)]}{13 + 8(T_d/\tau)}$	$\frac{4\,T_d}{11 + 2(T_d/\tau)}$

Table 1. [7]

From the obtained transfer function of the model, the system behaviour is observed by applying step input also known as step response test from which process gain, dead time and time lag values are obtained. Using these values we can implement Z-N and Cohen-Coon methods for deriving the PID parameters.

2.6 Analysing the transfer function

2.6.1 Step response test

A step response test is done to study the behaviour of the system. We can also obtain different system parameters like process gain and dead time of the system which will be used for obtaining the tuning parameters for the controller as discussed above.
Let the obtained transfer function be

$$T(s) = \frac{9.046\,s + 5.073}{s + 0.04371}$$

We formulate the function for step response such that

$$C(s) = R(s).T(s)$$

R(s) =1/s for step response test

Now we apply Inverse Laplace transform to the above system to convert it from frequency domain to time domain. [C (s) to C (t)]

$$L^{-1}[C(s)] = \frac{1}{s}L^{-1}[T(s)]$$

Now plotting the step response in time domain we get,

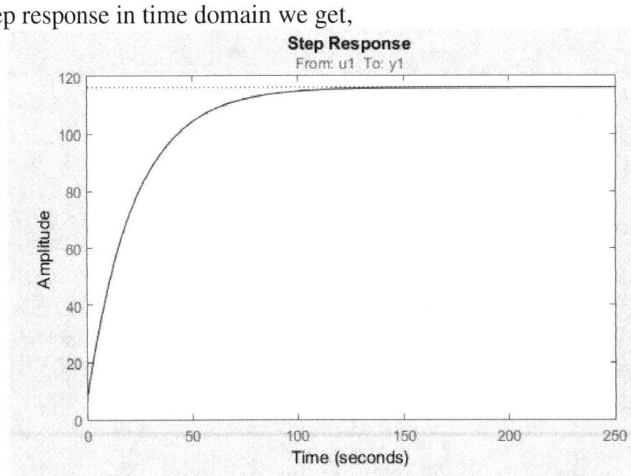

Fig.4 Step response of the obtained transfer function

From the step response test (Figure 4), we obtain the following process parameters

Process gain, $K = 116.03$
Time lag, $\tau = 22.836\ s$
Dead time, $T_d = 0.133\ s$

Substituting the above obtained values in the Table 1. We can find out the optimum PID parameters using both Zeigler-Nicholas and Cohen-Coon methods.

Table 2. PID parameter obtained from Z-N and Cohen-Coon methods

$\frac{Method}{Gain\ parameters}$	P (Controller gain)	I (Integral time)	D (Derivative time)
Zeigler-Nicholas method	1.775	0.266	0.0665
Cohen-Coon method	2.135	0.3265	0.04831

3. Future scope

The development of microcontrollers integrated in the control system allows the user to implement various tuning algorithms and to collect data points from the process and improve the process parameters further based on the data points [8]. Machine learning plays a vital role in analysing time series data. Using this feature we can train and develop a neural network which would be able to represent the nonlinear dynamic system (The furnace). Thus the trained model would be able to predict the future outcome of the system based on the past states the system has been through and the present input to the system. This is the key element in designing the Model Predictive Controller (MPC) for the furnace. MPC generates control actions by optimizing an objective function repeatedly over a predefined horizon [9]. The PID parameters obtained from this paper will be used to generate training data which will be used to train the Recurrent Neural Network (RNN). The training data obtained from a PID controlled plant is more efficient in training a Neural Network than the data obtained from open loop training [10].

4. Conclusion

The simulation driven way of obtaining transfer function can be helpful in designing control systems for the complex process where obtaining a proper mathematical model is difficult. As discussed above, the transfer function will be used for development of an intelligent PID controller based on

Machine Learning. Discontinuities can be a big challenge in the design of control systems. Machine learning is very effective at analysing discontinuities. This leads to precise control of furnace temperature. Critical manufacturing processes like investment casting, metal injection moulding and other thermal cycling processes like physical vapour deposition/chemical vapour deposition, e-waste processing, which require precise control of temperature are expected to be benefited by ML-integrated PID parameter auto-tuning and control.

Acknowledgement
Authors are thankful to Hon. Director, National Institute of Technology Delhi, Delhi, India.

Reference

1. Lawson D and Marion G, 2008. An Introduction to Mathematical Modelling, Glenn Marion, Bioinformatics and Statistics Scotland.
2. Hladky V andBielek V, 2014.Modelling and Control of Thermal System.Control Engineering, **12** (2).
3. Shankar SubramanianR. Unsteady Heat Transfer: Lumped Thermal Capacity Model, , Department of Chemical and Biomolecular Engineering , Clarkson University
4. Zhenhai D andLianyun S, 2012. Design of Temperature Controller for Heating Furnace in Oil Field, Dou Zhenhai and Sun Lianyun / Physics Procedia,**24**, 2083 – 2088.
5. Ang K H and Chong G, 2005.PID Control System Analysis, Design, and Technology, IEEE Transactions On Control Systems Technology, **13**(4).
6. ZiegleriJG, Nichols2 N B andRochester N Y, 1942. Optimum Settings for Automatic Controller, Transactions Of The A.S.M.E.
7. Control Loop Tuning - HVA 240 Principles of Process Control, http://h240.marcks.cc/downloads/10_tuning.pdf
8. O'DwyerA 2009.Handbook Of Pi And Pid Controller Tuning Rules, Page 4, 3rd edition, Published by Imperial College Press 57 Shelton Street Covent Garden London WC2H 9HE.
9. Yunpeng P and Wang J, 2012. Model Predictive Control of Unknown Nonlinear Dynamical Systems Based on Recurrent Neural Networks. IEEE Transactions On Industrial Electronics, **59**(8).
10. Draeger A, Engell S and Ranke H, 1995. Model Predictive Control Using Neural Networks", IEEE Control Systems, 0272- 1708/95/$04.000 1995IEEE

Analysis of Elastic Deflection of Single Point Cutting Tool and Optimization of Cutting Speed in Turning Operation by Using MADM Method

Saurabh Dewangan[1], Somnath Chattopadhyaya[2], Amar Sharma[1], Varun Chaudhary[1], Ritwik Rohan[1]

[1] Department of Mechanical Engineering, Manipal University Jaipur, Rajasthan, India, Pin-303007

[2] Department of Mechanical Engineering, IIT (ISM) Dhanbad, Jharkhand, India, Pin-826004

Corresponding author Email: saurabh22490@gmail.com

Abstract. In the present paper, an analysis was carried out to optimize the cutting parameter i.e. speed in turning operation. The turning was done on a cylindrical rod with diameter and length of 35 mm and 300 mm respectively. The actual length of cutting was 100 mm. Five different cutting operations were taken place with five different cutting speeds i.e. 45, 105, 160, 270, 630 (in rpm) respectively. The surface roughness of the job and forces involved during cutting were measured through profilo-meter and dynamometer respectively. Also, the tool life was calculated by using Taylor's tool life equation in each cutting condition. The elastic deflection of the tool in each cutting condition was simulated by using FUSION360 software. The optimum cutting speed was found out by applying multiple objective decision making (MADM) method.

1. Introduction

To maximize the production rate and to minimize the cost, an optimum cutting condition is always required. In every machining process, the optimization methods are necessary to find out the best set of cutting parameters which can save production time and production cost both. Turning is one of the oldest process of machining which is done on lathe machine. A manufacturing industry cannot be planned without a lathe machine. Even it is a very old process, still research is going on in this area. The day by day enhancement in the material's properties and analysis approach foster the researchers to do continue investigations [1-5]. Selection of the best set of cutting parameters, selection and enhancement of cutting fluid performance, reduction of tool wear, application of dry machining, etc. are some of the emerging field of research which are being focused now days. The selection of an optimum cutting parameter (i.e. speed, feed and depth of cut) depends on how the output parameters behave. These output parameters can include: surface finish of the work part, temperature generation on the tool, force and power consumption during machining, tool wear rate, vibration and chatter creation during cutting, etc. Many researchers have done a number of works related to turning processes and optimization work. Some of them are summarized here comprehensively.

Dewangan et al. (2019) reported that high cutting speed enhances the surface roughness as well as temperature on the tool tip. This high temperature might lead to plastic deformation of the tool tip. In addition, there was no proper relationship found out between feed rate and surface roughness in the experimental work [1]. It was noticed by Sumardiyanto et al. (2018) that the surface roughness was highly affected by the speed of the spindle. The machining was performed in S45-C steel. After speed, feed rate was found as second influencer and depth of cut is the least affecting parameter [6]. Graphite reinforced aluminium-8011 matrix metal matrix was taken under study by Anil et al. (2017) for checking the machinability. It was found that high cutting speed had drastically increased the surface finish whereas the depth of cut and feed rate had an adverse effect on surface finish. It was concluded that there was a requirement of high cutting force when feed and depth of cut were high [7]. Similar kinds of results were obtained by Kumar et al. (2012) also. After performing the turning operation in alloy steel on a CNC lathe machine, it was reported that roughness got increased as soon as the feed was increased. Also there was a directly proportional relationship between speed and surface finish [8]. A work was carried out by Ghani et al. (2004) to investigate the wear mechanisms in coated and uncoated carbide tools at different combinations of cutting parameters. The work piece material was AISI H13 which is usually applied for manufacturing tools. It was found by the experiment that the defects like cracks and mechanical breakage was prone in uncoated tools. However, the uncoated tools showed a uniform wear rate at low level of speed, feed and depth of cut [9]. The tool life was analysed according to the flank wear by Yong Huang et al. (2005). They found that as soon as the speed was increased, flank wear also got increased and thereby reduced the tool life [10]. Besides the experimental work, Sowjanya et al. (2015) performed finite element analysis to assess the tool life after turning operations. It was reported that lower cutting speed might increase the tool life to a great extent [11].

2. Experimental

The experimental work involves turning operation of AISI 1020- low carbon steel on a semi-automatic lathe machine. A single point cutting tool of high speed steel was selected for this work. The specification of the tool is given in Figure 1(a). The chemical constituents and mechanical properties of the work-piece is shown in Figure 1(b). Water-oil mixture was applied as a coolant during turning operations. The turning was carried out by varying the cutting speeds through changing the gear on head-stock. Five different cutting speeds were obtained in the lathe machine. They are: 45, 105, 160, 270, 630 (all in rpm). The feed rate and depth of cut were kept constant for all the cutting processes. The feed rate was selected as 0.04 mm/rev and depth of cut was fixed at 0.5mm. Prior to start machining, the single point cutting tool was used through a lathe dynamometer so that the cutting forces can be recorded by the data acquisition system. The cutting forces in three directions: opposite to cutting direction (main cutting force), opposite to feed direction (feed force) and opposite to depth direction (radial force) were recorded by the instrument. One by one, five different turning operations were carried out. After each cutting, the surface roughness of the job was measured by using surface roughness tester (or profilometer). The data obtained by the experiment is shown in Figure 1(c).

(a)

Specifications of the tool:

- Brand- JK- HSS Square Tool Bit Blanks
- Size: 3/8 × 6 inches
- Size: 10 × 150 mm
- Grade: KT-8 (M42)
- Hardness: RC 67

(b)

Chemical Properties of the proposed work material Mild steel

Element	C	Mn	Si	S	P	Fe
Weight %	0.18	0.90	0.40	0.04	0.04	Rest

Mechanical Properties of the proposed work material Mild steel

Mechanical properties	Hardness (BHN)	Yield Strength	Elongation	Poisson's Ratio	Machinability
Values	71	370 MPa	15%	29%	70%

(c)

Speeds	Forces(N)			Roughness		
	F_x	F_y	F_z	R_a	R_q	R_z
B-Low (45RPM)	32.69N	196.147N	218.54N	7.1585	8.786	40.855
B-High (270 RPM)	-62.759N	626.866N	297.51N	19.224	23.105	93.441
C-Low (105 RPM)	7.947N	88.141N	156.31N	7.896	23.105	93.441
C-High (630 RPM)	16.846N	126.144N	146.51N	7.778	6.642	32.076
D-Low (160 RPM)	8.702N	90.022N	75.46N	5.451	6.633	29.150

Figure 1. (a) Specification of tool; (b) Chemical and mechanical properties of the work-piece; (c) the experimental data

3. Modelling and Simulation

For modelling and simulation of elastic defelction of the tool during cutting, FUSION 360 software was used. The basic physical properties of the work-piece selected for modelling and simulation are provided in table 1. As per the values of forces found out by the dynamometer in three directions, the deflections are shown in table 2. A resultant deflection of 0.01821 mm was found maximum among all. This is corresponding to 270 RPM speed. The models and simulated diagrams are shown in Figure 2.

Table 1: Basic physical properties of the work-piece

Basic Thermal Properties	**Basic Mechanical Properties**	**Strength**
Thermal conductivity: 4.500E+01 W/(m-K) Specific heat: 0.480 J/(g.°C) Thermal expansion coefficient: 12.000 µm/(m.°C)	Young's modulus: 220.000 GPa Posson's ratio: 0.28 Shear Modulus: 81700.000 MPa Density: 7.850 g/cm^2 Damping coefficient: 0.00	Yield strength: 207.000 MPa Tensile strength: 345.000 MPa

Table 2: Different speeds, forces and corresponding deflections of the tool

Speeds	Forces in different directions(N)			Deflection(mm)
	F_x	F_y	F_z	In XYZ directions
45 RPM	32.69 N	196.147 N	218.54 N	0.00909 mm
270 RPM	-62.759 N	626.866 N	297.51 N	0.01821 mm
105 RPM	7.947 N	88.141 N	156.31 N	0.006237 mm
630 RPM	16.846 N	126.144 N	146.51 N	0.006065 mm
160 RPM	8.702 N	90.022 N	75.46 N	0.003378 mm

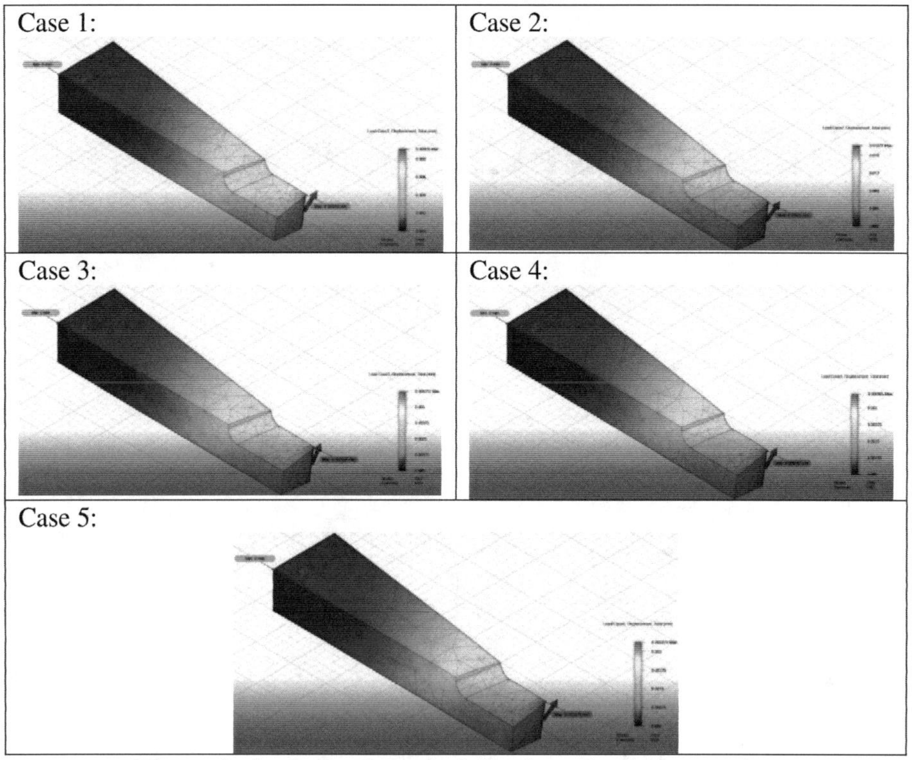

Figure 2. Analysis of elastic deflection of tool after turning

4. Tool life calculation

Tool life has been calculated for every cutting condition by using Taylor's tool life equation. As the feed rate and depth of cut were kept constant during turning, only speed were changed in each condition. $V_cT^n = C$ is the velocity dependent tool life equation where n and C are 0.15 and 180 respectively (for HSS tool and low-carbon steel workpiece). Hence the tool life at various speeds such as 45 rpm, 270 rpm, 105 rpm, 630 rpm and 160 rpm are 1.41×10^9 min, 9204.03 min, 5.005×10^6 min, 32.41 min, and 3.019×10^5 min respectively.

5. Combined MADM (TOPSIS+AHP) method

To find out an optimum cutting parameter, a combination of TOPSIS (Technique for Order Preference by Similarity to Ideal Solution) [12] and AHP (Analytic Hierarchy Process) [13] was used in this study. These are multiple attributes decision making methods. The theory behind combined MADM method has been referred from Rao (2008) [14, 17]. The author has used this method in other works too [15, 16]. As these methods are very common, a long explanation is given here. But, a short approach and steps to use combined MADM method are shown in Figure 3.

Steps to use TOPSIS and AHP together for decision making problem (All theory and explanations are referred from reference number [15], [16] & [17]

A detailed procedure for selection of the best alternative by using MADM method has been explained below. The MADM method discussed here is a combination of TOPSIS and AHP.

MADM 1: The first step involves determination of objective and selection of attributes which influence the objective directly or indirectly. The attributes may be quantitative and qualitative. The qualitative attributes must have to be converted into quantitative form.

MADM 2: In the second step, a decision matrix is formed which contain all the attributes in the quantitative form. Alternatives can be allocated in row of the matrix whereas the columns represent the attributes (According to below decision matrix).

$$D_{p\times q} = \begin{array}{c} \\ 1 \\ 2 \\ 3 \\ - \\ - \\ p \end{array} \begin{array}{c} \begin{matrix} 1 & 2 & 3 & - & - & q \end{matrix} \\ \begin{bmatrix} d_{11} & d_{12} & d_{13} & - & - & d_{1q} \\ d_{21} & d_{22} & d_{23} & - & - & d_{2q} \\ d_{31} & d_{32} & d_{33} & - & - & d_{3q} \\ - & - & - & - & - & - \\ - & - & - & - & - & - \\ d_{p1} & d_{p2} & d_{p3} & - & - & d_{pq} \end{bmatrix} \end{array}$$

Where,

$D_{p\times q}$= decision matrix with p= number of alternatives; q= number of attributes.

d_{mn}= d is n^{th} attribute for m^{th} alternative in decision matrix D.

MADM 3: In this step, normalization of above decision matrix is done by using given equation no. (2).

$$N_{mn} = \frac{d_{mn}}{\left[\sum_{m=1}^{p} d^2_{mn}\right]^{0.5}} \tag{2}$$

MADM 4: In this step, a relative importance of various attributes is found with respect to the objective. The comparison is done by the process explained in [30]. The relative importance of the attributes with each other is termed in qualitative manner, like, good/better/best/fair/excellent/etc. The qualitative judgement can be converted into some quantitative form as per the decision maker's choice. For example, in this work, four different numbers are defined by: 3=Fair; 5= Good; 7= Very good; 9= Excellent. Value '1' will come out when the attribute is compared with itself. Let us assume 'A' number of attributes. If attribute m is compared with attribute n, it is given a value s_{mn} which denotes relative importance of m with respect to n. In this way a square matrix $S_{a\times a}$ is formed. If $m=n$ then $s_{mn}=1$, hence all diagonal values are 1. Also $s_{nm}= 1/s_{mn}$.

$$S_1 = S_{a\times a} = \begin{array}{c} \\ 1 \\ 2 \\ 3 \\ - \\ - \\ A \end{array} \begin{array}{c} \begin{matrix} 1 & 2 & 3 & - & - & A \end{matrix} \\ \begin{bmatrix} 1 & s_{12} & s_{13} & - & - & s_{1a} \\ s_{21} & 1 & s_{23} & - & - & s_{2a} \\ s_{31} & s_{32} & 1 & - & - & s_{3a} \\ - & - & - & - & - & - \\ - & - & - & - & - & - \\ s_{a1} & s_{a2} & s_{a3} & - & - & 1 \end{bmatrix} \end{array}$$

Now, we have to find the relative normalized weight (w_{mn}) of each attribute by following these two steps one by one:

(i) To find out the geometric mean of each row and

(ii) Normalization of the geometric means of rows in the comparison matrix. This can be represented as

$$GM_m = \left[\prod_{n=1}^{A} s_{mn}\right]^{1/A} \tag{3}$$

And

$$w_n = \frac{GM_m}{\sum_{m=1}^{A} GM_m} \tag{4}$$

MADM 5: Formation of two new matrices S_3 and S_4: $S_3=S_1\times S_2$, and $S_4= S_3/S_2$, where $S_2= [w_1, w_2, w_3,......w_n]^T$ and $S_1= S_{a\times a}$

MADM 6: To find the average of matrix S_4 this becomes the maximum Eigen value λmax.

MADM 7: To find the consistency index (C.I.), this shows the deviation from ideal values. C.I. will be calculated as, C.I. = (λ_{max}- A)/(A-1). Small value of C.I. is favourable.

MADM 8: Random Index (RI) is to be calculated. It depends on number of attributes [22].

No. of attributes	3	4	5	6	7	8	9	10
R.I.	0.52	0.89	1.11	1.25	1.35	1.4	1.45	1.49

MADM 9: Finding consistency ratio CR=CI/RI. Low value (less than 0.1) of CR is acceptable.

MADM 10: To obtain the weighted normalized matrix. It can be calculated by multiplying the element of column of matrix N_{mn} with its respective weights w_n. Hence weighted normalized matrix (I_{mn}) is given by, $I_{mn}= w_n N_{mn}$

MADM 11: To obtain the ideal (I^+) and negative ideal (I^-) values. If number of alternatives is K and there are L attributes, then there will be L number of $K\times K$ matrices of judgement.

$$I^+ = \left\{(\sum_{m}^{max} I_{mn} / n \in N), (\sum_{m}^{min} I_{mn} / n \in N')/ m = 1,2...K\right\}$$
$$= \{I_1^+, I_2^+, I_3^+, ... I_L^+\} \tag{5}$$

$$I^- = \left\{(\sum_{m}^{min} I_{mn} / n \in N), (\sum_{i}^{max} I_{mn} / n \in N')/ m = 1,2...K\right\}$$
$$= \{I_1^-, I_2^-, I_3^-, ... I_L^-\} \tag{6}$$

Where, $N=(n= 1, 2, 3......L)/n$ is acceptable attributes and $N'=(n=1, 2, 3......L)/n$ is un-acceptable attributes. I^+ shows the higher values of acceptable attributes and too it is also showing lower values for un-acceptable attributes. Similarly, I^- indicates the lower value and higher value for acceptable and un-acceptable attributes respectively.

MADM 12: We have to find the separation of each alternative from the ideal solution. The following equations can be used for this purpose.

$$P_m^+ = \left\{\sum_{N=1}^{L} (I_{mn} - I^+_n)^2\right\}^{0.5} \tag{7}$$

$$P_m^- = \left\{\sum_{N=1}^{L} (I_{mn} - I^-_n)^2\right\}^{0.5} \tag{8}$$

Where, $m= 1, 2, 3........K$

MADM 13: For getting the best solution among all alternatives, we have to find the relative closeness of alternatives from the ideal solution. For this purpose, we have to use following equation.

$$C_m = \frac{P^-_m}{(P^+_m + P^-_m)} \tag{9}$$

MADM 14: In this step, we have to arrange the alternatives according to their preferences. The ranks are provided to each alternative and the decision is made on the basis of most favourable and least favourable solutions.

Figure 3. Steps to use combined MADM method using TOPSIS and AHP

Based on the experimental data, five different speeds are five alternatives. The force acting on the tool, surface roughness of the work piece and tool life are the three attributes based on these an optimum alternative has to be selected. The alternatives and attributes are given in table 3.

Table 3: Different alternatives (speeds) and attributes under study

S. No	Spindle Speed (RPM)	Force Exerted (N)	Surface Roughness	Tool Life(min)
1.	45	295.46	42.39	1.41×10^9
2.	270	696.70	98.15	9204.03
3.	105	179.26	96.57	5.005×10^6
4.	630	194.06	33.66	32.41
5.	160	75.46	30.38	3.019×10^5

By using the steps and formula given in Figure 3, all calculations have been done to take the experimental data. The calculation and analysis are shown in Figure 4.

Step I : The objective is to find the best speed for the turning operation.

For this Force Exerted (FE), Surface Roughness (SR) & Tool Life (TL) are taken as the 3 attributes.
There are 5 alternate attributes available.

Step II : Now we will form a decision matrix (D) which has all the data of observation.

$$D = \begin{bmatrix} 295.46 & 42.39 & 1.41 * 10^{\wedge}9 \\ 696.70 & 98.15 & 9204.03 \\ 179.26 & 96.57 & 5.005 * 10^{\wedge}6 \\ 194.06 & 33.66 & 32.41 \\ 75.46 & 30.38 & 3.019 * 10^{\wedge}5 \end{bmatrix}$$

Now we calculate normalised matrix by,

$$N_{mn} = \frac{d_{mn}}{\left[\sum_{n=1}^{p} d^2_{mn}\right]^{0.5}}$$

Here P = number of alternatives = 5

Now we calculate the normalized matrix

$$N = \begin{bmatrix} 0.371 & 0.2806 & 0.997 \\ 0.876 & 0.649 & 6.54 * 10^{\wedge} - 6 \\ 0.225 & 0.639 & 0.0035 \\ 0.244 & 0.222 & 2.92 * 10^{\wedge} - 8 \\ 0.0949 & 0.201 & 2.135 * 10^{\wedge} - 4 \end{bmatrix}$$

Step III :

1ST Column [Σd_{mn}^2]$^{0.5}$ = [631347.544]$^{0.5}$ = 794.573

2ND Column [Σd_{mn}^2]$^{0.5}$ = [22812.039]$^{0.5}$ = 151.036

3RD Column [Σd_{mn}^2]$^{0.5}$ = [1.988*10^{18} + 84714168.2 + 1050.40 + 2.505*10^{13} + 9.116*10^{10}]$^{0.5}$ =1.4135151397*10^{9}

Step IV : Now we will make the Importance matrix

Tool Life = 7, Force Exerted = 1, Surface Roughness = 5

$$Si = \begin{bmatrix} 1 & 5 & 7 \\ 1/5 & 1 & 5/7 \\ 1/7 & 7/5 & 1 \end{bmatrix}$$

Step V : We will calculate the Geometric Mean & normalized weight of each attribute.

$GM_{FE} = \sqrt[3]{1 * 5 * 7} = 3.271$

$GM_{SR} = \sqrt[3]{1/5 * 1 * 5/7} = 0.522$

$GM_{TL} = \sqrt[3]{1/7 * 7/5 * 1} = 0.584$

$$Wm = \begin{bmatrix} 0.747 \\ 0.119 \\ 0.133 \end{bmatrix} = S_2$$

To obtain matrix S_3 & S_4

$$S_3 = \begin{bmatrix} 2.273 \\ 0.363 \\ 0.406 \end{bmatrix}$$

$$S_4 = S_3/S_2 = \begin{bmatrix} 3.042 \\ 3.050 \\ 3.052 \end{bmatrix}$$

Step VI : Now we calculate max Eigen values λ_{max} i.e. average of S_4

λ_{max} = 3.042+3.050+3052/3 = 3.048

IOP Conf. Series: Materials Science and Engineering 1104 (2021) 012017 doi:10.1088/1757-899X/1104/1/012017

Step VII : Now since the number of attributes is 3, the Random Index (R.I) = 0.52

Now we calculate the Consistency Index (C.I)

$$C.I = \frac{\lambda max - A}{A-1} = \frac{3.048-3}{3-1} = 0.24$$

Here A = No. of Attributes

Now we will calculate Consistency Ratio [CR]

$$CR = \frac{CI}{RI} = \frac{0.024}{0.52} = 0.0461$$

CR ≤0.1, which is true here, hence the CR is favourable.

Step VIII : Now we will calculate weighted Normalised Matrix

$$I_{mn} = W_n.N$$

We will calculate I_{mn} by multiplying each column with respective element of $W_m(S_2)$

$$I_{mn} = \begin{bmatrix} 0.277 & 0.0333 & 0.1126 \\ 0.654 & 0.0772 & 8.69 * 10^{\wedge} - 7 \\ 0.168 & 0.0760 & 4.65 * 10^{\wedge} - 4 \\ 0.182 & 0.0264 & 3.883 * 10^{\wedge} - 9 \\ 0.0708 & 0.0239 & 2.83 * 10^{\wedge} - 5 \end{bmatrix}$$

Step IX : Now we have to obtain the ideal (I^+) & negative Ideal (I^-) values. If the no. of alternatives is k (=5) & L attributes (=3), then there will be L no. of k*k matrices of judgement. i.e. 3 matrices of 5*5

$$I^+ = \left\{ (\sum_m^{max} I_{mn} / n \in N), (\sum_m^{min} I_{mn} / n \in N') / m = 1,2...K \right\}$$

$$= \left\{ I_1^+, I_2^+, I_3^+, ...I_L^+ \right\}$$

$$I^- = \left\{ (\sum_m^{min} I_{mn} / n \in N), (\sum_i^{max} I_{mn} / n \in N') / m = 1,2...K \right\}$$

$$= \left\{ I_1^-, I_2^-, I_3^-, ...I_L^- \right\}$$

Since Force Exerted is non-beneficiary,

Hence $I_{FE}^- > I_{FE}^+$

So, $I_{FE}^- = 0.0708$, $I_{FE}^+ = 0.654$

For Surface Roughness (Beneficiary)

$I_{SR}^+ = 0.0772$, $I_{SR}- = 0.0239$

For Tool Life (Beneficiary)

$I_{TL}^+ = 0.1126$, $I_{TL}^- = 3.883*10^{-9}$

Step X : Now we calculate separation measures of each alternative by Euclidean distance by following equation.

$$P_m^+ = \left\{ \sum_{N=1}^{L} (I_{mn} - I^+_n)^2 \right\}^{0.5}$$

$$P_m^- = \left\{ \sum_{N=1}^{L} (I_{mn} - I^-_n)^2 \right\}^{0.5}$$

m = alternatives =5

$P_1^+ = \{(0.277 - 0.0708)^2 + (0.0333 - 0.0772)^2 + (0.1126 - 0.1126)^2\}^{0.5}$
$P_1^+ = \{0.0425 + 0.0019 + 0\}^{0.5} = 0.2107$

$P_1^- = \{(0.277 - 0.654)^2 + (0.0333 - 0.0239)^2 + (0.1126 - 3.883*10^{-9})^2\}^{0.5}$
$P_1^- = \{0.142 + 8.83*10^{-5} + 0.0126\}^{0.5} = 0.0239$

$P_2^+ = \{(0.654 - 0.078)^2 + (0.0772 - 0.0772)^2 + (8.69*10^{-7} - 0.1126)^2\}^{0.5}$
$P_2^+ = \{0.3317 + 0 + 0.0126\}^{0.5} = 0.586$

$P_2^- = \{(0.654 - 0.654)^2 + (0.0772 - 0.0239)^2 + (8.69*10^{-7} - 3.883*10^{-9})^2\}^{0.5}$
$P_2^- = \{0 + 0.00233 + 7.484*10^{-13}\}^{0.5} = 0.0482$

$P_3^+ = \{0.00944 + 1.89*10^{-9} + 0.0125\}^{0.5}$
$P_3^+ = 0.147$

$P_3^- = \{(0.168 - 0.654)^2 + (0.0706 - 0.0239)^2 + (4.65x10^{-4} - 3.883*10^{-9})^2\}^{0.5}$
$P_3^- = 0.488$

$P_4^+ = \{(0.182 - 0.0708)^2 + (0.0264 - 0.0772)^2 + (3.883*10^{-9} - 0.1126)^2\}^{0.5}$
$P_4^+ = 0.165$

$P_4^- = \{(0.182 - 0.654)^2 + (0.0264 - 0.0239)^2 + (3.883*10^{-9} - 3.883*10^{-9})^2\}^{0.5}$
$P_4^- = 0.4711$

$P_5^+ = \{(0.0708 - 0.0708)^2 + (0.0239 - 0.0772)^2 + (2.83*10^{-5} - 0.1126)^2\}^{0.5}$
$P_5^+ = 0.122$

$P_5^- = \{(0.0708 - 0.0654)^2 + (0.0239 - 0.0239)^2 + (2.83*10^{-5} - 3.883*10^{-9})^2\}^{0.5}$
$P_5^- = 0.583$

Step XI : Now we calculate Relative Closeness by

$$C_a = \{P_a^- / (P_a^+ + P_a^-)\}$$

$C_1 = \{P_1^- / (P_1^+ + P_1^-)\}$
$= \{0.0239 / (0.2107+0.0239)\}$
$= 0.1018$

$C_2 = \{P_2^- / (P_2^+ + P_2^-)\}$
$= \{0.0482 / (0.5864+0.0482)\}$
$= 0.0760$

$C_3 = \{P_3^- / (P_3^+ + P_3^-)\}$
$= \{0.488 / (0.147+0.488)\}$
$= 0.768$

$C_4 = \{P_4^- / (P_4^+ + P_4^-)\}$
$= \{0.4711 / (0.165+0.4711)\}$
$= 0.740$

$C_5 = \{P_5^- / (P_5^+ + P_5^-)\}$
$= \{0.583 / (0.122+0.583)\}$
$= 0.826$

So, $C_5 > C_3 > C_2 > C_4 > C_1$

Since C_5 is the highest

Hence 160 RPM is the Best Speed/Optimized Speed.

Figure 4. The steps for optimization method; result has been found out for an optimized speed

6. Conclusion

The present study deals with a modelling and simulation approach to find out the elastic deflection in the single point cutting tool used for turning operation. The turning process was performed on a low carbon steel work piece by taking five different speeds for cutting. For each speed, elastic deflection was analysed by FUSION 360 software. It was found that deflection (in x, y, z directions) was maximum for 270 rpm speed and it was minimum for 160 rpm. In addition, a combined MADM method was applied to find an optimum cutting speed on the basis of three different attributes i.e. surface roughness, tool life and forces acting on the tool during cutting. The entire process of optimization was explained comprehensively. It was reported that 160 RPM speed was found as an optimized speed on the basis of three attributes i.e. tool life, surface roughness and force exerted on the tool. Also, 160 rpm speed was found better than others from elastic deflection point of view. Hence, the simulation work and optimization result corroborate with each other.

References

[1] Dewangan S., Pattnayak A., Abhishek R., and Patel N. 2019 Some analysis into 0.18%-C steel work piece undergone through turning operation by using HSS [KT-8 (M42)] single point tool. AIP Conference Proceedings **2148**, 030030; https://doi.org/10.1063/1.5123952

[2] Dewangan S., Behera S., Chowrasia M.K., 2020 Comparative analysis into mechanical properties and microstructural attributes of quenched and tempered 0.2%-C steel. World Journal of Engineering 17(1), 127-133. https://doi.org/10.1108/WJE-11-2019-0327.

[3] Dewangan S., Mainwal N., Khandelwal M., Jadhav P.S., 2019 Performance analysis of heat treated AISI 1020 steel samples on the basis of various destructive mechanical testing and microstructural attributes. Australian Journal of Mechanical Engineering, 1-14,. https://doi.org/10.1080/14484846.2019.1664212

[4] Dewangan S., 2019 Critical observation and analysis into wear phenomena in WC-10%Co drill bits. AIP Conference Proceedings **2148**, 030033; https://doi.org/10.1063/1.5123955

[5] Dewangan S., Chattopadhyaya S., 2015 Critical Analysis of Wear Mechanisms in Cemented Carbide. JMEPEG 24:2628–2636. DOI: 10.1007/s11665-015-1546-6

[6] Sumardiyanto D., Susilowati S.E., Cahyo A., 2018 Effect of Cutting Parameter on Surface Roughness Carbon Steel S45C. Journal of Mechanical Engineering and Automation, 8(1): 1-6 DOI: 10.5923/j.jmea.20180801.01

[7] Anil K.C. et al., 2017 Effect of cutting parameters on surface finish and machinability of graphite reinforced Al-8011 matrix composite. IOP Conf. Ser.: Mater. Sci. Eng. **191** 012025

[8] Kumar S., Shetty A., Shetty A., Ananth K, Shetty H. 2012 Effect of spindle speed and feed rate on surface roughness of Carbon Steels in CNC turning. Procedia Engineering 38, 691 – 697

[9] Ghani J., Choudhary I.A., Masjuki H.H., 2004 Wear mechanism of TiN coated carbide and uncoated cermets tools at high cutting speed application", Journal of Materials Processing Technology, 1067-1073

[10] Huang Y., Liang S.Y., 2005 Effect of Cutting Conditions on Tool Performance in CBN Hard Turning", Journal of Manufacturing Processes, 7(1).

[11] Sowjanya S., Sreeramulu D., and Rao C. J., 2015 Experimental and numerical investigation of tool life of single point cutting tool during turning process. Indian Journal of Science and Technology, Vol 8(28), DOI: 10.17485/ijst/2015/v8i28/75726.

[12] Hwang C.L. and Yoon K.P., 1981 Multiple attribute decision making: methods and applications, *Springer*,New York.

[13] Saaty T.L., 1980 The analytic hierarchy process,*McGraw Hill*, New York.

[14] Rao R.V., 2008 Evaluating flexible manufacturing systems using a combined multiple attribute decision making method,International Journal of Production Research 46, 1975-1989.

[15] Dewangan S. and Chattopadhyaya S., 2014 Discussion on ranking the sawability of rocks using a combined multiple attribute decision making method, Applied Mechanics and Materials 592-594, 864-868.

[16] Dewangan S. and Chattopadhyaya S., 2016 Performance Analysis of Two Different Conical Picks Used in Linear Cutting Operation of Coal. Arab J Sci Eng 41:249–265. DOI 10.1007/s13369-015-1786-7

[17] Rao R.V., 2007 Decision making in the manufacturing environment. Springer series in advanced manufacturing, p. 31.

IOP Conf. Series: Materials Science and Engineering 1104 (2021) 012018 doi:10.1088/1757-899X/1104/1/012018

Characterization of dissimilar friction stir welded joints of aluminium alloys by simulation and experimentation

R Kumar[1*], A Kumar[1], S Chattopadhyaya[2], U Kumar[2], D Gope[2]

[1]Mechanical Engineering Department, Birla Institute of Technology, Mesra, Ranchi-835215, India
[2]Mechanical Engineering Department, Indian Institute of Technology (ISM) Dhanbad-826004, India

*Email:- ratan_876@yahoo.co.in

Abstract. The present work "Characterization of dissimilar friction stir welded joints of aluminium alloys by simulation and experimentation" is a combination of simulation and experimental work. In first part of the work friction stir welding (FSW) process is simulated in abacus software and then second part is based on the experimental work of friction stir welded joints of dissimilar aluminium alloys. In abacus, FSW process is well simulated. In this part heat generation profile during FSW is obtained in work-piece as well as in tool. In second part two alloys i.e. AA 5083 and AA 6061 is welded with friction stir welding process. Three process parameters were selected for welding. Three tool rotational speed is selected i.e. 500 rpm, 710 rpm and 900 rpm and one welding speed i.e. 25 mm/min. All the three welded samples were properly machined and different testing samples were prepared. Mechanical and micro-structural properties were studied on the variation of process parameters. From tensile test it can be concluded that with increase in tool rotational speed tensile strength increases. In all cases brittle fracture was observed. From microstructure study fine grains are observed in weld nugget zone.

1. Introduction

In the field of solid state joining process, Friction stir welding (FSW) has emerged rapidly as a reliable solution. It was developed and patented in the year 1991 at The Welding Institute (TWI) by Mr. Wayne Thomas. It has vast application in the field of automotive, railway, aerospace, maritime and lot more industries. As compared to other conventional fusion welding process this process it has lot of attractive advantages. Through this process, materials which are difficult to weld with conventional processes can be welded easily. Since it is a solid state joining process so, materials are joined below their melting point temperature. Therefore during this process defects like porosity, cracking, and difficulties raised due to second phase formation, embrittlement does not appear. During this process low distortion and low residual stresses are generated since it is a solid state joining process. It is observed that during process no filler material as well as no shielding gas is required, therefore it is an energy efficient process. Also during process no pollution, arc flash, spatter and fumes generated so it is an environment friendly process and also termed as green solid state joining process. Therefore, now these days, friction stir welding process is attracting the industry for its uses. [1]

Tang et al has performed the defect free friction stir welding for 316L stainless steel. They found excellent tensile properties for the welded samples and they are compared with the base metal. They also observed excellent bonding strength of weld interface [2]. Yan et al proposed a new calculating method based on fractal theory for frictional heat. They applied this method for the calculation of heat generated during FSW process. They correctly validated their results [3]. Zhang et al studied the effect of rotational speed on properties of dissimilar aluminium alloy welded joints. They observed that TMAZ on retreating side width is greater than that of advancing side width. They also concluded that with increase in rotational speed hardness value decreases. Dissimilar welded joints shows the mixed i.e. ductile brittle fracture mode [4]. Bertrand et al analysed the properties of two different alloy FSW joints i.e. AA2139 and AA7020. They concluded that these dissimilar alloy combination gives large range of process parameters fit for conservative mechanical properties of high performance structural components [5]. Ding et al studied the joint quality on friction stir welded joints due to ultrasonic vibration exerted at tool. They observed that the weld quality is improved due to ultrasonic vibration [6]. Cunha et al characterized the different properties of FSW GL E36 shipbuilding steel. They produced the joints at one rpm i.e. 500 rpm and at three different welding speed i.e. at 1, 2 and 3 mm/sec. They found that all tensile samples broke from the base material from which they concluded that joint achieved higher strength as compared to base metal [7]. Abrahams et al properties and microstructure of dissimilar FSW joints i.e. AA 5005 and AA 7075. They found greater influence of traverse speed as compared to rotational speed on mechanical properties. With increase in traverse speed micro-hardness and mechanical properties improves [8]. Pankaj et al studied the effect of pin profile and tilt angle of tool on FSW joints on brass. They tested samples for different mechanical properties. They found significant effect of pin profile and tilt angle on properties. They also correlated the mechanical testing results with micro-structural changes [9]. Donatus U et al investigated variations in stir zone and thermo-mechanically affected zone of dissimilar FSW joints of AA5083 and AA6082. They concluded that hardness profile changes in thermo mechanically affected zones of retreating side due to change in grain size [10].

From the above literature review it is found that researchers are working on friction stir welding of dissimilar alloys. Therefore this study is concentrated on FSW of dissimilar aluminium alloys. In this study modelling of the FSW process was done initially. Then, experimental was performed and welded joints mechanical and micro-structural properties were studied.

2. Abaqus modelling and simulation

At the start FSW tool model is drawn in abaqus software. Figure 1 shows the diagram of tool. Tool shoulder diameter is taken 30 mm and length of pin is taken 5.8 mm. Pin is selected as tapered profile pin. Length of shoulder is taken as 30 mm.

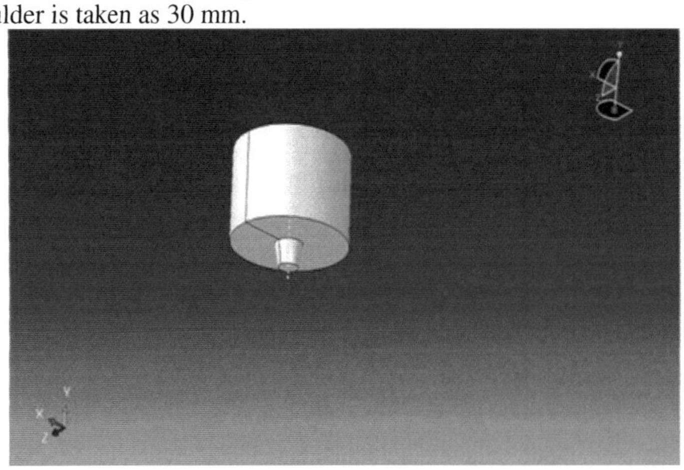

Figure 1. Modelling of tool

IOP Conf. Series: Materials Science and Engineering 1104 (2021) 012018 doi:10.1088/1757-899X/1104/1/012018

Two work pieces are selected of 6 mm thickness for welding simulation. First material is AA 5083 and second material is 6061 is selected. FSW tool and work-piece is assembled together as shown in figure 2. Figure 2 shows the assembly and meshing of tool and work piece together.

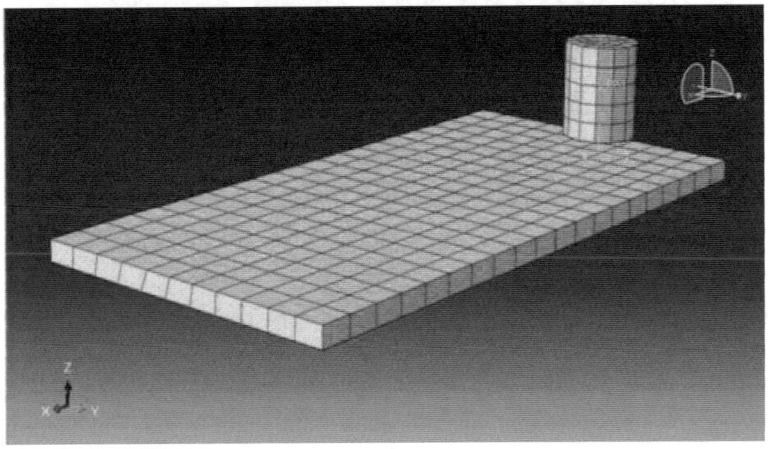

Figure 2. Meshing of tool and work-piece assembly

After meshing, simulation was done. Through simulation heat generation profile is generated. Figure 3 shows the simulation results. In which heat generation profile is shown. Maximum temperature generated is around the tool as shown in figure 3. This heat generated is due to the friction between tool and work-piece area.

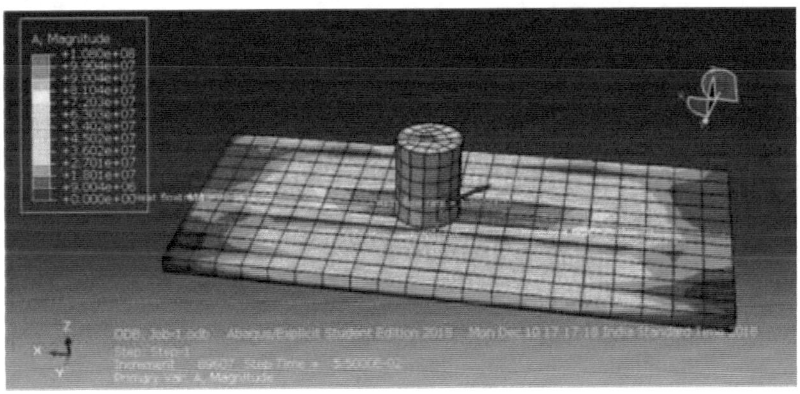

Figure 3. Heat generated profile

3. Experimentation work

Friction stir welding was done on a setup made in vertical milling machine as shown in figure 4. Two aluminium plates having thickness 6 mm is selected. Two grades of aluminium i.e. one is AA 5083 and second one is AA 6061 is selected. Friction stir welding was performed at three process parameters. During the welding, welding speed is kept constant and tool rotational speed is varied. Welding speed is kept constant at 25 mm/min and three tool rotational speed is selected i.e. 500, 710 and 900 rpm. FSW tool is made of SS 410 material. Tool diameter is 25 mm and pin length is 5.7 mm which is slightly lower than the thickness of material. Pin is tapered profile pin.

Figure 4. FSW set-up

3.1 Selection of samples

With friction stir welding three samples were prepared at three different process parameters. Detail of all process parameters are given in table 1.

Table 1. Samples process parameters

Sample No.	Material 1	Material 2	Tool rotational speed (rpm)	Welding speed (mm/min)
1.	AA 5083	AA 6061	500	25
2.	AA 5083	AA 6061	710	25
3.	AA 5083	AA 6061	900	25

4. Result and Discussions

4.1 Tensile test

The FSW welded samples were cut and machined to specified dimensions to prepare samples for tensile test. Tensile test samples were prepared as per ASTM E8 standard. All specimens were tested at strain rate of 5.00 mm/min. Figure 5 shows the three specimens prepared for all three different parameters.

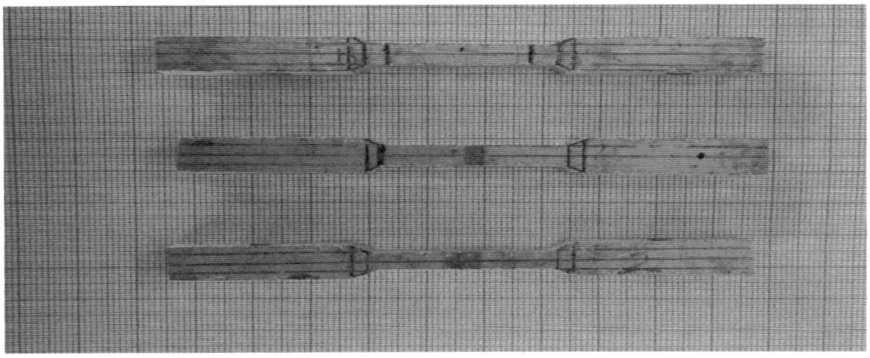

Figure 5. Tensile test specimen

All specimens were tested on UTM machines for the tensile test. Figure 6 shows the image of all samples after testing.

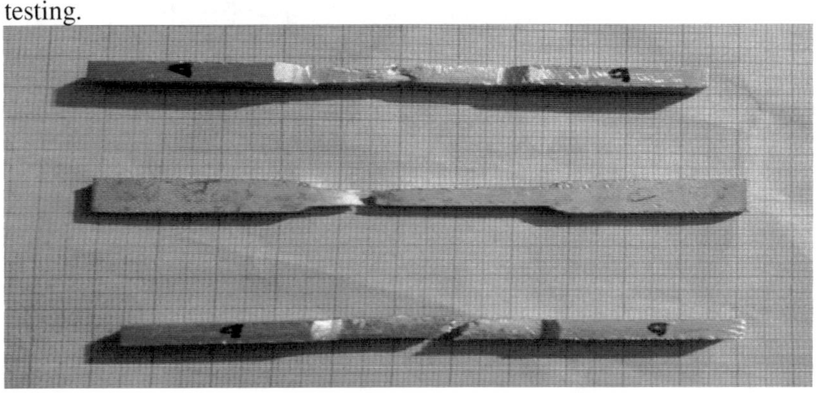

Figure 6. Fractures samples after tensile test

For all the three samples as shown in figure 6, it can be observed that samples were fractured during the tensile test is from middle portion, which is required for good results for UTM. All samples were fractured from the weld zone. So, from this observation it can be concluded that it is a ductile fracture. So FSW samples after welding shows the ductile phenomenon.
All the results obtained from UTM are tabled in table 2.

Table 2: Tensile test results of all samples

Sample no.	Ultimate tensile strength (MPa)	Percentage Elongation (%)
1.	103.08	9.7
2.	105.58	11.07
3	163.23	13.77

Figure 7 shows the stress-strain graph for sample 2 during testing.

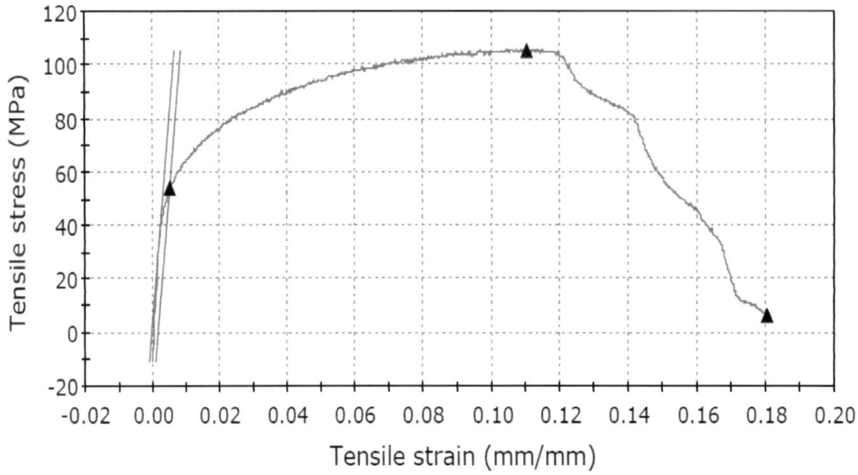

Figure 7. Stress-strain curve for sample 2

From figure 7, it is concluded that all FSW welded samples follows the ductile phenomenon during testing. For all three samples distinguished elastic region and plastic regions are observed from the stress-strain curve. For all test, clear ultimate tensile strength and clear fracture point is observed. In table 2 percentage elongations of all samples were shown, which also confirms the ductile phenomenon.

Figure 8 shows the variation of ultimate tensile strength for all three samples.

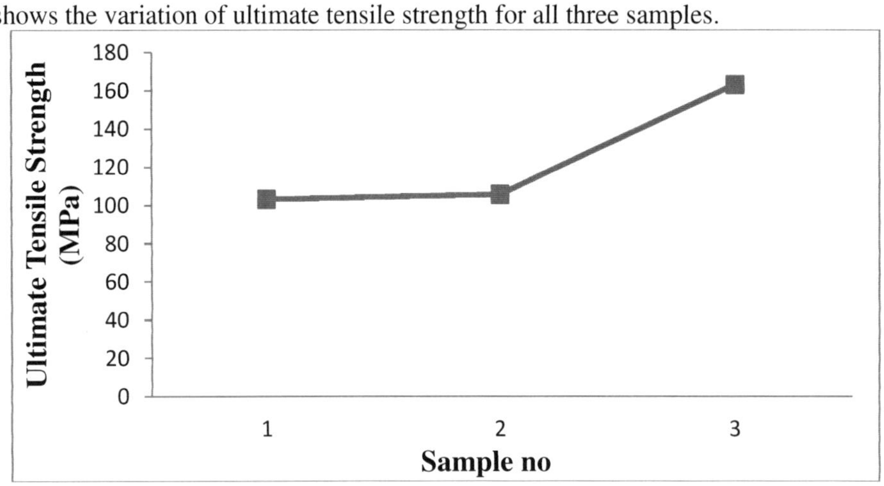

Figure 8. Variation of UTS for all samples

From figure 8 it can be observed that UTS is minimum for sample no 1 and maximum for sample no 2. So, UTS is increases from sample 1 to sample 3, i.e. with increase in rotational speed of tool ultimate tensile strength increases. As the increase in rotational speed, heat generated is increased. So, with increase in generated heat good weld quality is observed. Consequently with increase in rotational speed weld quality is increased. Standard UTS for AA 6061 is 115 MPa and for AA 5083 is 275 MPa. So, it is observed that UTS of welded sample is increased as compared to AA 6061 by 30% and become comparable to second aluminium alloy [11]. It is due to the grain refinement of the welded zone during friction stir welding. Since it is a solid state joining process so, materials are joined below the melting point of materials. Therefore grain refinement is better compare to other process due to formation of small heat affected zone.

4.2 Analysis of fractured surface

When samples are tested in UTM then fractured surfaces of samples are analysed in scanning electron microscope (SEM). After, tensile testing samples were cut in small size containing the fractured surface. Samples were studied under FESEM. Figure 9 shows the FESEM images of fractured portion. Figure 9 (b) is the magnified image of figure 9 (a) , likewise figure (c) is magnified image of (b) and (d) is magnified image of (c) and also (e) is magnified image of (d). From figure 9 clearly circular dimples are visible, which also confirms the ductile fracture of FSW samples during tensile test.

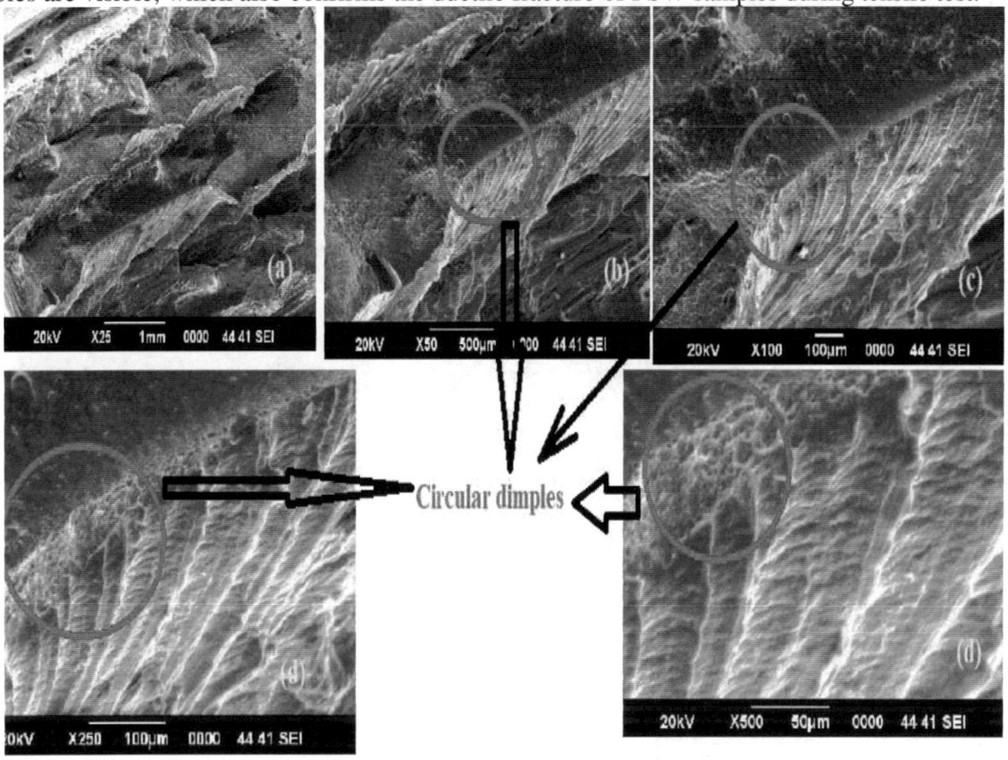

Figure 9. FESEM images of fractured portion

Figure 10 shows the FESEM image of other fractured part of the same sample. From the figure it is cleared that fracture is of ductile phenomenon. Figure (a), (b),(c) and (d) are the consecutive magnified images of the sample. Peaks and valleys type structure is observed in this image.

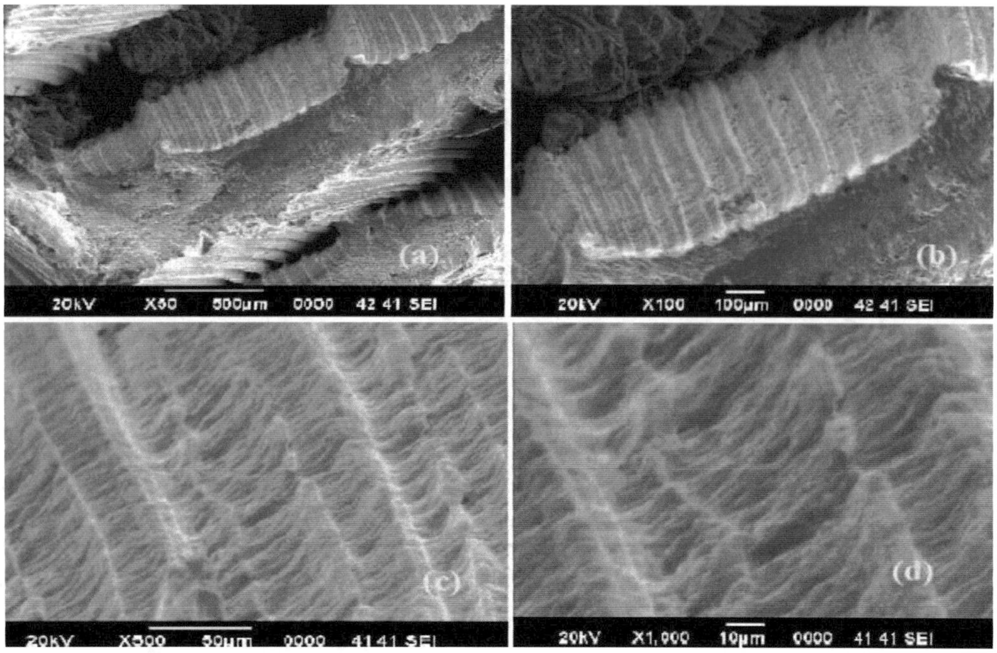

Figure 10. FESEM image of second fractured portion

4.3 Microstructure analysis

FSW samples were cut transversely along the weldment for the preparation of microstructure analysis samples. Samples were prepared on surface polishing machine and etched with keller's reagent (95ml water, 2.5ml HNO_3, 1.0 ml HF). In SEM analysis mixing of two different aluminium alloy can be observed. In figure 11, separating boundry layer between two materials can be observed.

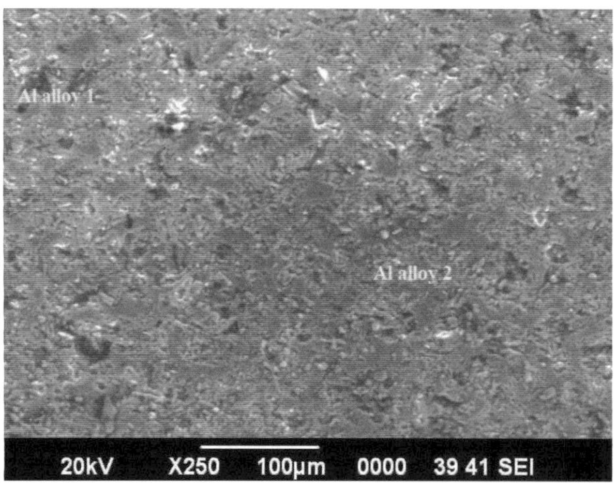

Figure 11. SEM image showing two different aluminium alloy

As shown in table 2 and figure 8, comparable tensile strength with parent material is observed. This can be explained with microstructure study. Fig 12 and 13 shows the equiaxed grains and fine grains of FSW samples. Since, during FSW materials are uniformly joined under frictional heat and pressure.

Materials are joined below melting point temperature. So, formation of second phase is absent here. Therefore fine grain and grain refinement can be achieved in weld zone. This type of grains can be observed in figure 12 and 13.

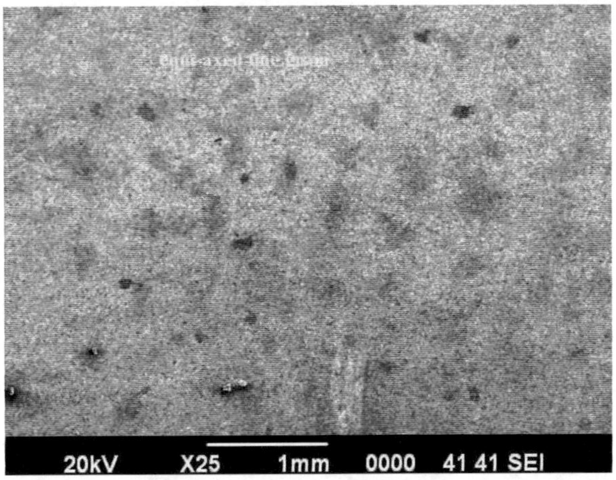

Figure 12. SEM images showing equiaxed grains

Figure 13. SEM images showing fine grains

Figure 14 shows the centre part of the welded samples. It is called weld nugget zone. In this part shoulder pin rotates and give stirring action to the material. Due to the movement of pin of tool this part has more refined grains as compared to other part of the welded samples. As in figure 14, small circular grains can be observed which is due to the heat generated and pressure applied by tool and stirring action of tool pin.

Figure 15 shows the SEM image of thermo mechanically affected zone (TMAZ). It is adjacent to weld nugget zone. It is not affected by the stirring of tool pin. Therefore very less circular grains are observed in this region.

Figure 14. SEM image of weld nugget zone

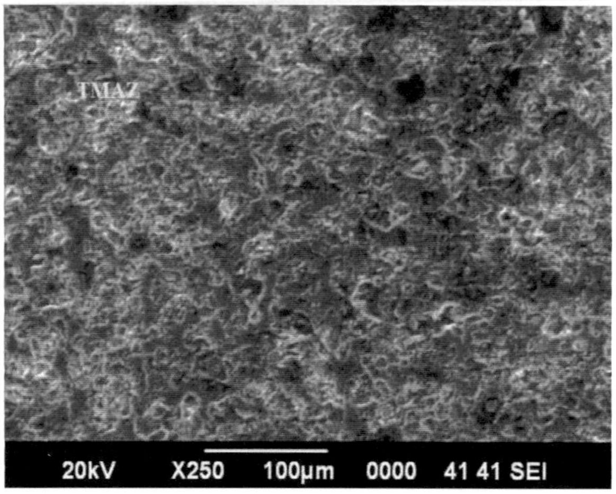

Figure 15. SEM image of TMAZ

5. Conclusion

Above study is based on the friction stir welding of dissimilar alloys. In this study at the staring FSW process is simulated on abacus software. After that experimentation work is done on two dissimilar alloy i.e. AA5083 and AA6061. After experimentation tensile strength, fracture study and micro-structural analysis is done on the FSW joints. On the basis of above study following conclusion is made.

i. FSW process can be well simulated on abacus software. As the time increases temperature generated in work-piece and tool is increased.

ii. From experimentation tensile strength of joint increases with the increase of rotational speed of tool.

iii. All fractures in tensile test follows the ductile phenomenon.

iv. From micro-structural study fine grains are observed in the nugget zone as compared to other weld zones.

References

[1] T. Wenshen, Y. Xinqi, L. Shengli, L. Huijun, *Microstructure and properties of CLAM/316L steel friction stir welded joints*, Journal of materials processing tech., 271 (2019), 189-201.

[2] F. Yan, Y. Zhang, X. B. Fu, J. H. Gao, Anew calculating method of frictional heat and its application during friction stir welding, Applied thermal engineering; 153 (2019), 250-263.

[3] Z. Chenghang, H. Guangjie, C. Yu, Z. Yulong, L. Qing, On the microstructure and mechanical properties of similar and dissimilar AA7075 and AA2024 friction stir welded joints :Effect of rotational speed, Journal of manufacturing processes, 37 (2019) 470-487.

[4] B. Remi, R. Hugo, T. Damien, Z. Yasser, F. Eric, B. Philippe, *Analysis of AA2XXX and AA7XXX friction stir welds*, Journal of materials processing tech. 271 (2019) 312-319.

[5] D. Wei, W. C. Song, Effect of ultrasonic vibration exerted at the tool on friction stir welding process and joint quality, journal of manufacturing processes 42 (2019) 192-201.

[6] G. Kumar, R. Kumar, R. Kumar, Optimization of process parameters of friction stir welded AA5082-AA7075 butt joints using resonance fatigue properties, Bulletin of Polish Academy of Sciences Technical Sciences ,68 (1) (2020), 99-108.

[7] C. Pedro , L. Guilherme, B. Luciano, R. Afonso, S. Jorge, M. Ricardo, P. Marcelo, *Effect of welding speed on friction stir welds of GL E36 shipbuilding steel*, J mater res technol 2019 8 (1), 1041-1051.

[8] R. Abrahams, J. Mikhail, F. Panahsadat, Effect of friction stir process parameters on mechanical properties of 5005-H34 and 7075-T651 aluminium alloys, Materials science & Engineering A 751 (2019) 363-373.

[9] P. Garg, R. Kumar, G. Amit, *Experimental investigation of mechanical and microstructural properties of FSWed Cu-Zn30 joints*, International journal of innovative technology and exploring engineering; 8 (3) 2019, 1-5.

[10] U. Donatus, G. E. Thompson, M. I. Momoh, N. B. Maledi, I. L. Tsai., Ferreira Raphael, Liu Z.; Variations in stir zone and thermomechanically affected zone of dissimilar friction stir weld of AA5083 and AA6082 alloys, Trans. Nonferrous met. Soc. China 28 (2018) 2410-2418.

[11] R. Kumar, B. Bora, S. Chattopadhyaya, G. Krolczyk, S. Hloch, *Experimental and mathematical evaluation of thermal and tensile properties of friction stir welded joint*, International Journal of Materials and Product Technology ,57 (2018), 204-229.

Influence of laser power and welding speed on dendrite structure growth of low power pulsed laser welded super alloy

Uday Kumar[1*], Somnath Chattopadhyaya[1], A K Das[1], Ratnesh Kumar[2], Shubham Sharma[3], Kaori Nagai[4]

[1]Department of Mechanical Engineering, Indian Institute of Technology (Indian School of Mines), Dhanbad India.
[2]Department of Mechanical Engineering, Birla Institute of Technology, Mesra India.
[3]Deptt. of Mechanical Engg., IK Gujral Punjab Technical University, Main Campus-Kapurthala, Jalandhar, Punjab, India.
[4]Associate Professor, PhD, Department of Architecture and Architectural Engineering, Nihon University, Japan.

[*]Email: kumaruday2404@gmail.com

Abstract. The influence of different welding parameters like welding speed and laser power on microstructure and the mechanical properties of low power pulsed Nd: Yag laser welded Nickel based super alloy inconel 617 sheets of thickness 1.5 mm has been investigated. In the present study, the welding of the specimens was done in a single pass aimed to full depth penetration. The microstructure of the various samples at fusion zone and at heat affected zone was characterized by field effect scanning electron microscope (FESEM) while the weld size of the welded metal at different parameters were measured using optical microscope (OM). The microhardness in the various zones for the welded sample has also been investigated. The microhardness of the welded samples was compared with base metal after experimental observations. The XRD technique was used to analyze the crystallography of the material at (2θ degree). The material has been used for the present study because of its exceptional properties of heat resistant and corrosion resistant. The Nd: YAG pulsed laser has been used because of its overlapping factor during welding that reduces porosity in the weld zone.

Keywords: Nd: YAG laser welding, Inconel 617, microstructure, FESEM, Optical microscope, microhardness, XRD.

1. Introduction

The material used for this experimental study is a Nickel based super alloy termed as Inconel 617 having the composition of group VIII elements in the periodic table i.e. Nickel (Ni), Cobalt (Co), Iron (Fe) with Nickel having maximum share of percentage to which other alloying elements are added(Ahmed et al., 2015). It is one of the best materials which has excellent strength, stability, creep-rupture strength and oxidation resistance at very high temperatures above 750ºC that makes it the most appropriate next generation material for high temperature power plants (Park et al., 2011) and (Viswanathan et al., 2013). Hot corrosion behaviour of Inconel 617 superalloy at temperature 700ºC, 800ºC and 900ºC for different exposure times to up to 100 hr has been studied (El-Awadi et al., 2016). Due to high oxidation resistance and corrosion resistance properties Inconel 617 has very wide range of applications in electric-resistance heater, gas turbines, petrochemical industries, aerospace industries and also as pressurized water reactors (PWR) in nuclear reactors (Yeh et al., 2014) and (Tung and Stubbins, 2012). The chemical composition of Inconel 617 is presented in table 1

Table 1 Chemical composition of Inconel 617

Elements	Ni	Cr	Mo	Fe	Co	Ti	Si	Al	C	O	Cl
Weight%	50.79	21.13	9.76	1.07	10.97	0.45	0.27	1.05	3.04	1.06	0.42

The laser welding technology is one of the most important metals joining technique because of its high peak power, precise welds with high aspects ratio, least thermal distortion, high welding velocity, design flexibility, low heat input and very high energy efficient process (Zhou and Tsai, 2008) and (Kazemi and Goldak, 2009). This technique is widely used in automotive, electric and electronics industries (Tzeng, 2000). The key feature of the laser welding technique is the ability to weld any material without using filler materials (Pang et al., 2008). The Nd: YAG pulsed laser welding has been chosen for the present work. This method allows more precise heat control as compared to other joining process. Nd: YAG laser welding of the material also reduces the region of heat affected zone (HAZ), residual stresses and discontinuities because of its overlapping factor(X.-L. Gao et al., 2014). Nd: YAG pulsed laser welding is featured by its periodic heating of the weld pool and its overlapping factor. Because of its high peak power density melting and solidification in the weld pool occur consecutively but solidification time is shorter as compared to other laser and conventional welding processes(Ventrella et al., 2011). There are a very few research works available on Inconel 617 material and especially with Nd: YAG laser welding process. The research work on Nd: YAG laser of other materials are available but very few research works are available for Inconel 617 material. The effect of overlapping factor on porosity, microstructure and mechanical properties of pulsed Nd: YAG laser welded Ti_6Al_4V sheet has been investigated(X. L. Gao et al., 2014). The fatigue damage evolution on pulsed Nd: YAG laser welded Ti_6Al_4V sheet has been analyzed(Liu et al., 2014). In the present study, a nickel based superalloy Inconel 617 has been welded using pulsed Nd: YAG laser system. The microstructure, microhardness and the other mechanical properties like tensile strength has been investigated experimentally for the single pass full depth penetration welding. The microstructure has been analyzed using field effect scanning electron microscope (FESEM) and optical microscope. The microhardness value has been taken using Vickers microhardness tester. Each and every observation has been analyzed in the weld zone and compared these values with the values obtained in the heat affected zone (HAZ) and base metal zone.

2. Experimental setup and procedure

A pulsed wave Nd: Yag laser welding was used to weld the Nickel based super alloy Inconel 617 sheets as specified in Table 2. The shielding gas used during the welding process was nitrogen gas. The schematic representation of Nd:Yag laser welding setup was presented in Fig.1.

Table 2 Nd:Yag setup specifications

Laser setup	Wave-length (in μm)	Max. Avg. Power (in Watt)	Max. Pulse energy (in Joule)	Pulse duration (in ms)	Max. Frequency (in Hz)	Focus Diameter (in mm)	Peak pulse Power (in KW)
Nd: Yag laser	1.06	600	100	0.5-20	1000	0.3-2.2	10

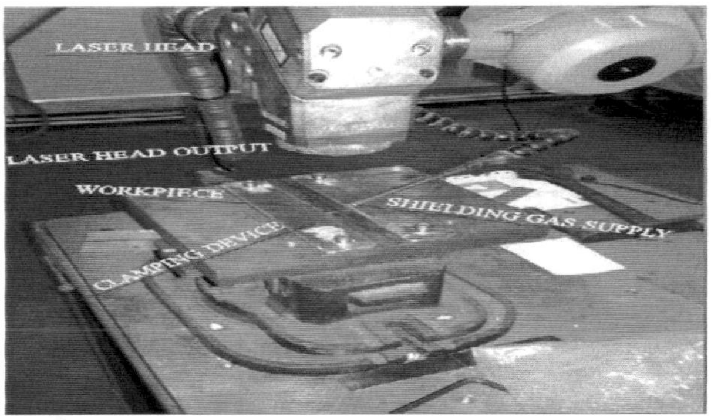

Fig.1 Setup of Nd: Yag laser welding

In the present study, a Nickel based super alloy Inconel 617 of thickness 1.5 mm thin sheet has been used as the work material. For better welding quality, two of the major welding parameter welding speed and laser were varied and rest of the parameters were kept constant.

Table 3 Fixed parameters for present study

Type of weld joint	Butt joint
Process parameters	Values
Laser beam dia. in mm	0.75
Focal position in mm	At the surface
Shielding gas pressure in litre/min.	5
Stand-off distance in mm	160
Pulse width in ms	2.1

Table 3 shows the parameters and their values remains fixed during the study while Table 4 shows the observed values recorded during the experiment for the combination weld parameters.

Table 4 Combination weld parameters

Sample	Power in watt	Speed in mm/s
K1	320	2.5
K2	340	2.5
K3	380	2.5
K4	380	3
K5	400	3

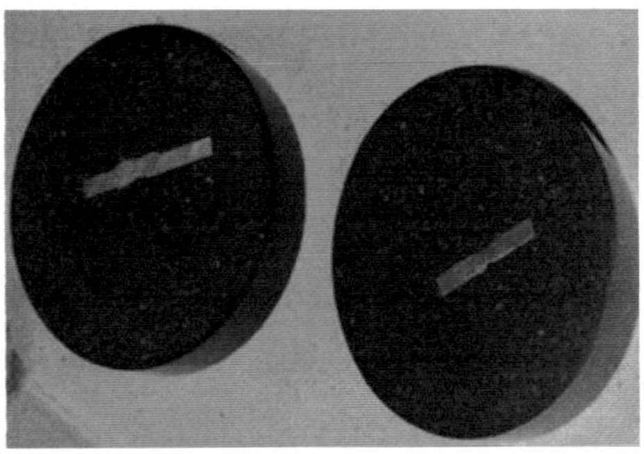

Fig.2 prepared mould specimen

Table 5 parameters applied for preparing mould

Temperature	160 °C
Pressure	100 Kg/cm^2
Time of heating	6 min.
Time of cooling	3 min.

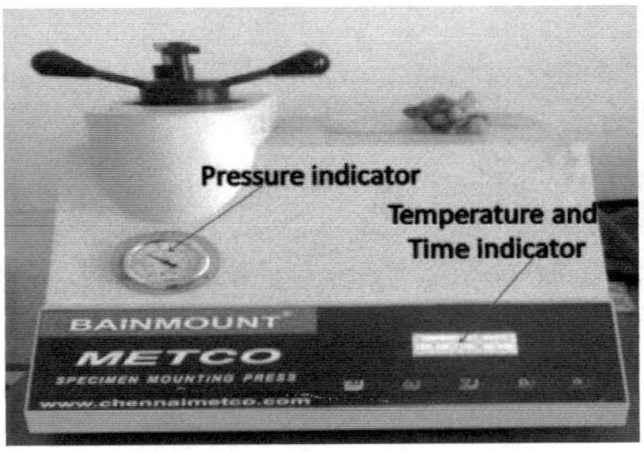

Fig.3. Specimen mounting press

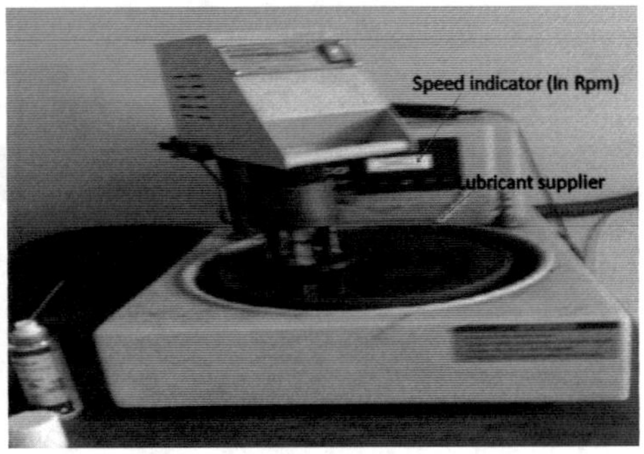

Fig.4. Machine for polishing mould specimen

The specimens were then prepared by grinding with different grits such as 400, 600, 800, 1000, 1200 and 1500 of SiC paper. Every sample were polished at 400, 600 and 800 grit sizes in the clockwise direction and for remaining grit size papers in the reverse direction. The grinding time for each sample for either rotating direction was 8 minutes. Etching was done for the polished surfaces for each sample at room temperature to see the images at various magnifications with the help of optical microscope and FESEM. The time of etching for each sample was 2 minutes for OM and 4 minutes for FESEM observations. The different reagents were used for getting images on OM and FESEM, glycergia reagent (a mixture of 10 ml glycerol, 10 ml hydrochloric acid and 5 ml nitric acid) was used for OM and a solution which was combination of hydrofluoric acid (HF), nitric acid (HNO_3) and distilled water was used for FESEM. Microhardness was measured at the interval of 200 μm distance which includes all the three zones in the sample i.e. base metal zone, HAZ and welding zone. The load applied on the sample was 0.1 Kgf and dwell time of 10s during the hardness testing.

3. Results and discussions

3.1. Microstructure and XRD analysis

The various region of a welded sample was shown in Fig.5 and the different regions i.e. base metal zone (BMZ), HAZ and FZ were indicated in the figure. The image shown was taken through optical microscope.

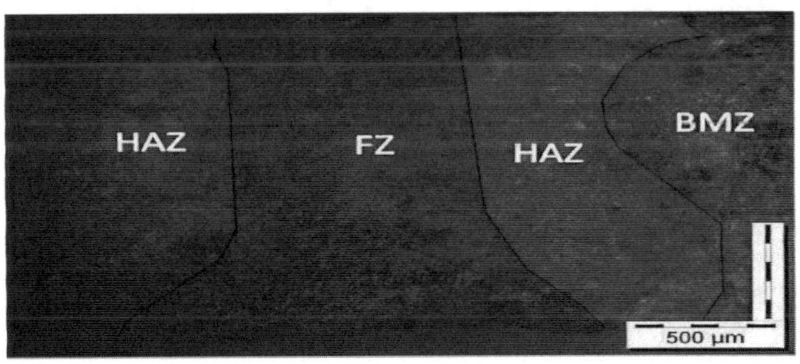

Fig.5. Represents all three zones of welded sample

The microstructure of the samples were analysed through FESEM as shown in Fig.6a, 6b and 6c that shows the microstructures of the specimens at various region i.e. base metal region, HAZ and FZ. The coarser grain size in the base metal region in which crystal orientation was not specified was observed while in the HAZ and FZ a fine equi-axed grain structure was observed due to fully dynamic recrystallization of material at elevated temperature. The change in microstructure was observed in the HAZ region due to recrystallization of the materials while it meets its austenitic temperature. The grain structure in the HAZ and FZ region of the samples were seems like branch of tree, known to be dendrite structure in terms of metallurgy(Kumar et al., 2017). These structures were growing significantly as molten metal freezes. The dendrite structure of the superalloy Inconel 617 was also observed by some other researchers(Ren et al., 2015). The surface energy present in the liquid solid interface is the main reason for the growth direction. The dendrite growth at HAZ was observed due to re-solidification process of the crystals when molecules were start to contract and tries to minimise its area of surfaces having maximum surface energy. The size of the grain largely related to the nucleation and the crystals growth in case of equi-axed grain structure. The dendritic growth of the crystals was shown in fig at HAZ while a fully developed dendrite structure was observed at FZ. Fine $M_{23}C6$ carbides, Ti(C, N) carbide and nitride also exist at FZ and crack at FZ was also observed in Fig.6d.

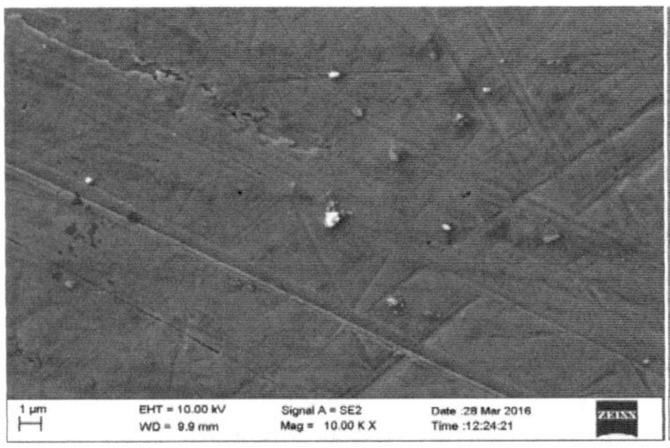

Fig.6a Base metal

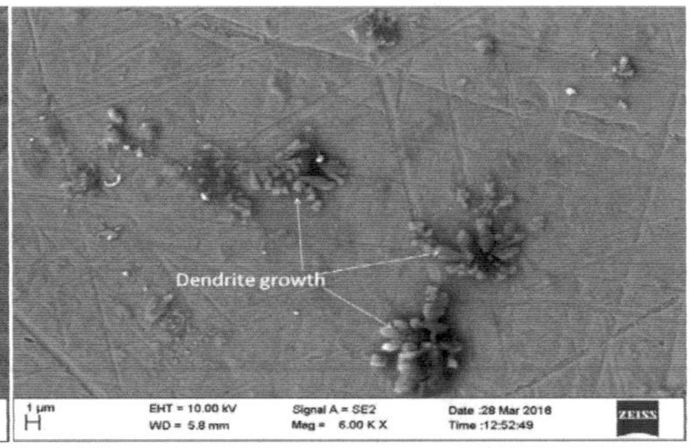

Fig.6b HAZ region

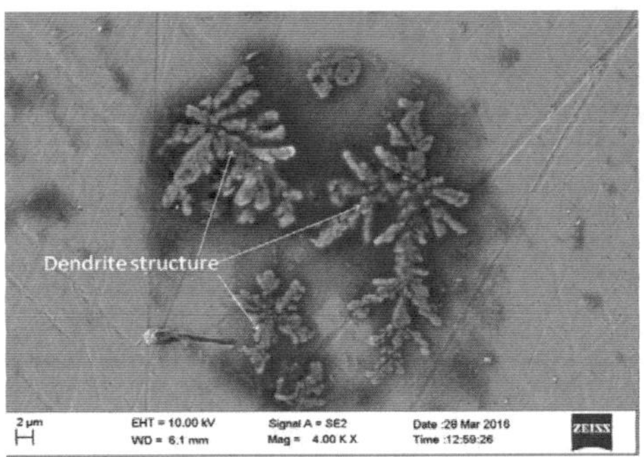

Fig. 6c FZ region

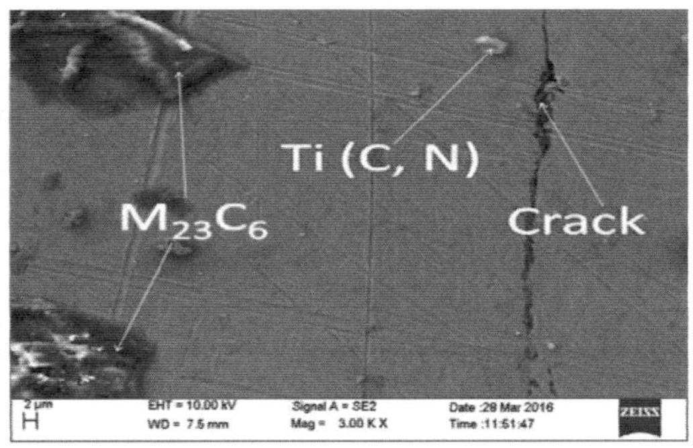

Fig.6d shows M23C6, Ti(C, N) and crack at FZ

The chemical composition of the material at base metal region and FZ region was shown in Fig.4a and 4b. A very little change was observed in the composition of sample at welded region in comparison with base metal. The composition of the sample was observed through EDAX analysis. The presence of nitrogen was observed as nitrogen gas was used as shielding gas which is very stable at room temperature but become unstable at elevated temperature. The increase in percentage of carbon element in the welded region shows the tendency of formation of carbide at elevated temperature of this superalloy.

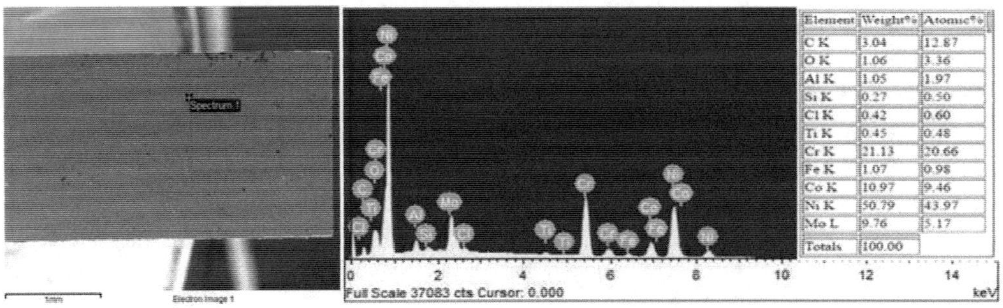

Fig.7a FESEM image and EDAX spectrum of the particles shown in square area in the Base Metal.

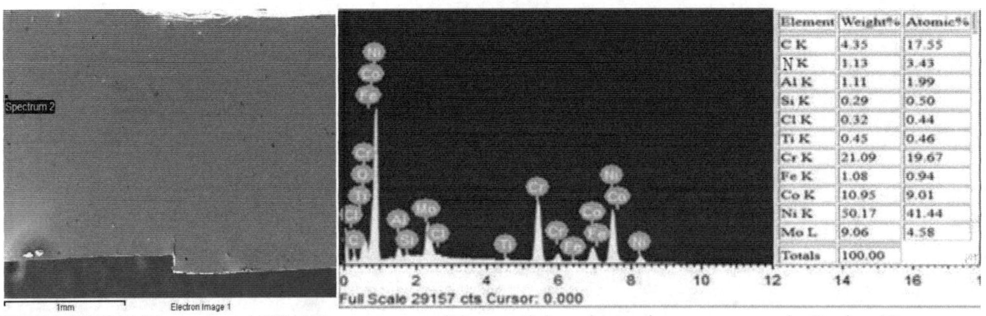

Fig.7b FESEM image and EDAX spectrum of the particles shown in square area in Fusion Zone

The XRD analysis was carried out to analyse the crystallographic structure of the samples at various positions (2θ°) as shown in fig. The analysis was done using X'pert Highscore plus software. The compounds like carbides, oxides and nitrides formed by the major elements such as Ni, Cr, Co, Mo etc. of the alloy were identified at the peak positions.

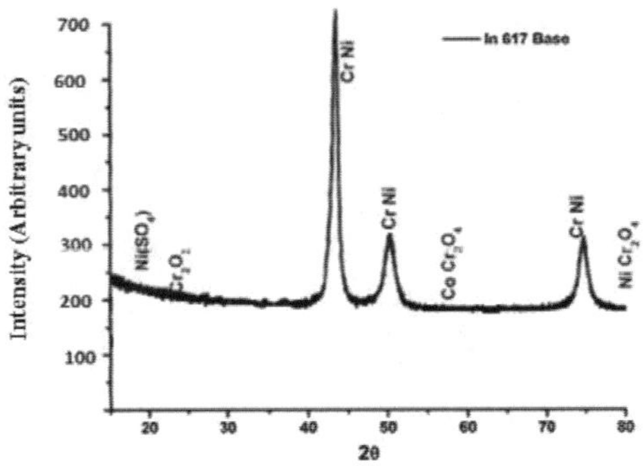

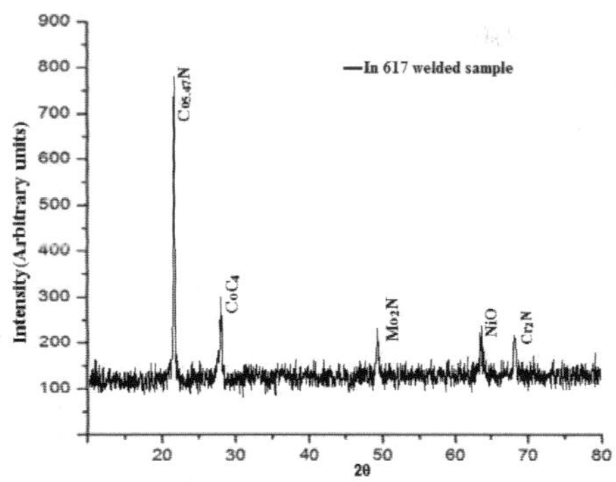

Fig.8a XRD graph at base metal region **Fig.8b** XRD graph at welded region

For both the regions, the positions ranged from 20° to 80° and the compounds formed by the major elements were identified at the peaks as shown in Fig.5a and 5b. For the BM region, oxides of the major elements were identified as some oxygen was present in the base metal but nitrides and carbides compounds were negligible as the carbon element was stable at room temperature. For welded region, nitrides of the major elements were identified and also there was some carbide compounds formation took place. As the nitrogen gas was used as shielding gas, the nitrogen molecules become unstable at elevated temperature that cause the bond dissociation of the nitrogen molecules which lead to the formation of nitride compounds such as chromium nitride (Cr_2N), Molybdenum nitride (Mo_2N). The formation of carbide compounds like cobalt carbide (CoC_4) of major elements was also observed in the weld zone. The domination of nitride and carbide compounds in the fusion zone shows anisotropic behaviour of the material in this zone as face centre cubic (fcc) structure of the crystals were observed along various directions.

3.2. Weld width

The images shown in fig. were taken through optical microscope and size of the weld width from bottom to top of the weldment was measured using software. The images shown of the welded samples clearly indicate the full depth penetration welding of the samples. The larger weld width on the top of the weldment as compare to bottom part of the weldment indicates the effect of heat input or laser power on the size of weld width.

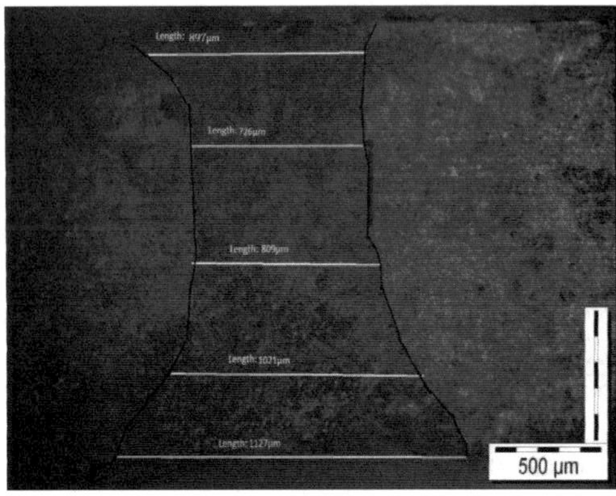

Fig.9a weldment at 320W, 2.5 mm/s

Fig.9b weldment at 340W, 2.5 mm/s

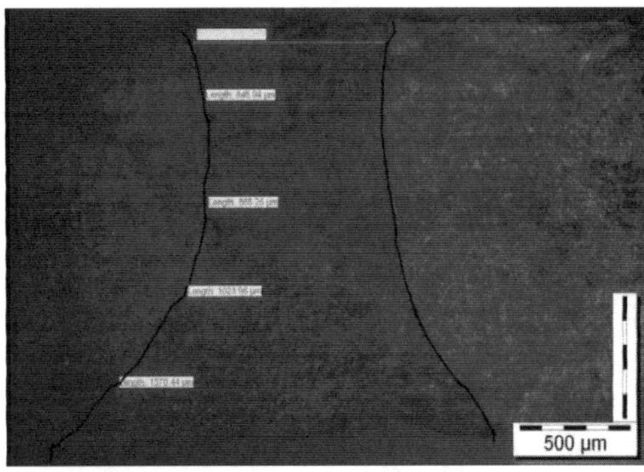

Fig.9c weldment at 380W, 3mm/s

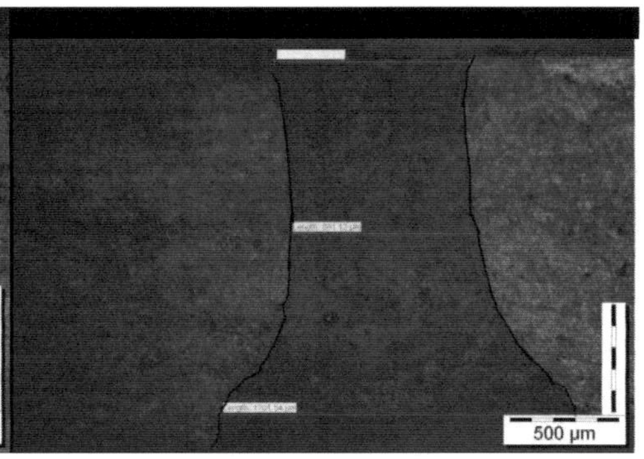

Fig.9d weldment at 400W, 3mm/s

The graph shown in fig.10 indicates the effect of laser power on the size of weld width which explains that the size of weld width increases as laser power increases. In case of welding, welding speed also plays a vital role, as the welding speed increases the size of weld width decreases therefore, various combination of power and speed was used and analyzed to get lower size of weld width and HAZ along with high quality of weldment. The value of the weld width for the 400W laser power was found to be highest in the top side of the weldment and lowest in the root side.

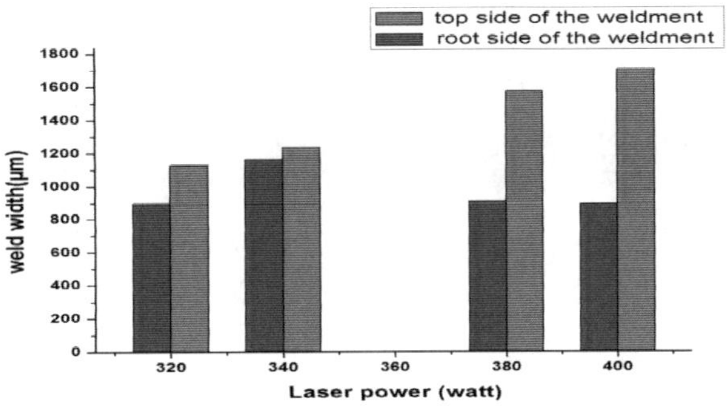

Fig.10. Represents bar chart between laser power and weld width

3.3. \Microhardness

The graph shown in Fig.11a and 11b represent the microhardness distribution of two different samples welded at same laser power but at different welding speed for various zones. In both the cases the maximum value of microhardness was observed in the FZ region while the least value was observed at BMZ region. The high value of microhardness in the weld zone was due to the presence of temperature gradient at FZ and HAZ which caused rapid cooling rate. At lower welding speed, due to preheating process of metal the temperature gradient was found to be low as compare to high welding speed. Therefore, high hardness value was observed in the FZ and HAZ region in case of high welding speed with same laser power. The high value of microhardness in the FZ region indicates increase in brittleness of the material when compare to original material. The rapid cooling rate during welding process was main cause of formation of hard and brittle structure at the weld zone which is also supported by the results of some other researchers (Casalino et al., 2015). At 380W laser power and 2.5mm/s welding speed the maximum value of microhardness at FZ was found to be 296.54HV while the maximum value for 324.26HV

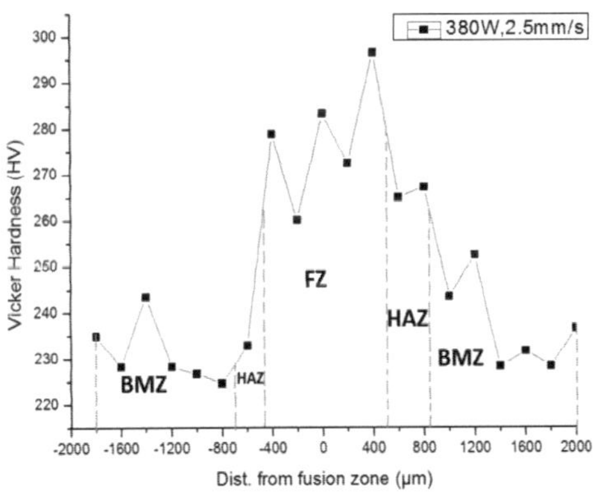

Fig.11a Microhardness of sample K3

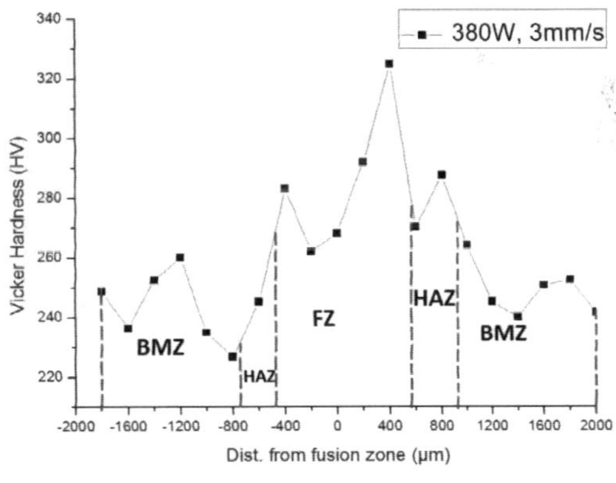

Fig.11b Microhardness of sample K4

4. Conclusions

The present work was mainly focused on the study of influence of laser power and welding speed on the microstructure, weld width and mechanical properties such as tensile strength and microhardness of the low power Nd:Yag laser welded Inconel 617 thin sheet of thickness 1.5mm. From the above results, we conclude that the weld quality was very good as the tensile strength of the welded specimen were greater than 95% of the tensile specimen of base metal. To ensure even greater weld quality to get the tensile strength of welded specimen upto or greater than 100%, the inert gas (like

Argon, Co_2 etc.) can be used as shielding gas for further research work so that the gas could not be entrapped within the specimen during welding and will not have any alter effect. The entrapped nitrogen was seen during EDAX analysis and also observed due to formation of nitride compound in the XRD analysis. The microstructure was identified as dendrite structure and the coarser grain structure of base metal and the fine grain structure in the weld zone of the sample was identified from the fractograph of the tensile specimen. The weld size was directly affected with heat input and the welding speed as the heat input increases the size of weld width also increases but low weld width was observed at high welding speed. The microhardness analysis confirmed the formation of hard and brittle fusion zone.

References

[1] Ahmed, G.M.S., Mohiuddin, M.V., Sultana, S., Dora, H.K., Singh, V.D., 2015. Microstructure Analysis and Evaluation of Mechanical Properties of Nickel Based Super Alloy CCA617. *Mater. Today Proc.* **2,** 1260–1269.

[2] Casalino, G., Mortello, M., Campanelli, S.L., 2015. Ytterbium fiber laser welding of Ti6Al4V alloy. *J. Manuf. Process.* **20,** 250–256.

[3] El-Awadi, G.A., Abdel-Samad, S., Elshazly, E.S., 2016. Hot corrosion behavior of Ni based Inconel 617 and Inconel 738 superalloys. *Appl. Surf. Sci.* **378,** 224–230.

[4] Gao, X.-L., Liu, J., Zhang, L.-J., Zhang, J.-X., 2014. Effect of the overlapping factor on the microstructure and mechanical properties of pulsed Nd:YAG laser welded Ti6Al4V sheets. *Mater. Charact.* **93**, 136–149.

[5] Gao, X.L., Zhang, L.J., Liu, J., Zhang, J.X., 2014. Porosity and microstructure in pulsed Nd:YAG laser welded Ti6Al4V sheet. *J. Mater. Process. Technol.* **214**, 1316–1325.

[6] Kazemi, K., Goldak, J.A., 2009. Numerical simulation of laser full penetration welding. *Comput. Mater. Sci.* **44,** 841–849.

[7] Kumar, P., Saw, K., Kumar, U., Kumar, R., Chattopadhyaya, S., Hloch, S., 2017. Effect of laser power and welding speed on microstructure and mechanical properties of fibre laser-welded Inconel 617 thin sheet. *J. Brazilian Soc. Mech. Sci. Eng.***17,** 734-738

[8] Liu, J., Gao, X.-L., Zhang, L.-J., Zhang, J.-X., 2014. A study of fatigue damage evolution on pulsed Nd:YAG Ti6Al4V laser welded joints. *Eng. Fract. Mech.* **117,** 84–93.

[9] Pang, M., Yu, G., Wang, H.H., Zheng, C.Y., 2008. Microstructure study of laser welding cast nickel-based superalloy K418. *J. Mater. Process. Technol.* **207**, 271–275.

[10] Park, Y.S., Ham, H.S., Cho, S.M., Bae, D.H., 2011. An assessment of the mechanical characteristics and optimum welding condition of Ni-based super alloy. *Procedia Eng.* **10,** 2645–2650.

[11] Ren, W., Lu, F., Yang, R., Liu, X., Li, Z., Elmi Hosseini, S.R., 2015. A comparative study on fiber laser and CO2 laser welding of Inconel 617. *Mater. Des.* **76**, 207–214.

[12] Tung, H.M., Stubbins, J.F., 2012. Incipient oxidation kinetics of alloy 617 and residual stress of the oxide scale formed in air at temperatures between 850 and 1000°C. *J. Nucl. Mater.* **424,** 23–28.

[13] Tzeng, Y.-F., 2000. Process Characterisation of Pulsed Nd:YAG Laser Seam Welding. *Int. J. Adv. Manuf. Technol.* **16,** 10–18.

[14] Ventrella, V.A., Berretta, J.R., De Rossi, W., 2011. Micro welding of Ni-based alloy Monel 400 Thin foil by pulsed Nd:YAG laser. *Phys. Procedia* **12,** 350–357.

[15] Viswanathan, R., Henry, J.F., Tanzosh, J., Stanko, G., Shingledecker, J., Vitalis, B., Purgert, R., 2013. U.S. Program on materials technology for ultra-supercritical coal power plants. *J. Mater. Eng. Perform.* **22,** 2904–2915.

[16] Yeh, T.K., Chang, H.P., Wang, M.Y., Yuan, T., Kai, J.J., 2014. Corrosion of Alloy 617 in high-temperature gas environments. *Nucl. Eng. Des.* **271,** 257–261.

[17] Zhou, J., Tsai, H.L., 2008. Modeling of transport phenomena in hybrid laser-MIG keyhole welding. *Int. J. Heat Mass Transf.* **51,** 4353–4366.

Latest trends in Additive manufacturing

M.V.D Sai Kalyan, Harish Kumar and Leeladhar Nagdeve

Department of Mechanical Engineering, National Institute of Technology Delhi, Sector A-7, Institutional Area, Narela, Delhi 110040, India

kalyan865@gmail.com

Abstract. Additive manufacturing is the most recent and the most revolutionary technology in the production industry. It is an unconventional method of manufacturing. In this manufacturing method, material is added one layer upon another layer to get the required shape and size of object. Most complex objects can be built very easily by this technology. Additive manufacturing has the capability to reduce the supply chain by reducing a number of processing stages of a product. Additive manufacturing technology developed a lot from the day 3D printing technique, introduced as Stereolithography in 1984 by Charles W. Hull. Today we have different types of 3D printing techniques such as polymer resin based printing, wire based printing(FDM) and powder based printing(SLS) and some hybrid Additive manufacturing like High speed sintering. Additive manufacturing has applications spread from Aerospace industry to Toy industry and from Biomedical industry to Construction field. Additive manufacturing leads to "Social manufacturing" which involves customers in all stages of product manufacturing. Most of the large scale industries are investing a large capital in R&D for this 3D printing. Many startups are entering into the market each day. Today bio-compatible kidneys , bones and even hearts are bio printed, multi-storey buildings have already been 3D printed and research works are in progress. Now researchers are stepping towards 4D printing which uses "Intelligent" materials which can be coded. Research work is being done on different Additive manufacturing techniques and different materials and process parameters. This paper is a review of Additive manufacturing and it's latest trends. The Trends related to printing technique as well as materials.

Keywords: 3D printing, Additive manufacturing, Stereolithography, Selective laser Sintering, Selective laser melting, Bio-printing, Fused deposition modeling.

1. Additive manufacturing

AM is similar to ink jet printing on paper but added with the 3rd dimension. The Additive manufacturing involves designing, slicing printing, printing the object, post processing and inspection and testing. In designing phase 3D models are created by using some computer aided software packages. These designs either be created manually on a CAD software or be generated automatically by using scanned data and digital image processing then by using some software packages. These CAD files are saved with Standard Tessellation Language (STL) having .STL extension. These .STL files then transferred to slicing software packages. Generally the 3D printer manufacturing companies will provide their own slicing software. Some online platforms are also providing these slicing software packages which can be compatible with the printers. This slicing software will slice the 3D model into a number of 2D planes and generate the G-codes and M-codes automatically for each layer and save this data in a format which can be read by the 3D printer. The 3D printer will read the data and will print the product in layer by layer manner.

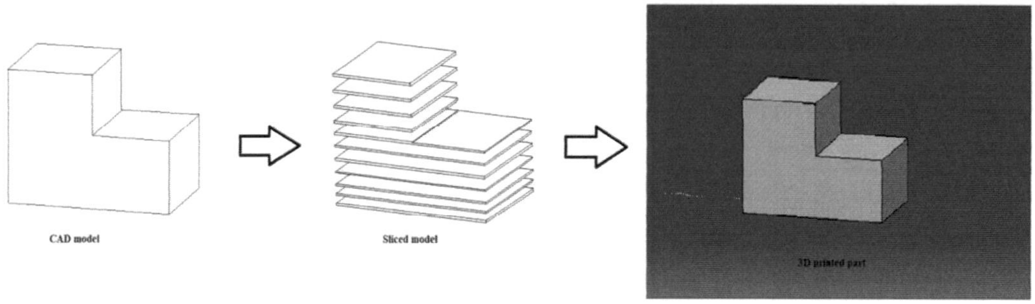

Figure 1. basic phases of Additive manufacturing

While printing, the printer will follow the instructions generated by the slicer software and material extruder, photo curing light, laser light and base table will move accordingly. There are different types of additive manufacturing techniques available based on material feeding and curing techniques from the beginning but only few of them are commercialized and became popular. After making products some post processing like finishing, cleaning, heat treatment etc. should be performed depending on the type of printing technology.

2. AM processes

Additive manufacturing is an umbrella term under which different manufacturing techniques are available. Additive manufacturing techniques are generally based on the form in which the material is being used. Different AM processes are based on material form are:

- Photopolymer resin
  - Stereolithography (SLA)
- Metal sheets
  - Laminated object manufacturing (LOM)
- Wire material
  - Fused deposition modelling (FDM)
- Powder material
  - Selective laser sintering (SLS)
  - Selective laser melting (SLM)
  - Direct metal laser sintering (DMLS)

However some hybrid techniques like Direct energy deposition, High Speed Sintering etc., are being developed for improving quality and to reduce time consumption. Out of different techniques, Fused deposition modeling (FDM) and Selective laser sintering/ selective laser melting (SLS/SLM) are most commonly used.

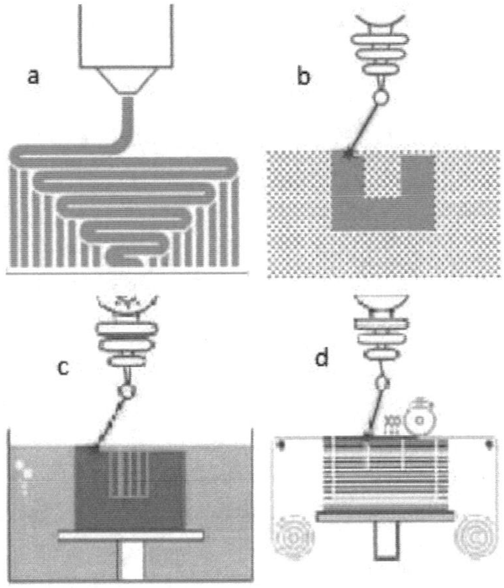

Figure 2. (a) Fused deposition modeling (b) Selective laser sintering / Selective laser melting (c) Stereolithography (d) Laminated object manufacturing [1]

2.1. Fused deposition modeling

FDM is the most commercially available printing technique. It is famous because of its low cost, ease of printing, flexibility in materials, ease of material availability and its applications. These printers use material in the form of wire. The wire will pass through the print jet. In the print jet hot extrusion of the material will take place and sends semi molten material out from the nozzle. Continuous supply of material is provided by the spool of material. The extruder will deposit the material which fuses later. The extruder will move in the contours defined by G-codes and M-codes generated by slicer software. After printing one layer, the extruder will print the next layer up on the previous layer of the component. Most of the FDM printers are available in the form of desktop printers and are very easy to use. Most of the FDM printers use plastic as a feed material however few models use metal wires also. The parts produced require less number of post processing operations so that we can find FDM printers even in some schools also [2].

Thermoplastic materials such as Acrylonitrile Butadiene Styrene (ABS) and its variations, nylon, ply-lactic acid, polystyrene, polycarbonates and thermoplastic urethane.MED610, a raw material that Stratasys provides is bio-compatible etc. are used as the feed material for the FDM printers. The printed part strength varies from material to material and it also depends on some process parameters. The process parameters include layer height, initial layer height, line width, wall thickness, top/bottom surface thickness, infill, printing nozzle temperature, build plate temperature and print speed. Here time and the quality of product both are contradicting each other. If we want a quality product we need to compromise with time vice versa., Multi color as well as multi material 3D printing is possible with this type of 3D printers [3].

Even though most of the FDMs are using plastics as a material, some FDM printers use metals and metal alloys as printing material. These 3D printers are named as FDMm (Fused Deposition Modeling metal printers). These FDMms may use different deposition techniques while printing. One of those techniques is by using a heater in the depression nozzle for melting the wired material which is the same as FDM for plastic material. Electrohydrodynamic redox printing (or EHD-RP) is another FDMm that utilises electrochemistry to synthesise metallic deposits. Here focussed electrohydrodynamic ejection of metal ions dissolved from sacrificial anodes and their subsequent reduction to elemental metals on the substrate [4]. FDM is mostly used in 3D printing technique. But this is limited to print polymer parts only while

Selective Laser Sintering can build both polymer as well as metallic parts Selective Laser Melting is a dedicated printing technique for metals only. In addition to prototyping, Additive manufacturing can compete with conventional manufacturing techniques in mass production and the metal additive manufacturing can help for mass manufacturing [5].

2.2. Selective laser sintering/ selective laser melting

Selective laser sintering uses material in powder form. Selective laser sintering or selective laser melting is the most dominating printing technique because it allows printing with metals. Metals have the capability to withstand loads like tensile, compressive, impact, fatigue, creep loads.SLS using powdered material which is cured by high intensity laser beam. Different powder manufacturing methods are available as in the case of powder metrology. Some binders are added to the powdered material to improve the properties. The powder material can be plastic as well as metal also, but the SLS/SLM preferred to use metals as the powders. While printing the part the material is spread on the printer bed with the help of a roller for one thin layer. Laser now focuses on the powder bed selectively in the path decided by G and M codes generated by the slicer software. This laser melts the powder material and fuses together. After sintering on one layer, the roller will move on the powder bed and spread the powdered material on the previous layer then the laser will cure the powder selectively. This process will be repeated until the completion of the entire product [6]. In this type of printing, The loose powder of the previous layer will act as the support material for cantilever parts. So, no need to use additional support material required as in the case of FDM. The loose powder after sintering can be reused for many numbers of cycles.

Different binding techniques are available in the powder bed additive manufacturing. Solid state sintering (SSS), uses heat energy between the temperature range from half of melting point temperature to the melting point temperature. The diffusion is the most important process in this SSS. It involves neck formation between the powder particles and sintering occurs by lowering the free energy between the powder grains. A wide variety of materials can be processed by this method. Chemically induced binding uses self binding property. In this binding method, particles are heated to higher temperatures so that disintegration of particles into elements occurs. The disintegrated element bonds with oxygen and forms oxides which acts as a binder. Liquid phase sintering can be a partial melting or a full melting process. Liquid phase partial melting can be classified into two groups based on usage of binder material. In one group, a binder material and a structure material are used. These binders and the structure materials can be in separate powder form or each grain of powder can be a composite of structure material and binder material or the structure material can be coated with a binder material. Here the heat source melts the binder material and this molten binder will bind the structure material together. In the second category of partial melting, binder material is not used. This non binder partial melting can be a single phase material partial melting. In this type the heat source is not sufficient to melt the structure material completely so that only the shell of powder grains can be melted. The core of the material remains unmelted. The molten material will combine the unmelted core materials. The non binder partial melting can also be a fusing powder mixture in which partially molten powders consists of multiple phases. In full melting, no binder material is used. Full melting technique is used for achieving full dense objects and the properties near to the traditionally manufactured products. Full melting can be a single component single material where powders are of single material or single component alloyed material where a single kind of grains consists of alloyed materials or fusing powder mixture [7].

Part produced by liquid phase partial melting will have less density, nearly half of the theoretical density because the structure material is not fused. These parts are called green parts. Here bonding strength is sufficient only for maintaining the shape. Crack free green parts can be built by SLS but strength is very less. So post processing operations are compulsory. The post processing has the objective to increase the density of the green part. This post processing involves debinding where bonding polymers are removed, Warm Isostatic Pressing (WIP) which will fuse the structure material by homogeneous pressure in all directions and pressure infiltration where another molten material is forced to enter and fill the internal pores [8]. This secondary processing may lead to shrinkage of the volume. The SLS is easy to control when compared to liquid phase full melting or Direct Metal Laser Sintering (DMLS) Selective Laser Melting (SLM) because in SLS structure material does not melt so that viscosity is more. But in DMLS, the structure material will also melted so viscosity is minimum. As viscosity is minimum and because of surface tension of liquid metals, it is difficult to control the metal flow during the printing process. As there is full melting of structure material, porosity of the final product is least and density of the part is maximum so no need to do post processing operations for increasing the density.

2.3. *Stereolithography*

Stereolithography is the pioneer of Additive manufacturing technology. This technology uses photo curable polymeric resin as a printing material. These photosensitive materials are cured by a photochemical process. Ultraviolet light is used as a source of energy and cures by cross linking the polymers with each other and converts liquid phase resin into solid phase layer by layer. The stereolithography can be either bottom down type or bottom up type. In bottom-down type of stereolithography, the base table moves down into the polymeric resin in steps wise for each layer of printing while the UV light source will focus on the vat of polymer. In the bottom-up layer, the UV light source will be inside of the resin while the base plate will move upwards.

There are three types of resin printing technologies available based on the method of light source used. In the first type it is the UV laser which is used to draw the predefined design or shape defined by slicing on the surface of the vat of polymer resin. In this laser intensity, hatch spacing and thickness of the layer will define the resolution of the printed part. The strength of the part printed in this way will depend upon the layer thickness. As the layer thickness is minimum, the strength will be maximum [9]. The second type uses Digital light processing (DLP) projector as the light source for curing the resin [10]. DLP projector consists of a digital mirror device (DMD) which has thousands of mirrors and each mirror has a size of order of red blood cells. Light is selectively directed using DMD and cures the whole layer at a time. Here light comes from a small source to cover a wide area. This has the chance of distorted pixels at their edges. For best resolution smaller parts are preferred by DLP type of 3D printers. In the third type, the light source is a liquid crystal diode (LCD) projector [11]. It uses an array of UV LCDs. Light from the flat LCD panel shines directly in a parallel fashion on the vat of polymer. Here it is similar to the DLP projector and prints the whole layer at a time. In LCD, light does not expand. So, pixel distortion is not an issue with LCD printing. The more pixels it has, the better print quality it gives.

2.4. *Laminated object manufacturing*

In LOM printing technique, parts are made by sticking sheet material one up on another. The required part of the sheet from continuously supplied sheet material is cut by a laser and this part of sheet is then made to stick by applying some adhesives, external pressure and by heating. The adhesive coating is applied to the bottom side of the sheet. A roller used for the hot-pressing process. Properties of the printed part can be varied by changing temperature, pressure values of the roller and by applying the pressure and temperature separately [12].

3. **Materials**

Based on the type of application and availability different materials are being used in Additive manufacturing. Mainly there are two phases of material used i.e liquid phase material and solid phase. The earliest is the pioneer material for the revolutionary 3D printing technology. The stereolithography uses material in liquid phase i.e thermosetting photocurable polymeric resin. The former one is in solid form which can be either wired form or powder form. Based on the printer manufacturers different materials available in each form of materials.

Table 1. Categories of materials available for Additive manufacturing.

Material type	Material form	Printing technique
1 Plastics	1.1 Photo curable resin	1.1.1 Stereolithography (SLA)
	1.2 wire	1.2.1 Fused deposition modeling
	1.3 Powder	1.3.1 selective laser Sintering
2 Metals	2.1 wire	2.1.1 Arc welding technique 2.1.2 Direct energy deposition
	2.2 powder	2.2.1 selective laser sintering 2.2.2 Selective laser melting
	2.3 sheets	2.3.1Laminated object manufacturing
3 Ceramics	3.1 Powder	3.1.1Selective laser sintering
4 Cement	4.1 Powder	4.1.1 Binder jetting
	4.2 Cement slurry	4.2.1 Fused deposition technique
5 Bio-ink	5.1 Resin	5.1.1 stereolithography
	5.2 wire	5.2.1 Fused deposition modeling

4. Processing parameters to control

There are many parameters that can decide the quality of the final product. Generally the quality of the final product depends on the design data generated by the designer, material properties and machine parameters. Different parameters influencing the quality of FDM, SLA and SLS/SLM printing techniques are mentioned in table 2. In the market various varieties of materials are available and each material property will vary from one manufacturer to another manufacturer. It is because of a lack of standardization in making of the materials. Several initiatives have been taken by several organizations around the world and several institutions like International Organization for Standardization (ISO),

Additive manufacturing society of india (AMSI), American Society for Testing and Materials (ASTM), with the help of projects and technical groups focused on the standardization of AM [13]. The influence of material properties can be eliminated by standardization of powder material. While manufacturing a product by 3D printing machine we are available with many process parameters.

Table 2. influencing parameters on product quality [18][19][20]

Material properties (SLS/SLM)	Process parameters (SLS/SLM)	Process parameters (FDM)	Process parameters (SLA)
Viscosity	Laser power	Layer thickness	Liquid resin depth
Surface tension	Scan speed	Build orientation	Curing time
Particle size	Scan path	Raster Angle/ Raster orientation	Post curing time
Particle distribution	Raster angle	Air Gap	Layer thickness
Particle shape	Scan vector length	Extrusion temperature	Part orientation
Absorptivity	Laser type	Print speed	Curing temperature
Reflectivity	Hatch spacing	Infill pattern	Temperature of previous layer
Thermal conductivity	Print bed temperature	Infill density/ Interior infill percentage	Photo initiator concentration
Specific heat	Layer thickness	Nozzle diameter	Intensity of UV light
Emissivity	Melt pool size	Raster width	Absorptivity of photo initiator
Melting temperature	Intensity of heat source	Number of contours	Fill spacing
Component ratio	Scan radius	Contour width	Fill cure depth
Chemical composition	Gas flow	Contour to contour Air gap	Line width
Powder density	Orientation of the part		Hatch spacing
Powder porosity	Pre heating		Hatch overcure
	Cooling rate		Border overcure
	Wavelength of laser		
	Machine		

These parameters will vary from machine to machine. By controlling these machine parameters we can control the quality of the final product according to our requirement. But it is difficult and very costly to choose each and every parameter available for controlling. The process parameters are to be selected based previously available data on their influence on the final quality of product. Design of experiments (DOE) are experimentation techniques for optimization of process parameters. Different design of experiment techniques like One-factor at a time, Taguchi, ANOVA, Genetic algorithms, Artificial neural networks (ANN) etc., are available to develop the process parameters and optimization of process parameters. For a specific property optimization, many experiments are needed in a trial-and-error approach that will lead to a high cost of the SLS process. One factor at a time is one approach. It reduces the number of experiments. But the problem is that it will not consider mutual interaction of parameters.

Full Factorial and fractional factorial mutual interaction of parameters is considered. Aboma Wagari Gebisa et. al., [14] have used five process parameters: air gap, raster width, contour number, raster angle and contour width for investigation of tensile properties of ULTEM 9085 in the FDM process with the help of full factorial design of the experiment method. Raster angle, road width, air gap, Layer thickness, number of contours and build orientation as primary parameters in investigation of dynamic mechanical performance of FDM/ABS printed parts through ANOVA design technique [15]. Deng X, Zeng Z, Peng B, Yan S, Ke W [16] have used Layer thickness, Print speed, Infill density/ percentage and Extrusion Temperature as process parameters and used Taguchi as experimentation technique. Gebisa0F A.W., Lemu H.G have used Air Gap, Raster angle/Raster Orientation, Raster width, Number of contours/Shells and Contour width/Shell width as process parameters and used ANOVA as experimentation method for optimising Tensile Strength and Tensile strain [17].

5. Opportunities

Additive manufacturing in its early days didn't find new opportunities because of patent restrictions. But after expiry of patents it is growing opportunities into many fields. From the past two decades AM spreading its applications to

5.1. Manufacturing

Prototyping in one of the product development processes. After conceptualization of market need, before manufacturing a component in full scale a scaled component will be produced. This scaled product is called a prototype. The prototype helps in analysing the part and for communicating the idea of the product to the customers. As on spot manufacturing is possible with Additive Manufacturing, It has application in prototyping. Additive manufacturing technology is rapidly growing in prototyping stages. This additive manufacturing avoids tooling like moulds, cutting tools etc., and can easily produce very intricate shapes easily. So, the number of assemblies in a component can be reduced which leads to an increase in rate of production as well as reduces production time. Instead by a big manufacturing plant only one desktop 3D printer can build a required component. So that it reduces manufacturing plant space. Multi Material additive manufacturing is a recently growing technique where more than one material can be used for tailored properties of the final part.

5.2. Supply chain management

Additive manufacturing leads to drastic changes in Supply chain management. It will lead to mass customization of components. Social manufacturing [21]is possible with AM so that people can get which model or component they want instead of choosing from the available components. Social manufacturing or on demand manufacturing is one of the opportunities from AM. It will help in reduction of inventory cost of parts. Social manufacturing enhances the “human-machine interaction” quality. A personalized product according to the customer’s need can be designed online, during the design stage of a product, a product can be tested and its user experience can be achieved virtually with 3D model display. It helps user’s desires to be simulated and product consumption is then increased. Whenever a product is required, customers just open the product data file and press the “PRINT” button, and then the product can be printed out [22]. The social manufacturing model will help individuals with the 3D printers to participate in manufacturing the parts that are needed by the customers [23]. Democratization of manufacturing has been taking place where people can print the parts of their choice. Today most of the schools had their own FDM printers. Many households prefer to purchase FDM printers for making products like spare parts and toys for their children at their home of their choice. For the cinema industry these 3D printers are the must. The manufacturing industry especially in the design phase these FDM printers are used for prototyping. Some FDM printers are already installed in the International Space Station for making spare parts because it is difficult to make parts by other traditional manufacturing methods. These FDM printers are also used in some rescue operations at remote locations where it is difficult to find any spare parts. In addition to printed parts people are willing to purchase these 3D printers which became a new market opportunity to the printer manufacturers.

5.3. *Aerospace*

AM to revolutionize the aerospace manufacturing industry, with impacts to production lead time, acquisition cost, and performance envisioned. In general aerospace uses very complex structures. Each component of an airplane is a combination of multiple intricate components. Additive manufacturing helps in reducing part count per component. General Electric company used AM for manufacturing of Fuel nozzles for jet engines. In general the fuel nozzle has 20 parts. They have redesigned the fuel nozzle and reduced the part count from 20 to a single component. GE included 20 components into a single component and printed as a single part. so that they reduced the inventory by 95% and 5 times more durable [24]. Additive manufacturing can print very complex shapes very easily. It helps in producing lighter parts with similar strength that produced by conventional manufacturing methods. Reduction of weight is very useful in the aerospace industry because it will increase the efficiency of the engine. Rib-Web Structural Components, Turbine Engine Cases, Engine Blades and Vanes are examples of AM printed parts in the Aerospace industry [25].

5.4. *Bio printing*

5.4.1. Tissue printing. The application of AM in medicine is named as Bio-printing. Human organs are difficult to create manually by cad modeling, we use CT or MRI scan. MRI (magnetic resonance imaging) uses powerful magnetic fields and radio frequency pulses to produce detailed pictures of organs and other internal body structures. CT scans utilize X-rays to produce images of the inside of the body. These images then are digitally processed. These digital processed models are then modified by using CAD software. Then as usual sliced and then transferred to the printer. In tissue printing, the material used is Bioink. Bioink is a solution of a biomaterial or a mixture of several biomaterials in the hydrogel form, usually encapsulating desired cell types. This bioink can be crosslinked or stabilized during or immediately after bioprinting to generate the final shape, structure, and architecture of the designed construct [26]. Bio-ink is a material which can be compatible with that particular person's body. Stem cells from the person for whom the part is being produced will be collected and these cells will be developed in the laboratory. These bio-inks will be converted into wire form. The printing can be extrusion of wire based material. After printing the part will be kept in an incubation where cell development will take place. Bioprinting is able to make functional tissue constructs to replace injured or diseased tissues. It is a relatively new approach that provides high reproducibility and precise control over the making of parts.

5.4.2. Medical implants. Medical implants which are used in replacement of broken bones musculoskeletal systems. These parts are of irregular and very intricate structures. 3D printing techniques such as SLS/SLM has been extensively used for manufacturing these parts. Materials that are biocompatible and stable over a long period of time have been used in implant printing. Pure metals can't be used as material for body organs because pure metals can degrade easily. So, metal alloys are used for printing bone implants.Most of the bones are manufactured from surgical grade 5 titanium alloy(Ti-6Al-4V). Some ceramics and Polymers are the mostly used alloplastic materials [27]. Some implanted alloys are generally fabricated with Additive manufacturing from stainless steels (SS), Co/Cr-based alloys, titanium and its alloys [28], [29]. Some shape memory alloys like NiTi are also used. Shape memory allows NiTi implants to be initially deformed by refrigeration and then return to their predefined shape at body temperature which is a useful property for the deployment of stents [30].

5.5. *Food printing*

Food manufacturing with customized shape, color, flavor, texture, and even nutrition is possible with the Food printing process. Three-dimensional (3D) food printing can revolutionize the food making process. By this method food products can be customized and fabricated to meet individual needs. It enables us to control the amount of printing material and nutrition content. Three types of printing materials (i.e., natively printable materials, non-printable traditional food materials, and alternative ingredients) and two types of recipes (i.e., element-based recipe and traditional recipe) have been used for customized food fabrication. Essentially, 3D food printing provides an engineering solution for new food product development, and a potential machine to reconfigure a customized food supply chain [31]. Simplifying supply chain, and broadening of the available food material are opening the scope of food printing. Currently, 3D printing is being applied in food areas such as elderly food, sweets food,military and space food etc., [32].

5.6. *Construction printing*

Nowadays for construction of a building at least takes 6 months and complex shapes are difficult to build by conventional construction techniques. 3D construction printing gives greater freedom and flexibility in architecture. 3Dcp can build the same construction in 2 to 3 days. This printing technique is named as Material Deposition Method (MDM). The printer can be a computer controlled 4 axes gantry type or six axes robot type. Out of the two techniques, 4 axis gantry is mostly used and for complex shape printing, 6 axis robot printer is most preferred. Contour crafting is one of the MDM techniques where extruder is the combination of nozzle and trowel. The trowel guides the printed material to create an exceptionally smooth and accurate surface. The Trowel can be deflected at different angles to create various non-orthogonal structures. Concrete mixture passes through the pipes to the extrusion nozzle then extruded out in a pre-decided manner.

5.7. *Printed electronic circuit boards*

Additive manufacturing technology has been successfully utilized in multi-layeredPCB circuit manufacturing, which can reduce the lead time from the PCBdesign to prototype PCBs. Currently used subtractive or hybrid processes that involve wet chemical processes can be replaced by direct ink printers to obtain a rigid PCB.Fabrication of a double-sided PLC is also possible with 3D PCB printing. This technology involves the deposition of conductive materials in a highly viscous liquid state. High viscosity helps to retain their shape after deposition. Highly viscous silver nanoparticle (AgNP) ink was used as the printing material which has excellent conductivity and better mechanical reliability. The printing was done through a small nozzle of ink extrusion printer. The PCB layout can be designed using conventional PCB softwares. The nozzle size, ink feed rate, printing speed, and other parameters were adjusted to obtain an optimal printing quality. With a minimum pin-to-pitch printing resolution of100mm, our 3D PCB fabrication method satisfies most conventional PCB design rules and modern integrated circuit pack-aging requirements [34].

5.8. *Textile printing*

Because of it's zero wastage property it has many opportunities in apparel, fashion designs and jewelry making. Similar to architecture in construction, these apparels, fashion design and jewelry require more degrees of freedom in making complex parts. 3D printing technology is widely used in these fields [33].

6. **Future application research gap**

As additive manufacturing is a recently growing technology, There have been many limitations to overcome and many applications to be explored. Following are the some of the future and ongoing developments and research gaps in the field of AM:

6.1. *Hybrid 3D printing*

This EHD-RP allows multi-metal and alloy printing from a single nozzle. Another technique in metal deposition is the use of Gas Metal Arc Welding [35]. Here instead of a print jet and electric heater, material is melted because of the arc. The controlled deposition of material achieved by movement of electrode and base table by the instructions of G-codes and M-codes. Here inert gas is provided to separate the printing area from the atmosphere. It is very difficult to control metal in its molten state. So we have to compromise with the resolution and dimensional accuracy of the final product. Finally the Direct Energy Deposition (DED) method. This is the most recent and the fastest 3D metal printing process. Here we may use a laser, plasma or electron beam as an energy source for melting the feed material and then the molten metal can be deposited wherever it is required according to the codes generated by the slicer software. But the major problem with the DED is that we have to compromise with the resolution of the final object produced.

High Speed Sintering is the combination of binder jetting and melting. Time consumption in the case of SLS and SLM is more because the laser has to travel the path generated by the slicer. In some machines more than two lasers are used to minimize the time to trace the path. In HSS lasers need not to trace the path instead Radiation Absorbing Material (RAM) will print on the desired area and then sintered using an inexpensive infrared lamp. This binder will help in binding the powders as well as to concentrate and absorb more amount of the heat when compared to the support material to melt that required portion [36], [37].

MIT has initiated the concept of 4D printing.4D printing is a recent development of 3D printing, shows promising capabilities and broad potential applications. It completely depends on mathematical modeling and design, smart materials and 3D printers. 4D printing shows advantages over 3D printing in several aspects.4D printing can make dynamic structures with adjustable functionality, shapes and properties. This capability can be achieved by appropriate combination of smart materials in the three-dimensional space. Mathematical modeling is required for the design of the distribution of multiple materials in the structure. There are minimum two stable states available in a 4D printed structure, and the structure can shift from one state to another under the corresponding stimulus [38], [39].

6.2. In-situ monitoring

As of now we can collect the data related to in situ-measurements or online measurements of the part while it's being printed with the help of feedback system by implementing in situ infrared (IR) thermography high speed camera and digital CCD cameras AM technologies[40], [41]. Works are in progress in-situ monitoring which enables the data collected from the sensors through the feedback system to analyse and compare with the target design data. If there are any deviations from the targets the system will be able to set the process parameters such as laser power, hatch spacing, speed etc inorder to control the quality of the part whiler it has been printing. So that it can assure more reliability and quality in the product.

6.3. Topology optimization

AM can also produce topology optimized objects. Topology optimization is a numerical approach that identifies where material should be placed in a given domain. For a given set of loads and constraints, topology optimization can facilitate minimal material usage of weight or uniform stress distribution. Macro structure topology optimization considers that a structure is composed of a single homogeneous material and that material is either present or absent in each part of the design domain. Aerospace and automotive industries [42] where weight reduction can lead to substantial energy savings over the life of the product, topology optimization maximizes efficiency. Topology optimization has also been application in biomedical implants [43], investment casting processes, and more [44]

6.4. *smart manufacturing*

A 3D printer generates a large amount of data during printing a part. This data is related to the materials, process parameters and its influence on output quality. Machine learning can use this data for choosing suitable material for a particular application and this machine learning can even set the process parameters for the required set of applications [45]. 3D printing along with information technology and Artificial intelligence will be a key player for smart manufacturing or industry 4.0.

6.5. *Standardization*

Assurance of quality is a major issue with Additive manufacturing. Lack of standardization in material and machine is one of the main factors of the problem while the gap in the simulation part is another important factor. Different standards development organizations (SDOs) are working on developing standardized tests for measurement of the key mechanical properties including tensile strength and fatigue resistance etc., [46].

References

[1] P. Ahangar, M.E. Cooke, M.H. Weber, D.H. Rosenzweig, 2019, Current biomedical applications of 3D printing and additive manufacturing, J. Appl. Sci. 9 (2019) 1713.
[2] Babak Mohajeri et al., Shift to Social Manufacturing: Applications of Additive Manufacturing for Consumer Products.
[3] San Draw, Kai-Jui at al., 2017, Color or multi-material three-dimensional (3D) printing.
[4] Alain Reiser et al., 2019, Multi-metal electrohydrodynamic redox 3D printing at the submicron scale.
[5] Mohsen Attaran, 2017,The rise of 3-D printing: The advantages of additive manufacturing over traditional manufacturing.
[6] Carl R. Deckard, 1986,Method and apparatus for producing parts by selective sintering.
[7] Prof. Dr. Ir. J.P. Kruth et. al., 2004, Binding Mechanisms in Selective Laser Sintering and Selective Laser Melting.
[8] Khuram Shahzada, 2013, Additive manufacturing of alumina parts by indirect selective laser sintering and post processing.
[9] Chockalingam, K., Jawahar, N. and Chandrasekhar, U. (2006), "Influence of layer thickness on mechanical properties in stereolithography", *Rapid Prototyping Journal*, Vol. 12 No. 2, pp. 106-113. https://doi.org/10.1108/13552540610652456
[10] Quanyi Mu, Lei Wang, Conner K. Dunn, Xiao Kuang, Feng Duan, Zhong Zhang, H. Jerry Qi, Tiejun Wang, 2017, Digital light processing 3D printing of conductive complex structures, *Additive Manufacturing,* Volume 18, Pages 74-83,ISSN 2214-8604,
[11] Wang, Daniel Zongjie & Martin, Nicholas & Hini, Dela & Mills, Barry & Kim, Keekyoung. (2017). Rapid Fabrication of Multilayer Microfluidic Devices Using the Liquid Crystal Display-Based Stereolithography 3D Printing System. *3D Printing and Additive Manufacturing*. **4**. 10.1089/3dp.2017.0028.
[12] Liao, Y., Li, H. & Chiu, Y. 2006, Study of laminated object manufacturing with separately applied heating and pressing. *Int J Adv Manuf Technol* **27**, 703–707. https://doi.org/10.1007/s00170-004-2201-9
[13] Monzón, M.D., Ortega, Z., Martínez, A. *et al.* 2015, Standardization in additive manufacturing: activities carried out by international organizations and projects. *Int J Adv Manuf Technol* **76**, 1111–1121. https://doi.org/10.1007/s00170-014-6334-1.
[14] Gebisa, Aboma & Lemu, Hirpa. (2019). Influence of 3D Printing FDM Process Parameters on Tensile Property of ULTEM 9085. *Procedia Manufacturing*. **30**. 331-338. 10.1016/j.promfg.2019.02.047.
[15] Mohamed, O.A., Masood, S.H., Bhowmik, J.L. *et al., 2016* Effect of Process Parameters on Dynamic Mechanical Performance of FDM PC/ABS Printed Parts Through Design of Experiment. *J. of Materi Eng and Perform* **25**, 2922-35. https://doi.org/10.1007/s11665-016-2157-6
[16] Deng X, Zeng Z, Peng B, Yan S, Ke W. 2018 Jan 30, Mechanical Properties Optimization of Poly-Ether-Ether-Ketone via Fused Deposition Modeling. Materials (Basel);11(2):216. doi: 10.3390/ma11020216. PMID: 29385756; PMCID: PMC5848913.
[17] A.W. Gebisa0F, H.G. Lemu, Influence of 3D Printing FDM Process Parameters on Tensile Property of ULTEM 9085. 14th Global Congress on Manufacturing and Management (GCMM-2018). *Procedia Manuf.*, **30** (2019) 331–338.
[18] Männel, M. J., Selzer, L., Bernhardt, R., Thiele, J., *Adv. Mater. Technol.* 2019, **4**, 1800408. https://doi.org/10.1002/admt.201800408
[19] Ankita Jaisingh Sheoran, Harish Kumar, 2020, Fused Deposition modeling process parameters optimization and effect on mechanical properties and part quality: Review and reflection on present

research, *Materials Today: Proceedings*, Volume **21**, Part 3, Pages 1659-1672, ISSN 2214-7853, https://doi.org/10.1016/j.matpr.2019.11.296.

[20] H.S. Cho, W.S. Park, B.W. Choi and M.C. Leu, 2000, Determining Optimal Parameters for Stereolithography Processes via Genetic Algorithm, *Journal of Manufacturing Systems* Vol. **19**/No. I,

[21] Babak Mohajeri et al.,Shift to Social Manufacturing: Applications of Additive Manufacturing for Consumer Products.

[22] G. Xiong et al., "From mind to products: towards social manufacturing and service," in *IEEE/CAA Journal of Automatica Sinica*, vol. **5**, no. 1, pp. 47-57, Jan. 2018, doi: 10.1109/JAS.2017.7510742..

[23] Markko Hamalainen, Jesse Karjalainen, 2017, Social manufacturing: When the maker movement meets interfirm production networks, *Business Horizons*, Volume **60**, Issue 6, Pages 795-805, ISSN 0007-6813, https://doi.org/10.1016/j.bushor.2017.07.007.

[24] Marie Ann McMasters, Michael A. Benjamin, Alfred Mancini, Steven Joseph lohmueller, Fuel Nozzle, United States Patent, 2009

[25] P.A. Kobryn, N.R. Ontko, L.P. Perkins and J.S. Tiley, 2006, Additive Manufacturing of Aerospace Alloys for Aircraft Structures ,

[26] P. Selcan Gungor-Ozkerim, Ilyas Inci, Yu Shrike Zhang, Ali Khademhosseini, and Mehmet Remzi Dokmeci , 2018, Bioinks for 3D Bioprinting: An Overview, *Biomater Sci. 2018 May 1; **6**(5): 915–946.* doi: 10.1039/c7bm00765e

[27] Michael Larsen, Ruchi Mishra, Michael Miller, David Dean, 2015, Bioprinting of Bone, Editor(s): Anthony Atala, James J. Yoo, *Essentials of 3D Biofabrication and Translation*, Chapter **17**, Pages 293-308,

[28] P. Laheurte, F. Prima, A. Eberhardt, T. Gloriant, M. Wary, E. Patoor, 2010, Mechanical properties of low modulus beta titanium alloys designed from the electronic approach, *J.Mech. Behav. Biomed. Mater.* **3** (8) (2010) 565–573.

[29] Kong, Decheng. (2018). Bio-functional and anti-corrosive 3D printing 316L stainless steel fabricated by selective laser melting.

[30] Stoeckel D, Pelton A, Duerig T. 2003, Self-expanding nitinol stents: material and design considerations. Eur Radiol. 2004 Feb;**14**(2):292-301. doi: 10.1007/s00330-003-2022-5.

[31] Sun, J., Zhou, W., Huang, D. *et al.* 2015,An Overview of 3D Printing Technologies for Food Fabrication. *Food Bioprocess Technol* **8**, 1605–1615 . https://doi.org/10.1007/s11947-015-1528-6

[32] Zhenbin Liu, Min Zhang, Bhesh Bhandari, Yuchuan Wang, 2017, 3D printing: Printing precision and application in food sector, *Trends in Food Science & Technology*, Volume **69**, Part **A**, Pages 83-94, ISSN 0924-2244, https://doi.org/10.1016/j.tifs.2017.08.018.

[33] Pasricha, A., Greeninger, R. 2018, Exploration of 3D printing to create zero-waste sustainable fashion notions and jewelry. *Fash Text* 5, 30 (2018). https://doi.org/10.1186/s40691-018-0152-2

[34] Yue Dong, Chao Bao and Woo Soo Kim, 2018, Sustainable AdditiveManufacturingof Printed CircuitBoards, Joule**2**, 579–582, https://doi.org/10.1016/j.joule.2018.03.015

[35] Anzalone, Gerald & Zhang, Chenlong & Wijnen, Bas & Sanders, Paul & Pearce, Joshua. (2013). A Low-Cost Open-Source Metal 3-D Printer. *IEEE Access*. **1**. 803-810. 10.1109/ACCESS.2013.2293018.

[36] Thomas, Helen R. Hopkinson, Neil Erasenthiran, Poonjolai, 2006, High Speed Sintering - Continuing research into a new Rapid Manufacturing process.

[37] Adam Ellis, Christopher J. Noble, Neil Hopkinson, 2014,High Speed Sintering: Assessing the influence of print density on microstructure and mechanical properties of nylon parts, *Additive Manufacturing,* Volumes **1–4**, Pages 48-51, ISSN 2214-8604, https://doi.org/10.1016/j.addma.2014.07.003.

[38] Farhang Momeni, Seyed M.Mehdi Hassani.N, Xun Liu, Jun Ni, 2017, A review of 4D printing, *Materials & Design,* Volume **122**, Pages 42-79, ISSN 0264-1275, https://doi.org/10.1016/j.matdes.2017.02.068.

[39] Pei, Eujin. (2014). 4D Printing: dawn of an emerging technology cycle. *Assembly Automation*. **34**. 310-314. 10.1108/AA-07-2014-062.
[40] Rodriguez, Emmanuel & Mireles, Jorge & Terrazas, Cesar & Espalin, David & Perez, Mireya & Wicker, Ryan. (2015). Approximation of absolute surface temperature measurements of powder bed fusion additive manufacturing technology using in situ infrared thermography. *Additive Manufacturing*. 5. 10.1016/j.addma.2014.12.001.
[41] Sarah K. Everton, Matthias Hirsch, Petros Stravroulakis, Richard K. Leach, Adam T. Clare, 2016, Review of in-situ process monitoring and in-situ metrology for metal additive manufacturing, *Materials & Design,* Volume **95**, Pages 431-445, ISSN 0264-1275, https://doi.org/10.1016/j.matdes.2016.01.099.
[42] Rozvany, G. I. N, 2009 A Critical Review of Established Methods of Structural Topology Optimization, *Structural and Multidisciplinary Optimization*, **37(3)** pp. 217-237.
[43] Cansizoglu, O., Harrysson, O. L., West II, H. A. 2008, Applications of Structural Optimization in Direct Metal Fabrication, *Rapid Prototyping Journal*, **14(2)** pp. 114-122.
[44] Mary Kathryn Thompson, Giovanni Moroni, Tom Vaneker, Georges Fadel, R. Ian Campbell, Ian Gibson, Alain Bernard, Joachim Schulz, Patricia Graf, Bhrigu Ahuja, Filomeno Martina, 2016, Design for Additive Manufacturing: Trends, opportunities, considerations, and constraints, *CIRP Annals*, Volume **65**, Issue 2, Pages 737-760, ISSN 0007-8506, https://doi.org/10.1016/j.cirp.2016.05.004.
[45] Ivanna Baturynska, Oleksandr Semeniuta, Kristian Martinsen, 2018, Optimization of Process Parameters for Powder Bed Fusion Additive Manufacturing by Combination of Machine Learning and Finite Element Method: A Conceptual Framework, *Procedia CIRP*, Volume **67**, Pages 227-232, ISSN 2212-8271, https://doi.org/10.1016/j.procir.2017.12.204.
[46] Seifi, M., Gorelik, M., Waller, J. *et al.* 2017, Progress Towards Metal Additive Manufacturing Standardization to Support Qualification and Certification. *JOM* **69**, 439–455. https://doi.org/10.1007/s11837-017-2265-2

Development and analysis of mathematical model of capillary for conventional refrigerants R22, R427A, R407C, R32, R410A

Animesh Pandey, Akash, Deshdeep Gambhir and Naveen Solanki

Maharaja Agrasen Institute of Technology, Delhi-110086, India

E-mail:pandey.animesh@yahoo.com

Abstract. The Vapor Compression Refrigeration System (VCRS) uses a circulating refrigerant as the medium which absorbs and removes heat from the space to be cooled and subsequently rejects that heat elsewhere. Capillary tubes are commonly used as expansion device, serving almost all small refrigeration systems to reduce the pressure and temperature of the refrigerant and to regulate the flow of refrigerant to the evaporator. Objective of this research is to develop an analytical method for computing the length of capillary tube required for a specified reduction in temperature and to determine various thermodynamic and transport properties for refrigerant fluids and mixtures viz. R22, R427A, R407C, R32, R410A, available in REFPROP database and perform a comparative analysis. For achieving this objective, a mathematical model is developed using the integration of MATLAB and REFPROP software programs.

KEYWORDS: Refrigerant, Vapor Compression Refrigeration System, Capillary, MATLAB, REFPROP

NOMENCLATURE:

VCRS= Vapor Compression Refrigeration System
T = Temperature, *°C*
L = Capillary tube length, *m*
D = Inner Diameter (I.D.) of capillary tube, *m*
A = Cross-sectional area of inside of capillary, m^2
w = Mass flow rate, *kg/s*
V = Velocity, *m/s*
v = Specific volume, m^3/kg
h = Enthalpy, *kJ/kg*
p = Pressure, *Pa*
f = Friction factor, dimensionless
u = Viscosity, *Pa-s*
Re = Reynolds Number, dimensionless
ΔL = Incremental length, *m*
x = Dryness fraction, dimensionless

SUBSCRIPTS:
f = Saturated Liquid
g = Saturated Vapor
in = Inlet
out = Outlet
m = Mean
1 = Point-1
2 = Point-2

1. INTRODUCTION

Refrigeration is recognized as an imperative method for improving human beings' living standards since early twentieth century. Capillary tubes are widely used as expansion devices for small-scale VCRS These tubes are long circular copper pipes measuring 0.5-2 mm in diameter and 1-6 m in length. The capillary tube is a thermodynamic throttling device through which refrigerant flows and experiences pressure as well as temperature drop. Capillary tubes have found universal acceptance because of their low-cost and reliability.

Even though capillary tubes appear quite simple in construction but they are actually very complex expansion devices. At tube entrance, the refrigerant undergoes slight pressure drop, mainly due to abrupt change in cross-sectional area. The pressure then decreases along the tube length in a linear manner due to wall friction. This type of regime applies until the flow reaches saturated conditions. From this flash point, where vapor first appears, the pressure drop per unit length increases as the end of tube is reached. This happens because the pressure drop in the two-phase flow region corresponds to both flow friction and the increase in momentum of the flow media [1].

Since 1940, many theoretical, experimental and numerical research works have been reported on the flow characteristics of refrigerant in a capillary tube, capillary tube dimensions and geometry and their effect on the performance of VCRS. Bolstad and Jordan [2] conducted first comprehensive- theoretical and experimental work on metastable flow in capillaries. Cooper et al. [3], proposed analytical design of capillary tube based on Fanno flow (adiabatic flow through constant cross-section) of refrigerant with chocked flow condition. The aforementioned researches involved complicated calculation procedures and the validity of their results are restricted to refrigerants R12 or R22. Li et al. [4] experimentally investigated the metastable phenomenon and under-pressure of vaporization in an adiabatic capillary tube. Lin et al. [5] examined the friction coefficient through capillary tube and proposed a new method of dynamic viscosity calculation.

The literature survey reveals that most of the previously undertaken studies are focused on modeling of capillary for a limited variety of refrigerants and involve complex computational techniques. Accordingly, the present study focuses on developing an analytical model of capillary which is valid for all the refrigerant fluids and mixtures available in REFROP database and the comparative analysis of the refrigerants viz. R22, R427A, R407C, R32, R410A,

2. SYSTEM DESCRIPTION AND MODELING

2.1 DESCRIPTION

The VCRS is the most popular system type mainly due to its simplicity and compactness. It consists of four main components, i.e. Compressor, Condenser, Expansion device and Evaporator. Heat is absorbed from the cooled room/cabin at the Evaporator while heat is rejected to the ambient at the Condenser. Power is consumed mainly to operate the Compressor. Refrigerant flows through the components continuously while undergoing various thermodynamic processes.

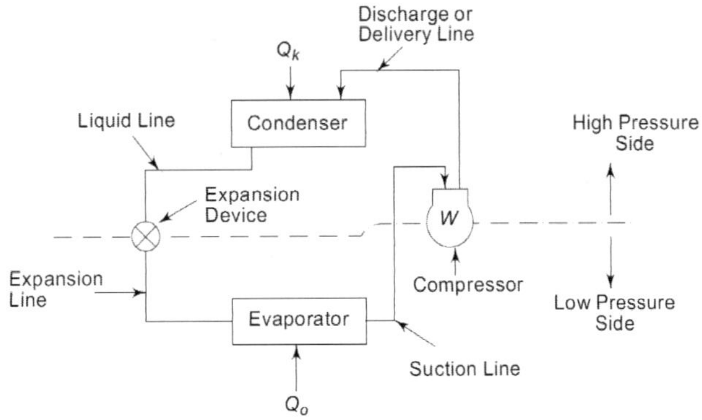

Figure 1. Schematics of vapor compression refrigeration system. [7]

2.2 MODELING

A mathematical model of capillary tube is developed on MATLAB software in order to analytically compute the pressure drop and the cumulative length corresponding to a set of specified operating conditions of the VCRS by determining the states and conditions across successive incremental lengths of the tube.

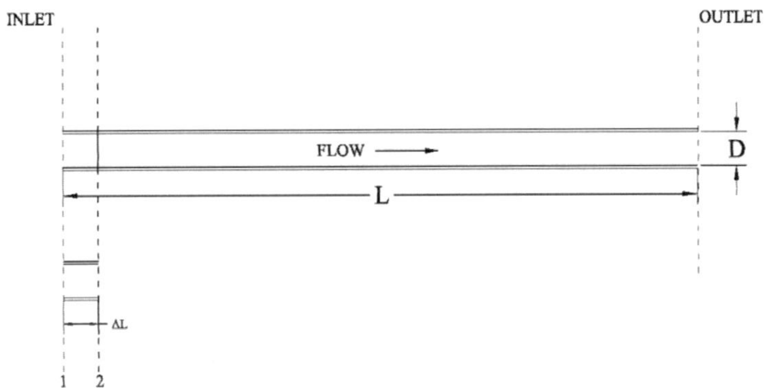

Figure 2. Incremental length of capillary tube.

The fundamental equations applicable to the Control Volume (C.V.) bounded by points 1 and 2 (figure 2) are:

CONSERVATION OF MASS: -

Continuity Equation for the flow of refrigerant through the C.V.: -

$$\frac{A\,V_1}{v_1} = \frac{A\,V_2}{v_2} = w \tag{1}$$

$$\Rightarrow \frac{w}{A} = \frac{V_1}{v_1} = \frac{V_2}{v_2} = \text{Constant} \tag{2}$$

CONSERVATION OF ENERGY: -

ASSUMPTIONS: -

a) Heat and Work interactions through the system are negligible, $Q,\ W_{C.V.} \cong 0$
b) Potential Energy change is negligible, $\Delta P.E. \cong 0$

Steady Flow Energy Equation (S.F.E.E.) for throttling of the refrigerant through the C.V.: -

$$1000\,h_1 + \frac{V_1^2}{2} = 1000\,h_2 + \frac{V_2^2}{2} \tag{3}$$

CONSERVATION OF MOMENTUM: -
Differential force applied to the element because of drag and pressure difference on opposite sides of the element accelerates the fluid, i.e.

$$[(p_1 - p_2) - f_m \frac{\Delta L}{D} \frac{(V_m)^2}{2\,v}]\,A = w(V_2 - V_1) \tag{4}$$

$$\text{where, } f_m = \frac{f_1 + f_2}{2} \text{ and } V_m = \frac{V_1 + V_2}{2} \tag{5}$$

APPROXIMATION: For Reynolds Number (R_e) in the lower range of the turbulent region;

$$f = 0.33\,R_e^{0.25} \tag{6}$$

$$\text{where, } R_e = \frac{V\,D}{u\,v} \tag{7}$$

From equation (2)

$$\because \frac{V}{v} = \frac{w}{A}$$

$$\therefore f_m \frac{\Delta L}{D} \frac{(V_m)^2}{2\,v} = f_m \frac{\Delta L}{D} \frac{V_m}{2} \frac{w}{A} \tag{8}$$

Substituting (4) in (3)

$$[(p_1 - p_2) - f_m \frac{\Delta L}{D} \frac{V_m}{2} \frac{w}{A}]A = w(V_2 - V_1) \tag{9}$$

As the refrigerant flows through the capillary, its pressure and saturation temperature progressively drop whilst the vapor fraction increases. At any point,

$$\begin{aligned} h &= h_f(1-x) + h_g x \\ v &= v_f(1-x) + v_g x \\ u &= u_f(1-x) + u_g x \end{aligned} \tag{10}$$

CALCULATION OF INCREMENTAL LENGTH: -
The objective of the analytical calculation method is to determine the incremental length for a given reduction in saturation temperature of the refrigerant. The state conditions at inlet of the capillary i.e. Point-1 are pre-requisite for the calculation and are thereby required to be provided by the user as input data. For an arbitrarily selected temperature at Point 2, the state conditions at Point 2 and the incremental length *(ΔL)* are computed as follows: -

i. Select T_2, determining the temperature drop per iteration.
ii. To facilitate the computation, equations for properties of saturated refrigerant $(p_2, h_{f_2}, h_{g_2}, v_{f_2}, v_{g_2})$ are imported from REFPROP database.
iii. Calculate the vapor fraction (x_2) of the refrigerant by simultaneously solving equation (2) and (3) and substituting the determined parameters.
iv. With the value of x_2 known, h_2, v_2 and V_2 are computed.
v. Compute R_{e_2} using u_2 from equation (10) and f_m, f_2 *from* equation (5) and (6) respectively.
vi. Substitute V_m from equation (5) in equation (9) to compute (ΔL).

CONTINUATION TO SUCCEEDING INCREMENTS: -
The conditions computed at Point 2 become the entering conditions for the next increment of length with subsequent drop in saturation temperature and the procedure is repeated until the required (user specified) outlet temperature (T_{out}) at the end of the capillary is obtained.
MATLAB program executes a loop wherein the computed outlet parameters of the previous iteration become inlet parameters for the subsequent iterations. The incremental length computed in an iteration of the loop is cumulated with the outcomes of the previous loop iterations to determine the cumulated length of the capillary, thereby calculating the total length of the capillary tube required for the specified reduction in temperature (i.e. up to T_{out}).

3. MODEL VALIDATION

In order to validate the developed mathematical model, the system conditions and parameters required as user input data are taken in accordance with Stoecker W.F., Jones J.W. [6] as follows:-

Table 1. Input conditions.

REFRIGERANT	R22
Diameter of capillary tube *(D), m*	0.00163
Mass flow rate of refrigerant through the capillary *(w), kg/s*	0.010
Temperature at inlet of capillary (T_{in}), °C	40
Temperature required at outlet of capillary (T_{out}), °C	5
Temperature drop per iteration $(T_2\text{-}T_1)$, °C	1

Table 2 presents the computed capillary tube calculations for the specified input conditions and parameters as follows: -

Table 2. Capillary tube calculations for R22.

Position	Temperature (°C)	Pressure (kPa)	Vapour Fraction, x	Specific Vol(m^3/kg)	Enthalpy (kJ/kg)	Velocity (m/s)	Incremental Length(m)	Cumulative Length of Capillary (m)
1	40	1533.579	0	0.000886	249.646552	4.246399	0	0
2	39	1496.505	0.007889	0.000886	249.644129	4.782965	0.242568	0.242568
3	38	1460.097	0.015635	0.000998	249.641304	5.341161	0.211277	0.453845
4	37	1424.346	0.023244	0.001115	249.638034	5.921851	0.185394	0.639239
5	36	1389.246	0.03072	0.001236	249.634274	6.525937	0.163686	0.802924
6	35	1354.788	0.038069	0.001362	249.629975	7.15436	0.145264	0.948188
7	34	1320.965	0.045295	0.001493	249.625085	7.808101	0.129476	1.077665
8	33	1287.770	0.052401	0.001629	249.619543	8.488185	0.115829	1.193493
9	32	1255.195	0.059391	0.001771	249.613287	9.19568	0.103944	1.297437
10	31	1223.233	0.06627	0.001919	249.606248	9.931701	0.093526	1.390963
11	30	1191.876	0.073039	0.002072	249.59835	10.69741	0.084343	1.475305
12	29	1161.117	0.079704	0.002232	249.589511	11.49402	0.076206	1.551511
13	28	1130.949	0.086265	0.002398	249.579642	12.32281	0.068963	1.620474
14	27	1101.364	0.092727	0.002571	249.568644	13.18509	0.062491	1.682966
15	26	1072.356	0.099091	0.002751	249.556413	14.08223	0.056687	1.739653
16	25	1043.917	0.10536	0.002939	249.542832	15.01568	0.051464	1.791116
17	24	1016.041	0.111537	0.003133	249.527777	15.98694	0.04675	1.837867
18	23	988.719	0.117622	0.003336	249.511109	16.99756	0.042485	1.880351
19	22	961.947	0.123619	0.003547	249.492682	18.04916	0.038616	1.918967
20	21	935.715	0.129528	0.003766	249.472332	19.14346	0.035098	1.954065
21	20	910.018	0.135351	0.003995	249.449884	20.28220	0.031894	1.98596
22	19	884.848	0.14109	0.004232	249.425147	21.46723	0.02897	2.01493
23	18	860.200	0.146746	0.00448	249.397912	22.70046	0.026297	2.041226
24	17	836.065	0.152319	0.004737	249.367954	23.98389	0.023849	2.065076
25	16	812.437	0.157811	0.005005	249.335027	25.31957	0.021606	2.086682
26	15	789.310	0.163222	0.005283	249.298865	26.70967	0.019546	2.106227
27	14	766.676	0.168553	0.005574	249.259176	28.15640	0.017653	2.12388
28	13	744.530	0.173804	0.005875	249.215648	29.66210	0.015911	2.139791
29	12	722.864	0.178975	0.00619	249.167937	31.22916	0.014307	2.154097
30	11	701.672	0.184066	0.006517	249.115676	32.86007	0.012828	2.166926
31	10	680.948	0.189077	0.006857	249.05846	34.55741	0.011464	2.17839
32	9	660.684	0.194007	0.007211	248.995857	36.32384	0.010206	2.188596
33	8	640.875	0.198856	0.00758	248.927394	38.16212	0.009043	2.197639
34	7	621.513	0.203623	0.007963	248.852561	40.07509	0.007969	2.205607
35	6	602.593	0.208307	0.008363	248.770807	42.06567	0.006975	2.212582
36	5	584.108	0.212906	0.008778	248.681536	44.13686	0.006057	2.218639

IOP Conf. Series: Materials Science and Engineering 1104 (2021) 012021 doi:10.1088/1757-899X/1104/1/012021

In accordance with the specified conditions of Stoecker W F, Jones J W [6] the cumulative length of the capillary tube required for the specified reduction in temperature of refrigerant R22 is provided to be 2.118 m. Results of the developed mathematical model of capillary computed the corresponding cumulative length required as 2.2186 m (table 2), which is found to be at a deviation of +4.749%.

4. RESULTS AND DISCUSSION

The analytical method developed and validated for refrigerant R22 for the aforementioned input conditions is further utilized to compute the various thermodynamic properties for refrigerants R427A, R407C, R32, R410A and the summary of the obtained capillary tube calculations are tabulated upto evaporator temperature of 5°*C*, as follows: -

Table 3. Comparison of capillary tube calculations for R22, R427A, R407C, R32, R410A.

TEMPERATURE (°C)	PROPERTY	REFRIGERANT DESIGNATION				
		R22	R427A	R407C	R32	R410A
40	Pressure (kPa)	1533.5797	1678.057	1749.0458	2478.3132	2425.6463
	Vapour Fraction, x	0	0	0	0	0
	Specific Vol(m^3/kg)	0.000886	0.000944	0.000937	0.00112	0.001025
	Enthalpy (kJ/kg)	249.64655	260.53821	261.24466	275.61142	267.54842
	Velocity (m/s)	4.246399	4.525085	4.488475	5.366171	4.913408
	Incremental Length(m)	0	0	0	0	0
	Cumulative Length of Capillary (m)	0	0	0	0	0
35	Pressure (kPa)	1354.7885	1473.8956	1535.8079	2189.8337	2144.5419
	Vapour Fraction, x	0.038069	0.049072	0.047033	0.041293	0.053925
	Specific Vol(m^3/kg)	0.001362	0.00153	0.001512	0.001573	0.001435
	Enthalpy (kJ/kg)	249.62998	260.51554	261.22266	275.59274	267.53297
	Velocity (m/s)	7.15436	8.112556	8.008598	8.133898	7.419271
	Incremental Length(m)	0.145264	0.178771	0.190027	0.233396	0.250969
	Cumulative Length of Capillary (m)	0.948188	1.087224	1.152844	1.453403	1.554987
30	Pressure (kPa)	1191.8762	1288.9355	1342.6623	1927.5067	1888.8126
	Vapour Fraction, x	0.073039	0.093338	0.089517	0.077735	0.100521
	Specific Vol(m^3/kg)	0.002072	0.002415	0.002381	0.002249	0.002047
	Enthalpy (kJ/kg)	249.59835	260.46988	261.17844	275.55964	267.50572
	Velocity (m/s)	10.697413	12.53535	12.352434	11.504452	10.466511
	Incremental Length(m)	0.084343	0.121006	0.128985	0.145414	0.158018
	Cumulative Length of Capillary (m)	1.475305	1.79507	1.906597	2.336465	2.510231
25	Pressure (kPa)	1043.9176	1121.9488	1168.3268	1689.6211	1656.775

	Vapour Fraction, x	0.10536	0.13362	0.128223	0.110377	0.141595
	Specific Vol(m^3/kg)	0.002939	0.003506	0.003453	0.003072	0.00279
	Enthalpy (kJ/kg)	249.54283	260.3867	261.09788	275.50398	267.46008
	Velocity (m/s)	15.015689	17.986248	17.711296	15.609938	14.17123
	Incremental Length(m)	0.051464	0.085331	0.091077	0.093659	0.102843
	Cumulative Length of Capillary (m)	1.791116	2.287135	2.431541	2.898142	3.124184
20	Pressure (kPa)	910.01839	971.7613	1011.5732	1474.5658	1446.8635
	Vapour Fraction, x	0.135351	0.170474	0.163677	0.139943	0.178336
	Specific Vol(m^3/kg)	0.003995	0.004851	0.004776	0.004076	0.003694
	Enthalpy (kJ/kg)	249.44988	260.2433	260.95892	275.41334	267.38606
	Velocity (m/s)	20.282203	24.704426	24.323027	20.614449	18.677907
	Incremental Length(m)	0.031894	0.060992	0.065148	0.061186	0.067941
	Cumulative Length of Capillary (m)	1.98596	2.637368	2.805525	3.263303	3.52766
15	Pressure (kPa)	789.31027	837.24717	871.22228	1280.8129	1257.6026
	Vapour Fraction, x	0.163222	0.204239	0.196201	0.166935	0.211547
	Specific Vol(m^3/kg)	0.005283	0.006508	0.006408	0.0053	0.004795
	Enthalpy (kJ/kg)	249.29887	260.00447	260.72722	275.26879	267.26849
	Velocity (m/s)	26.70967	32.984159	32.481049	26.721727	24.16631
	Incremental Length(m)	0.019546	0.043535	0.046533	0.039953	0.044942
	Cumulative Length of Capillary (m)	2.106227	2.887758	3.073073	3.502073	3.794688
10	Pressure (kPa)	680.9483	717.32541	746.14048	1106.9054	1087.5901
	Vapour Fraction, x	0.189077	0.23506	0.225937	0.191689	0.241763
	Specific Vol(m^3/kg)	0.006857	0.008549	0.008422	0.006796	0.006138
	Enthalpy (kJ/kg)	249.05846	259.6161	260.3498	275.04154	267.08433
	Velocity (m/s)	34.557413	43.182209	42.542364	34.184085	30.859659
	Incremental Length(m)	0.011464	0.030687	0.032834	0.025685	0.029373
	Cumulative Length of Capillary (m)	2.17839	3.06536	3.263004	3.656831	3.970335
5	Pressure (kPa)	584.10873	610.95711	635.23724	951.44802	935.48505
	Vapour Fraction, x	0.212906	0.262876	0.252831	0.214403	0.269318
	Specific Vol(m^3/kg)	0.008778	0.01106	0.010902	0.008626	0.007776
	Enthalpy (kJ/kg)	248.68154	258.99607	259.74602	274.68785	266.79868
	Velocity (m/s)	44.136867	55.720358	54.931118	43.312122	39.033736
	Incremental Length(m)	0.006057	0.021118	0.022638	0.015926	0.018646
	Cumulative Length of Capillary (m)	2.218639	3.188974	3.395391	3.754527	4.083453

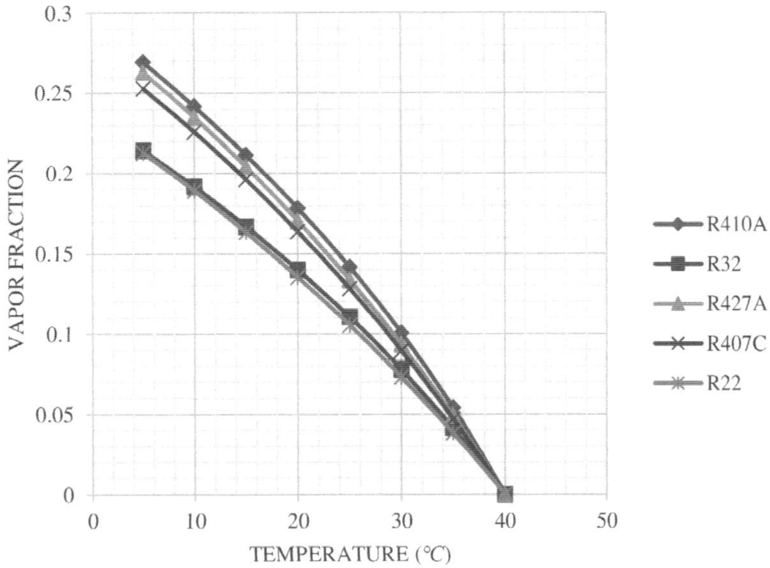

Figure 3. Variation of vapor fraction with temperature for R22, R427A, R407C, R32, R410A.

From figure 3, we can observe that vapor fraction at the outlet of capillary is minimum in case of R22.

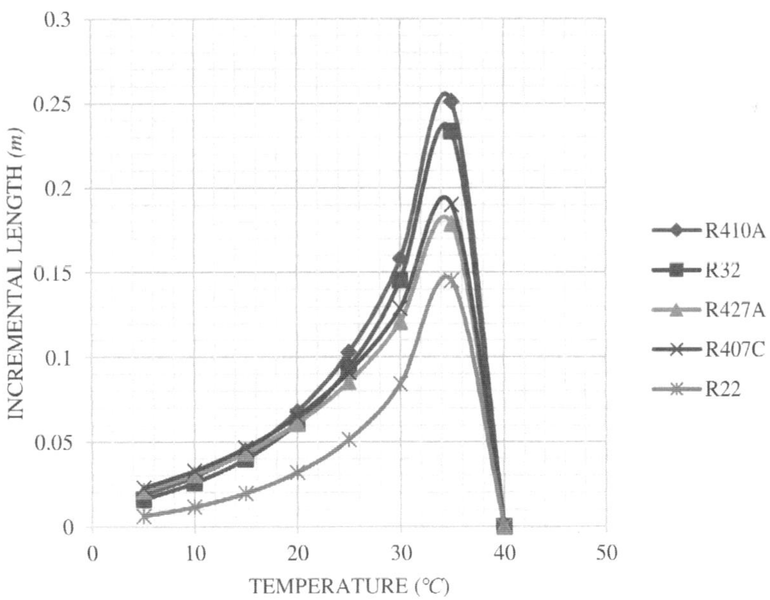

Figure 4. Variation of incremental length with temperature for R22, R427A, R407C, R32, R410A.

From figure 4, we can infer that the incremental length of capillary per unit drop of temperature for R22 is minimal.

IOP Conf. Series: Materials Science and Engineering 1104 (2021) 012021 doi:10.1088/1757-899X/1104/1/012021

Table 4 shows a continuation of table 3, wherein the evaporating temperature is further lowered to -15℃

Table 4. Continuation of Capillary Tube Calculations.

TEMPERATURE (°C)	PROPERTY	REFRIGERANT DESIGNATION				
		R22	R427A	R407C	R32	R410A
0	Pressure (kPa)	497.98789	517.14294	537.46284	813.10126	800.00107
	Vapour Fraction, x	0.23457	0.287395	0.276615	0.235147	0.294366
	Specific Vol(m^3/kg)	0.01112	0.014139	0.013948	0.010865	0.009779
	Enthalpy (kJ/kg)	248.09804	258.0225	258.79552	274.14158	266.35872
	Velocity (m/s)	55.812751	71.076676	70.131426	54.483817	49.025987
	Incremental Length(m)	0.002398	0.013975	0.015033	0.00918	0.011175
	Cumulative Length of Capillary (m)	2.237379	3.272296	3.48487	3.812934	4.153201
-5	Pressure (kPa)	421.80172	434.91955	451.80501	690.57636	679.90099
	Vapour Fraction, x	0.253769	0.30806	0.296762	0.25385	0.316886
	Specific Vol(m^3/kg)	0.013968	0.01789	0.017667	0.013604	0.012228
	Enthalpy (kJ/kg)	247.20594	256.52043	257.32448	273.30361	265.68518
	Velocity (m/s)	69.994724	89.755497	88.659426	68.150018	61.242274
	Incremental Length(m)	-0.000076	0.008674	0.009395	0.004498	0.005947
	Cumulative Length of Capillary (m)	2.241557	3.325622	3.542436	3.844104	4.192647
-10	Pressure (kPa)	354.78599	363.35386	377.28328	582.63242	573.99372
	Vapour Fraction, x	0.270013	0.324027	0.312467	0.270285	0.33666
	Specific Vol(m^3/kg)	0.01741	0.022413	0.022163	0.01695	0.015219
	Enthalpy (kJ/kg)	245.86134	254.25135	255.09403	272.02771	264.66058
	Velocity (m/s)	87.111777	112.22386	111.00176	84.830578	76.156608
	Incremental Length(m)	-0.001721	0.004799	0.005276	0.001259	0.002297
	Cumulative Length of Capillary (m)	2.235963	3.356864	3.576521	3.856393	4.210903
-15	Pressure (kPa)	296.19698	301.53383	312.9398	488.0745	481.13195
	Vapour Fraction, x	0.282604	0.334202	0.322673	0.284032	0.353243
	Specific Vol(m^3/kg)	0.021532	0.027785	0.027517	0.021019	0.018861
	Enthalpy (kJ/kg)	243.87098	250.91485	251.80009	270.1037	263.11464
	Velocity (m/s)	107.56007	138.80635	137.51098	105.09162	94.295785
	Incremental Length(m)	-0.002768	0.002046	0.002347	-0.000949	-0.000222
	Cumulative Length of Capillary (m)	2.224011	3.372203	3.593695	3.855714	4.214447

From table 4 we can observe that when the evaporating temperature is lowered beyond a certain stage, the incremental length of the capillary tube required to achieve a unit drop in temperature is obtained to be negative. Thus, an unrealistic situation is developed, known as Choked Flow. The Choked Flow condition develops in long capillary tubes, wherein pressure drop due to friction results an increase in the specific volume and velocity of the refrigerant, until Sonic velocity (Mach No.=1) is attained.
The developed impractical situation can be scrutinized using Fanno line, shown on the enthalpy-entropy diagram of figure 5, where the enthalpy decreases with corresponding increase in entropy as the refrigerant flows through the tube. At the sonic velocity, the Fanno line demands decrement in entropy which will be a violation of the Second Law of Thermodynamics for this adiabatic process. Nature regulates the flow rate so that the sonic velocity occurs at the very exit of the tube.

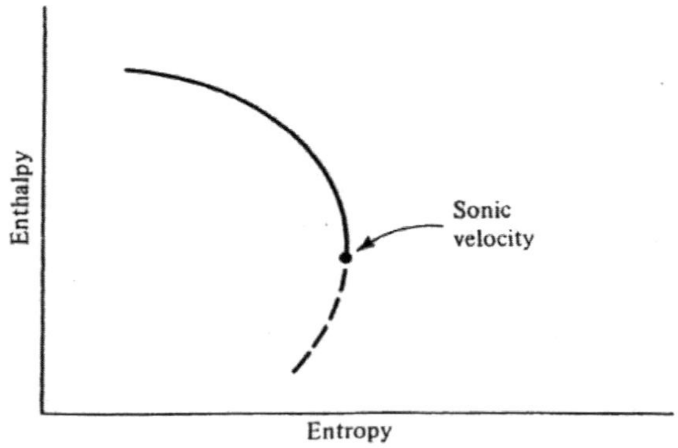

Figure 5. Fanno line showing choked flow conditions. [6]

5. CONCLUSION
From table 3 and figure 4, we can conclude that near the evaporator, the increments of length required for unit drop in saturation temperature becomes progressively smaller.

The comparative analysis of various refrigerants used in VCRS viz. R22, R427A, R407C, R32 and R410A is conducted under same set of conditions. Various thermodynamic properties are compared to determine the appropriate refrigerant for the given system under the specified conditions.

From the results and discussions, it can be concluded that

- Length of the capillary required is minimal for R22 refrigerant.
- Vapor fraction of the refrigerant at the outlet of capillary is also minimum in case of R22 (figure 3).
- Enthalpy at the inlet of evaporator is observed to be minimum in case of R22, which indicates greater refrigeration effect and hence, higher C.O.P. of the system.

Thus, the analysis converges to the conclusion that although R22, an HCFC refrigerant, is harmful for the environment and contributes to Climate Change and the depletion of the Ozone Layer (Global Warming Potential *(G.W.P.)* = 1810 and Ozone Depletion Potential *(O.D.P.)* = 0.055), but for the scope of comparative analysis, is most efficient and effective for the given system model.

REFERENCES
[1] Mezavila M M and Melo C 1996, Capheat: A Homogeneous Model to Simulate Refrigerant Flow Through Non Adiabatic Capillary Tubes.
[2] Bolstad M M and Jordan R C 1948, “Theory and Use of the Capillary Tube Expansion Device,” Refrigerating Engineering, 519-552.
[3] Cooper L, Chu C K and Brisken W R 1957, “Single Selection Method for Capillaries Derived from Physical Flow Conditions,” Refrigerating Engineering, 37-107.

[4] Li R Y, Lin S L, Chen Z Y and Chen Z H 1990, "Metastable Flow of R12 Through Capillary Tubes,"*Int. J. Ref.*, **13**, 181-186.
[5] Lin S, Kwork C C K, Li R Y, Chen Z H and Chen Z Y 1991, "Local Frictional Pressure Drop During Vaporization of R12 Through Capillary Tubes," *Int. J. Multiphase Flow*, **17**, 95-102.
[6] Stoecker W F and Jones J W, *Refrigeration & Air Conditioning*, 2nd Edition, Tata McGraw Hill, New Delhi 1982.
[7] Arora C P , *Refrigeration and Air Conditioning*, 2nd Edition, Tata McGraw Hill.

IOP Conf. Series: Materials Science and Engineering 1104 (2021) 012022 doi:10.1088/1757-899X/1104/1/012022

Effects of Modulus of Elasticity on airflow in Human Respiratory airway

Rakesh Kumar Shukla[1*], Sandeep Tiwari[2], Ashutosh Kumar[3]
Department Of Mechanical Engineering, Krishna Engineering College, Ghaziabad India

Abstract—In this work, Computational analysis of air flow in human respiratory tract model is done by the use of CFD technique. Fluid Structure Interaction (FSI) method is used on the third-to-sixth generation simplified respiratory tract model for computational study. The modulus of Elasticity is varied in a range of 1 to 5 MPa and its effect on flow-physics is captured inside human airway model. The variation of velocity and its effect on deformation has been captured for both rigid and flexible model and its analysis is carried out. The result from simulation shows that as the Modulus of elasticity increases, wall deformation decreases.

Keywords—Compliant human airway model, Computational Fluid Dynamics (CFD), Fluid Structure Interaction (FSI), Modulus of elasticity (E).

I. INTRODUCTION

THE human respiratory model is similar to the Y-shaped bifurcated duct. Its main function is to exchange oxygen and carbon dioxide between the lungs and atmosphere. The human respiratory tracts are flexible in nature. However most of the previous CFD studies considered it as rigid wall [1-3].

Koombua et al. [4] discussed CFD study of flexible human respiratory wall considering it homogeneous/isotropic having density 1365.6 kg/m^3 and Young's modulus of elasticity of 5.8 MPa. Third to fifth generation simplified model was considered for the study. It was found that wall elasticity affected the fluid pressure and wall strain distribution while no significant change is observed on flow velocity and wall shear stress (WSS). Later, Koombua and pidaparti [5] varied value of elasticity in their study and noted corresponding effects. Fluid-Structure Interaction (FSI) study of rigid and compliant CT scan model was discussed by Xia et al.[6]. It is noted that the respiratory wall deformation increases with the increase in Reynolds number. In case of compliant wall, the peak wall shear stress is found five times greater than that of the rigid wall.

The present study is concentrated to see the effects of Young's modulus of elasticity on the deformation of respiratory wall. FSI method in Ansys-Fluent-Academic [7] has been used to capture and study the wall deformation. The velocity contours of the different cross section have been obtained to see the effects of wall elasticity.

II. MODEL OF HUMAN RESPIRATORY TRACT

Third to fifth generation simplified model of human respiratory model is taken in the present study. The dimensions of the respiratory model is adopted from Koombua et al.[4]. For the study of compliant airway, both of the domain namely solid and fluid are constructed in Ansys-Workbench and is shown in the Fig.1. Thickness of the solid domain (in green colour) is taken as 5 mm, however the fluid domain (in yellow colour) is created using 'fill-up option available in Ansys-Workbench.

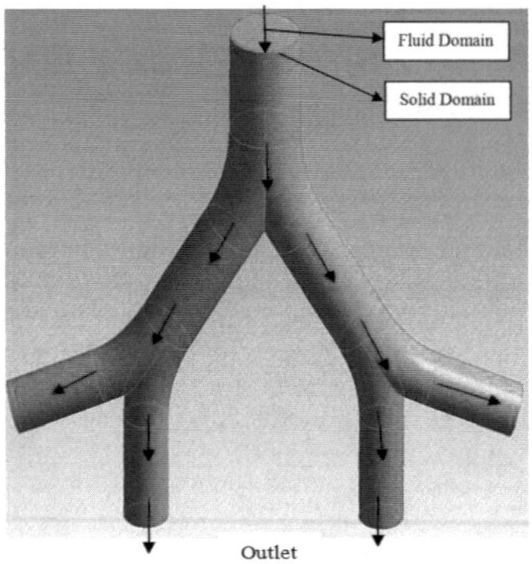

Fig.1: Fluid and Solid Domain of Third to Fifth generation Human respiratory Model (Adopted From Koombua et al.[4]

III. Grid Generation

Grid generation of both the fluid and solid domain is done using the tetrahedral elements because it provides minimum skewness at the curved surfaces. The size of the tetrahedral elements is taken from 0.25 to 0.35 mm.

IV. Boundary Conditions

Parabolic inlet velocity profile (Re=1935) and outlet pressure is applied at the inlet of third generation bronchus and outlet of fifth generation bronchus respectively. The inlet velocity is applied corresponding to the inlet mass flow rate 60 L/min at the inlet of trachea. The properties of air at normal conditions are considered in the present study i.e., density is 1.19 kg/m^3 and dynamic viscosity is 1.82× 10^{-5} kg/ms.

V. Numerical Simulation

The commercial CFD packages 'Ansys-Fluent 19.0'(Acedemic) and 'Ansys-Structural 19.0'(Acedemic) have been used for fluid-structure interaction (FSI) study. The different terms of the transport equation were discretized using second-order upwind scheme. A convergence criterion is taken as 10^{-4}. Two-way coupling method has been used to study the effects of fluid on wall and vice versa.

V. GRID INDEPENDENCY

Grid independency test is performed to verify the computational accuracy of the airway model. Initially 16710 elements was generated which was further refined six times. Grid is found independent for 1152284 numbers of elements. Table 1 shows how the grid independency is achieved.

Table 1: GRID INDEPENDENCY

Case	Number of elements	Successive increase in elements	Avg. Velocity*
1	16710		3.880
2	78906	372.2%	3.900
3	132867	68.4%	3.810

IOP Conf. Series: Materials Science and Engineering 1104 (2021) 012022 doi:10.1088/1757-899X/1104/1/012022

4	511681	285.1%	3.820
5	1152284	125.2%	3.840
6	2113594	83.4%	3.842

* computed at the bifurcation junction.

VI. CFD Model Validation

The present computational results were validated with the results reported by Koombua et al. [4]. The velocity contours for the two results is shown in the Fig. 2. The maximum local velocity is computed 10.19 m/s against the velocity 10.17 m/s reported in Koombua et al. [4].

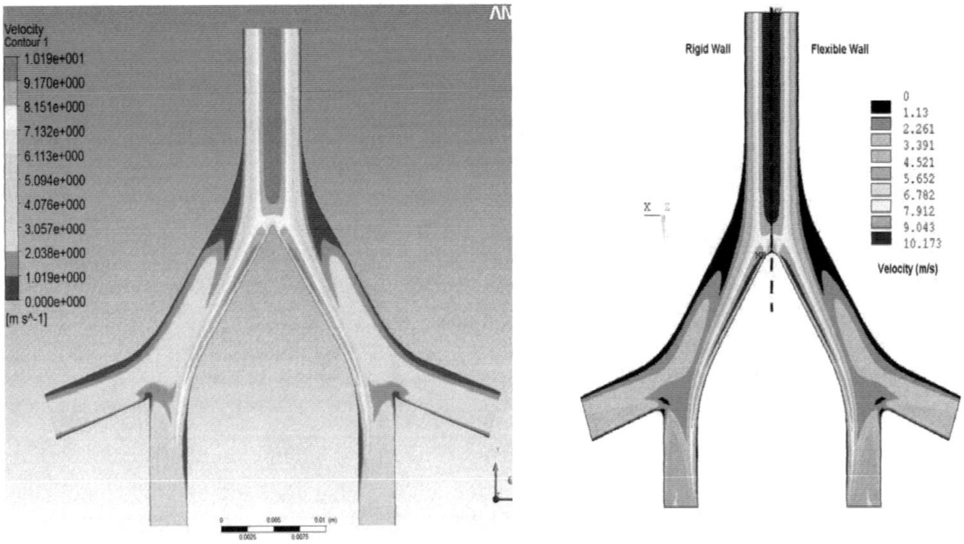

Present Result (Vmax= 10.19 m/s, Vmin = 0 m/s)

Koombua et al. (2008) (Vmax= 10.17 m/s, Vmin = 0 m/s)

Fig.2: CFD Validation of the Present Model

VII. Results and Discussion

A. Velocity Distribution:

The velocity contours corresponding to the Young's modulus of elasticity equals to 0.4 MPa are shown in Fig. 3. Three different horizontal planes are located at the inlet and bifurcation junction is plotted in the following figure. Since parabolic profile is given at the inlet of the third generation bronchus, maximum velocity is therefore, obtained in the center of the bronchus.

However, as airflow moves towards the bifurcation (divider), velocity distribution is skewed towards the inner wall at the fourth generation bronchus due to centrifugal force.

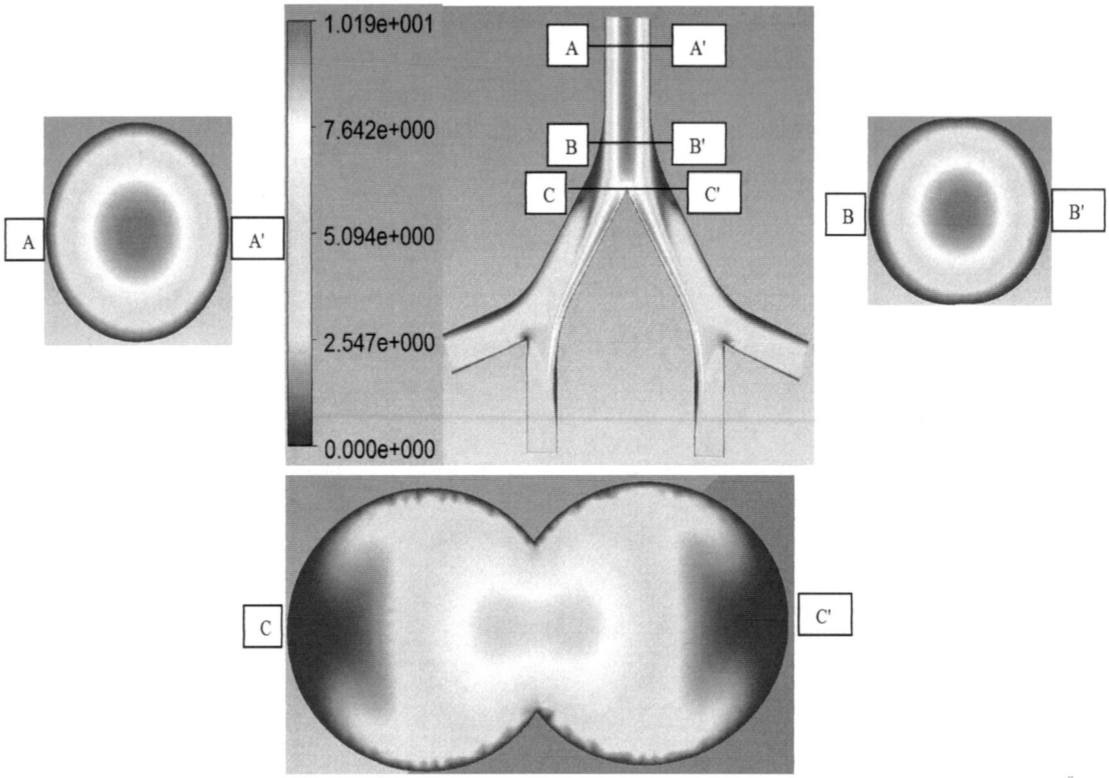

Fig.3: Velocity Contours of the vertical and horizontal cross sectional planes

B. Total Deformation

Total deformation (deflection) is computed for different range of Young's modulus of elasticity (E). The total deformation contour for elasticity 1.0 MPa and 0.4 MPa is shown in Fig. 4. Maximum deformation is observed at the bifurcation junction. Higher deformation is obtained for $E = 0.4$ MPa as compared to that for $E = 1.0$ MPa.

It is found that as the Young's modulus of elasticity increases, total deformation decreases. Total deformation is computed as 7.46×10^{-6} and 18.73×10^{-6} m corresponding to Young's modulus of elasticity 1.0 MPa and 0.4 MPa respectively(Fig.5).

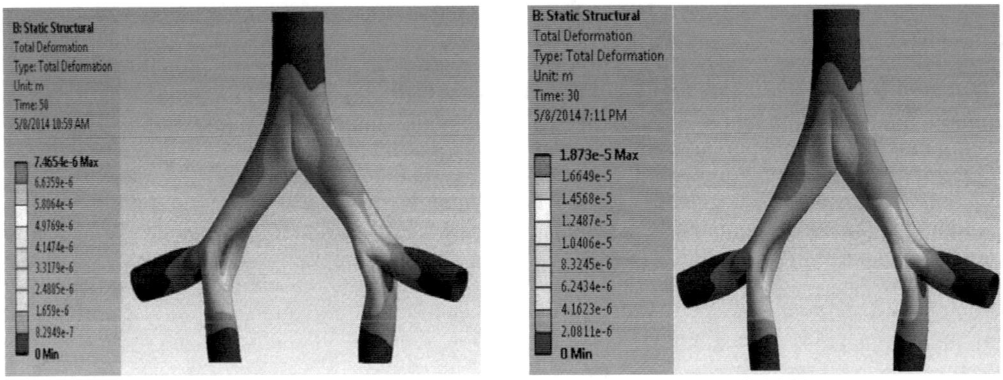

E= 1.0 MPa E= 0.4 MPa

Fig.4: Distribution of Total Deformation

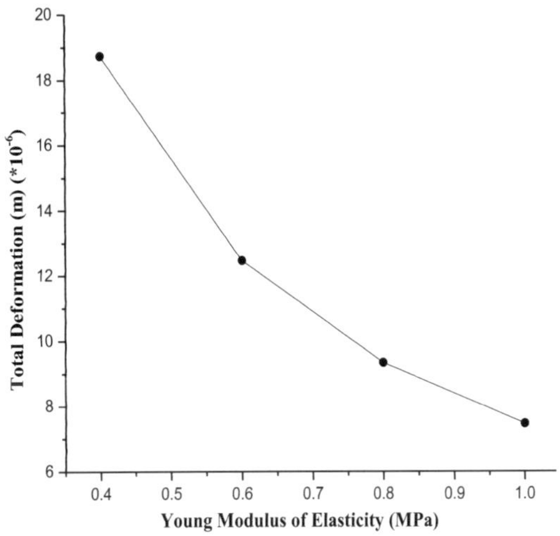

Fig.5: Elasticity versus total deformation

VIII. CONCLUSIONS

The airflow study in the compliant human respiratory model has been carried out for different range of Young's modulus of elasticity. The conclusions are given below:

1) It is found that as the value of Elasticity increases, total wall deformation decreases.
2) Fluid-structure interaction (FSI) method helps to find out the realistic flow physics in human airways.
3) The present study helps to correlate the medical aspects with the engineering solution.

REFERENCES

[1] A.S. Green. Modelling of peak-flow wall shear stress in major airways of the lung.Journal of Biomechanics. Vol. 37, 2004. pp. 661-667.

[2] X.Y.Luo, J.S. Hinton, T.T.Liew, K.K.Tan,. LES modelling of flow in a simple airway model.Medical Engineering & Physics. Vol. 26, 2004 pp. 403–413.

[3] V. K. Srivastav, A. R. Paul, A. Jain. Effects of Cartilaginous Rings on Airflow and Particle Transport through Simplified and Realistic Human Upper Respiratory Tracts. Acta Mechanica Sinica. (Springer). Vol. 29, Issue 6, 2013, pp 883-892.

[4] K. Koombua and R. M. Pidaparti. Inhalation Induced Stresses and Flow Characteristics in Human Airways through Fluid-Structure Interaction Analysis. Hindawi Publishing Corporation. Modelling and Simulation in Engineering. Volume 2008, Article ID 358748, 8 pages.

[5] K. Koombua, R. M. Pidaparti P. W. Longest and K. R. Ward. Computational Analysis of fluid characteristics in rigid and flexible human respiratory airway models. Engineering applications of computational fluid mechanics Vol. 2, 2008, pp. 185-194.

[6] G U. Xia, Merryn, Awhai, H. T., Eric, Hoffman, A., Lin., C. L..Airway wall stiffening increases peak wall shear stress: A fluid–structure interaction study in rigid and compliant airways. Annals of Biomedical Engineering, Vol. 38, 2010, pp. 1836-1853.

[7] Ansys-Fluent-15.0 User's Guide.

Mechanical Characterization & Phase Evaluation of ODS Aluminum Alloy

Satyajeet Kumar*[1], Jay Shankar Kumar[2], Sandeep Tiwari[1], Sachendra[1],Om prakash[1], Shailesh Kumar Singh[3]

[1]Krishna Engineering College, Ghaziabad, U.P, India

[2]Technocrats Institute of Technology, Bhopal, M.P, India

[3] Dronacharya Group of Institutions, Greater Noida, U.P, India

*E-mail: **Satyajeetkumar59@gmail.com**

Satyajeet.kumar@krishnacollege.ac.in

Abstract. Nano-oxide dispersed aluminum alloys are widely used as structural components in different fields in aerospace, defence, transport and automobile industries. It is observed that the hardness of these Aluminium alloys (65-100 HV) could be significantly enhanced by uniform dispersion of nanometric Y_2O_3 and TiO_2 in high-Al matrix in presence of other alloying elements like Fe and Ti. In the present work, synthesis of Al (70%), 22 Fe (22 %), Ti (8 %) , Al (69 %), Fe (22 %), Ti (8 %) with 1% nano-Y2O3 and Al (69%) , Fe (22%), Ti(8%) with 1% nano-TiO_2 dispersed particles through mechanical alloying has been carried out. These alloys have been named as alloy A, B & C respectively on the basis of their weight percentages Subsequent consolidation of this alloy has been carried out through conventional sintering at 1000°C for 1 hour. X-ray diffraction study (XRD), field emission scanning electron microscopy (FESEM) has been used to analyse the phases evolved during various stages of milling. The crystallite size and lattice strain were analyzed by Williamson-Hall method and finally, the structures and Phases are correlated. Sharp peaks with high intensity reveal the presence of amorphous and crystalline phases. For alloy B, intermetallic peaks like Al3Fe, andAl3Ti, Y2O3 and TiO2 for alloy C has been observed after sintering.

Keywords: *XRD, FESEM, Phase evolution, intermetallic, sintering, crystallite size and lattice strain*

1. Introduction

Al based alloys find its application in many areas like defence, aerospace, transport and automobile sector. The desirable properties including reduced density, low cost, and low elastic modulus makes Al alloys better to use. Al alloys also find its application on the basis of some other important properties like elastic limit, strength to weight ratio, corrosion & wear [1].

As compared to structural material (steel), Aluminum is sixty percent lighter and has superior strength to weight ratio. Both impact toughness and corrosion resistance behaviour can be possessed by Al in normal atmospheric condition. They can be used because of their better thermal and electrical conductivity. The method of fabrication also makes aluminium superior to use. In order to fulfil the requirements and challenges like greater efficiency, reduced emissions and environmental impact of the aerospace industries, the performance & characteristics of these materials must be improved [2].

Alloying of elements is one of the most common methods to enhance the strength of aluminium .properties of materials. Factors such as formability, cost, reusable-ability, increased toughness, and easy availability are considered while selection of alloying element especially for dynamic structure. In Today's world, Both Al and Mg alloys are widely used for structural materials in surface transportation and aviation. Mechanical alloying assumes further significance owing to the scope of developing heterogeneous microstructures comprising amorphous and nano crystalline phases either by controlled amorphization (during milling) or by subsequent partial crystallization of the amorphous precursor (by annealing), and thereby, attaining a novel composite microstructure with superior mechanical properties[3,4]. The Mechanical strength of Al-alloys can be enhanced by a heat treatment process called aged hardening. Yield strength of structural Al alloys can reach up to 600 MPa [3] by age hardening heat treatment process[5] .The general procedures to produce Al-based alloys are rapid quenching such as melt spinning[6,7] or with the help of mechanical alloying [8] but these processes restrict the sample size to slightly thin robins or in the shape of powder. Hence it is required to synthesize amorphous based bulk samples. It is very difficult for alloying elements to combine through traditional melting method .Hence it can be done with the help of mechanical alloying [9]. Intermetallics possess outstanding elevated temperature properties such as high strength with increased stiffness, better corrosion /oxidation resistance. These properties can be observed because of its systematic & ordered nature of existence. Reduced dislocation motion and low diffusivities are the two important attributes of intermetallics. Intermetallic compounds generally show brittle characteristics and having high melting point.

Oxide dispersion strengthening (ODS) alloys are produced by solid state and other powder processing techniques. One such example is Mechanical alloying (MA). The constituents of powder can be in the primary, intermetallic or pre alloyed state. Both the metallic and oxide (Y_2O_3 or TiO_2) powders are mixed together and mechanically alloyed by using a high energy ball milling.

The repeated impacts during the process like cold welding, fracture, and re-welding of mixed powder resulting in a very fine and homogenized mixture of constituents. The shearing action leads to the growth of new fragmented alloyed particles.

2.0 Materials and Method

2.1 Raw Material

Elemental powders of aluminium (Al), iron (Fe), titanium (Ti), oxides of yttrium and titanium oxide (Y_2O_3 & TiO_2) are used to prepare the desired alloys. These can be tabulated as shown in Table1.

Alloy	**Al (wt. %)**	**Fe (wt. %)**	**Ti (wt. %)**	**Y_2O_3 (wt. %)**	**TiO_2 (wt. %)**
Alloy A	70	22	8	-	-
Alloy B	69	22	8	-	1
Alloy C	69	22	8	1	-

Table 1: Composition (initial) of the powder mix for mechanical alloying/milling

2.2 Mechanical Alloying and Process variables:

In the present study, alloy A, B and C were subjected to high energy ball milling in a multivial Fritsch Pulverisette planetary ball mill .The ball mill consists of a hardened steel container of maximum capacity (500 ml) & steel balls having diameter 10 mm. For proper mixing, the ball is to powder ratio was kept 10:1 in terms of weight. Milling operation was carried out at 300 rpm with subsequent time interval of 30 minutes. Prior to milling, Samples were taken out after a time interval of 0h, 10h, 20h, 30h, 40 and 50 h for characterization (FESEM and XRD). The various milling parameters considered in study are mentioned below as shown in table 2.

Parameter	operational condition
Types of mill	Planetary mill
Milling container	Chrome steel
Milling speed	300 rpm
Grinding media	Toluene
Ball diameter	10mm
Ball to powder weight ratio	10:1
Time of milling	50hr

Table 2: Milling process parameter

2.3 Compaction and Sintering

The compaction experiments were executed to make a cylindrical sample. In order to make Al based alloy specimen, cylindrical punch and die made of stainless steel was used. Compacts of 15 mm diameter were made with the punch and die. Maximum of 450MPa pressure of load was applied on it very slowly. Compacts were cast out from the die in the same direction as the compression.

The process of sintering is carried out at high temperature (1000^0 C) but below the melting of the materials being sintered. The actual value of sintering temperature of most of the materials ranges between 70-80 % of their melting temperature. In order to produce a pore free body densification of the material is done which eliminates the pores because of neck formation.

2.4 X- Ray Diffraction (XRD) study

In order to find the phases present in Aluminium alloy X-pert MPD X-ray diffractometer is used. The purpose of XRD study is to identity the different phases and its evolution associated with different milling hours. . Phase evaluation can also be analysed after sintering processes. The variation in peaks and correlation with the standard specimen were analyzed with the help of (JCPDS) data file. Additional software was used to identify the crystal structure constituent phases. X-ray diffraction were recorded in the scanning range from 30° to 90° with an accelerating voltage of 40kV using Cu $K\alpha$ (λ=1.542Å) and current of 35 mA. Estimation of crystallite size and lattice strain was approximated by using Williamson-Hall Plot.

2.5 Field emission scanning electron microscope (FESEM):

In order to investigate the polished surface of Al-based alloys pellets specimen were carried out field emission scanning electron microscope

3.0 Result and Discussion

3. 1 Phase Evolution of Alloys during Mechanical Alloying:

Milling of Al (70%), 22 Fe (22 %), Ti (8 %), alloy A powders at different stages of milling time in hrs (0- 50hrs) can be seen through the series of XRD pattern (figure-1). During manual blending for 0 h, only the peaks of Aluminium, iron and Titanium were observed. Furthermore, as the milling time progresses, presence of high intensity peaks or sharp peaks can be observed. Broadening in XRD peak reveals that the powders are disjointed, cold weld and rewelded . This leads to the formation of amorpohous/ nanocrystalline phases. Shifting & Formation of new intermetallic (Al_3Fe, andAl_3Ti)

IOP Conf. Series: Materials Science and Engineering 1104 (2021) 012023 doi:10.1088/1757-899X/1104/1/012023

peaks can be seen after 30 h of milling. This is due to the reaction of stoichiometric ratio of all the metallic elements (Fe, Al & Ti) in course of milling.

3.2 Phase Evolution of Alloys after Sintering:

Fig. 2 shows the XRD pattern of alloy A, B and C respectively. Sintered samples at 1000°C for 1 hour in an inert (Ar) atmosphere were investigated .It reveals that the presence of FCC-Al peak is predominant peak and also some intermetallic peaks like Iron aluminide (Al_3Fe) and Al_3Ti, Yttrium oxide (Y_2O_3) for alloy B and Titanium oxide (TiO_2) for alloy C respectively.

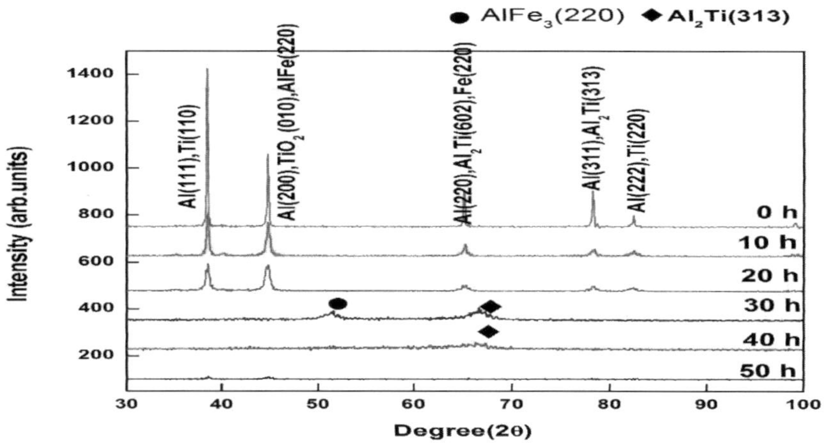

Figure 1

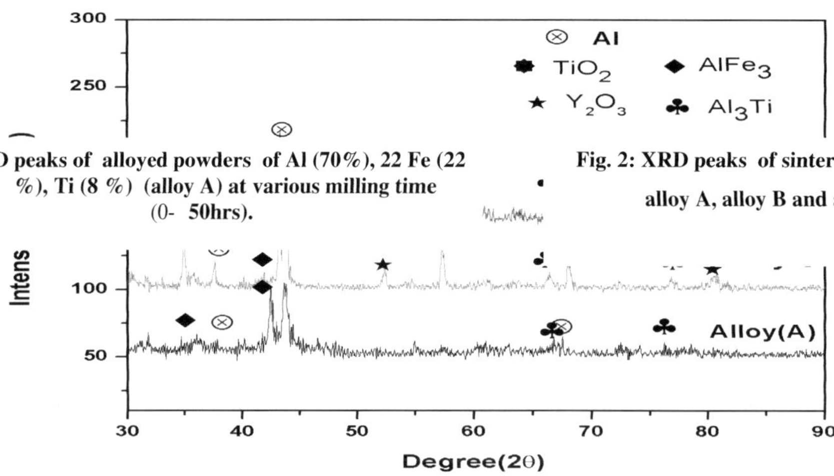

Fig .1: XRD peaks of alloyed powders of Al (70%), 22 Fe (22 %), Ti (8 %) (alloy A) at various milling time (0- 50hrs).

Fig. 2: XRD peaks of sintered compacts of alloy A, alloy B and alloy C.

3.3

Fig

tha

Figure 2

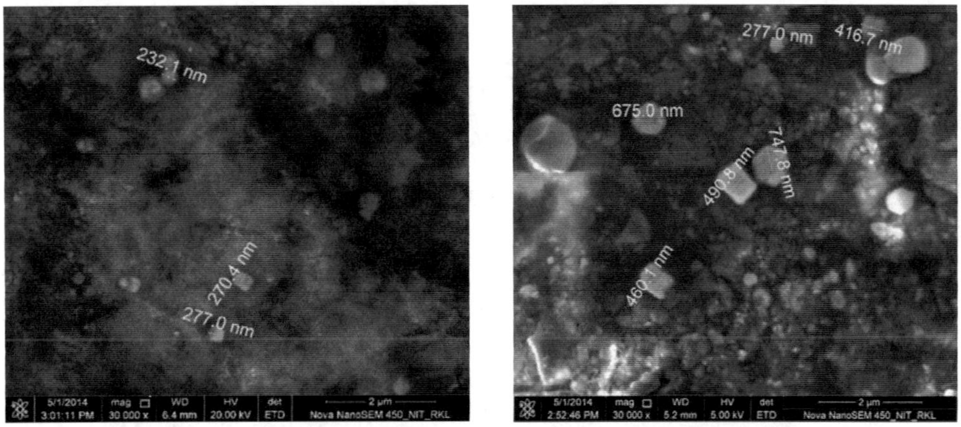

Fig.3: FESEM micrographs of sintered pellets of alloy B and alloy C respectively.

Fig. 4 shows the dispersion of nano-Titanium oxide (TiO_2) in aluminum matrix of alloy C. The dispersoids and intermetallic phases improve the creep resistance property of the aluminum alloy.

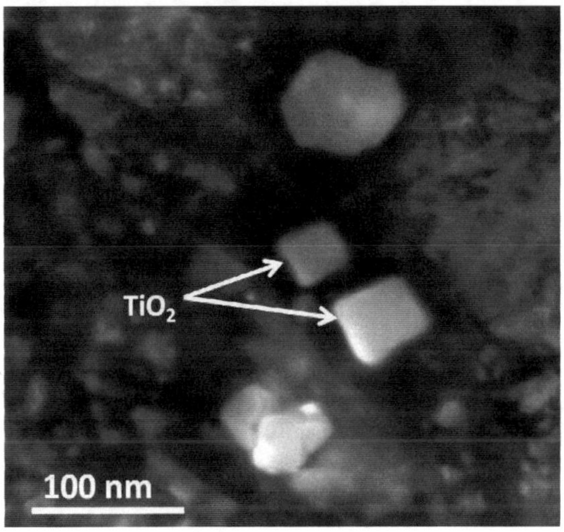

Fig.4: Nano-TiO_2 in Aluminium matrix

3.4 Determination of Crystallite Size and Lattice strain

Crystallite size and lattice strain of milled powders (0 h, 10 h, 20 h, 30h, 40h and 50 h) were determined with the help of William-Hall equation. As per earlier discussion, the XRD peaks get broaden (Fig.1) during milling this is due to the fact that various factors such as instrumental error, crystallite size and micro strain have been influenced during high energy ball milling. The broadening of peaks was used to determine the particle size and residual strain of the material. This broadening is because of the small crystal size and was evaluated through Scherrer's formula (Eq. .1).

The crystallite size is calculated by Scherrer's formula as follows:

$$d = \frac{0.9\lambda}{\beta_g cos\theta} \Rightarrow \beta_g = \frac{0.9\lambda}{dcos\theta} \quad (1)$$

Where, d = crystallite size; λ= the wavelength of the X-ray (0.1541nm); β_g = Full width half maximum intensity (FWHM and θ = angle of diffraction.

The lattice strain is calculated by the given formula:

$$\eta = \frac{\beta_s}{4tan\theta} \tag{2}$$

$$\Rightarrow \beta_s = 4\eta tan\theta \tag{3}$$

Where, η = Lattice strain; β_s = Full width half maximum intensity (FWHM) due to micro strain and θ = angle of Diffraction

$$\beta cos\theta = \frac{0.9\lambda}{d} + 4\eta sin\theta \tag{4}$$

We can calculate the crystallite size and residual strain with the help of Equation (4) called Williamson-Hall equation.

Milling Time	Crystallite Size (nm)	Lattice strain(%)
0 h	247	0.15
10 h	230	0.18
20 h	182	0.23
30 h	78	0.89
40 h	62	1.23
50 h	32	1.32

Table 3: Summary of crystallite size and lattice strain of alloy A: calculated by William –Hall equation at various stages of milling (0- 50 hrs).

It is found (Table-3) that there is a gradual decrement in crystallite size from 247 to 30 nm as milling time increases and residual strain increases from 0.15 to 1.32 (%) as milling time increases. A plot between crystallite size and lattice strain vs milling time has been shown in figure.5. It can be observed from the graph that with an increase in milling time the crystallite size decreases. Also there is an increase in lattice strain with increase in milling time.

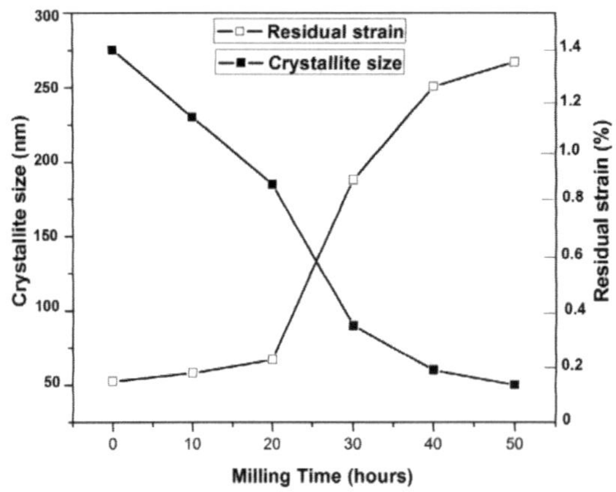

The present study investigates the characterization of ball-milled Aluminium alloy powder under various process parameters. Evolution of phases through XRD techniques and effect of milling on various Al based alloys (A, B and C) have been examined. The subsequent key findings obtained from the experimental results are mentioned. XRD pattern of different alloy with successive time of milling were investigated. FESEM micrographs of sintered pellets confirm that the average grain size of alloy B and alloy C was 249.67 nm and 510.84 nm respectively. For alloy C, dispersion of nano-TiO_2 in aluminum matrix is observed. These dispersoids and intermetallic phases improve the creep resistance property of the aluminum alloy. The analysis of particle size revealed that there is a decrement in particle ranges from 55 μm to 5 μm during 50 h of milling. The variation of crystallite size with milling time is linearly independent. Also there is an increase in lattice strain with increase in milling time. During manual mixing of elemental powders, XRD peaks of only Al, Fe & Ti were observed. With an increase in milling time from 0hr to 50 hrs, pointed and elevated XRD pattern were observed. The lengthening and sharpening of XRD pattern confers that the elemental powders are disjointed, cold weld and rewelded which leads to the formation of amorphous and nano crystalline phases. Shifting and formation of new XRD pattern as intermetallics phases (Al3Fe and Al3Ti) can be observed after 30 hrs of milling.

References

[1] Prashanth KG, Scudino S, Murty BS, Eckert J. ,2009, Crystallization kinetics and consolidation of mechanically alloyed Al70Y16Ni10Co4 glassy powders, Journal of Alloys and Compounds.

[2] Satyajeet Kumar, Shailesh Kumar Singh,,Jayshankar Kumar, Qasim Murtaza, 2018, Synthesis and Characterization of Al-alloy by Mechanical alloying", Elsevier: Materials Today Proceedings.

[3] C. Suryanarayana, 2001, Mechanical alloying and milling, Progress in Materials Science .

[4] Murthy B.S: et al. , 1998, Novel materials synthesis by mechanical alloying/milling. International Materials Reviews .

[5] Kawamura. et. al Y., Mano. H., Inoue. A., Nanocrystalline aluminum bulk alloys with a high strength of 1420 MPa produced by the consolidation of amorphous powders, Scripta Mater., 44 (2001) 1599-1604.

[6] S.C. Tjong, Haydn Chen,2004, Materials Science and Engineering .

[7]Lee W, Kwun SI. ,1996, J Alloys and Compounds

[8] Gaber A., Gaffar M.A., Mostafa M.S., Abo Zeid E.F.,2007, Precipitation kinetics of Al–1.12 Mg2Si–0.35 Si and Al–1.07 Mg2Si–0.33 Cu alloys, J. Alloys Compd.

[9] C. C. Koch, 1997, Synthesis of nanostructured materials by mechanical milling: problems and opportunities, Elsevier: Materials Today

A Comparative Study of hand operated Water Purifier

Anil Kumar[1], Shriram Hegde[2], and Shailesh Kumar Singh[3]
[1]Department of Mechanical Engineering, Krishna Engineering College, Ghaziabad, India.
Corresponding author Email: anilkumar.aman4892@gmail.com
[2]Department of Applied Mechanics, Indian Institute of Technology, Delhi
[3]Department of Mechanical Engineering, Dronacharya Group of Institutions, Greater Noida, India

Abstract
The present work shows the comparative analysis of various designs of portable water purifier without consuming electric power. Four different combinations of reverse-osmosis membrane, alum, potassium permanganate, sodium tri chlor has been used for obtaining optimized value of total dissolved solids (TDS) of the known sample of water. In the subsequent stages of the project the various components of the product will be designed and fabricated. The quality of the water filtered by the prototype device depends upon the membrane (filter) it is using so the proper designing and selection of the filter is also a very crucial task. The specialty of the product is that it is a compact product hence can be easily carried to anywhere. Its operation is very easy so it can be operated by anyone with ease.

1. Introduction

Since water is a necessity of life, it is important to drink water free from contaminants. The quality of usable water is deteriorating with time. It may be due direct disposal of industrial and household waste into the water bodies or the increasing level of pollution, which directly affect people's health and hygiene. There is an urgent need to address this problem at all possible levels to ensure the availability of safe drinking water. To treat water and make it worth drinking, different processes are used and one of the best treatment methods is reverse osmosis (RO). In this process of micro filtration, the water goes through semi permeable membranes with minute pores and becomes free of contaminants. The impurities left behind are flushed out [1].

The reverse osmosis in the treatment of drinking water is a useful technique. It is suggested to do pre-filtration before going for the reverse osmosis process. It occurs when the water is moved across the membrane against the concentration gradient, from lower concentration to higher concentration. To illustrate, imagine a semipermeable membrane with fresh water on one side and a concentrated aqueous solution on the other side. If normal osmosis takes place, the fresh water will cross the membrane to dilute the concentrated solution. In reverse osmosis, pressure is exerted on the side with the concentrated solution to force the water molecules across the membrane to the fresh water side [2].

The reverse osmosis systems that have a five stage systems. The first stage is the removal of sediments like clay, silt and stones from the water. The second stage of the Reverse osmosis treatment is the usage of carbon filter. The carbon filter is used to remove the chlorine and other harmful chemicals that enter the water sources. In third stage, water is passed from a dense and compacted carbon filter. The water that we get may have some unpleasant characteristics and this third step helps in the removal of all such characteristics. All the contaminants left in the water are removed at this stage and water becomes almost clean. In the fourth step, water passes through the membrane and all the heavy metals present in the water are removed. In this last stage, the bacteria, chlorine, and bad odor are removed from water. After water passes from this stage, it comes out of the faucet and is perfect for consumption [3].

There are various water treatment processes that can be implemented. The processes like chlorination, sand filtration, coagulation, Flocculation, sedimentation plus filtration, microfiltration, ultrafiltration, nano filtration. Chlorination is for removal of odor and some particles only. Sand filtration is a natural process which purifies the water in the natural bodies of water. Coagulation is addition of alum to the water to settle down the particles which are there in on the surface of water. Sedimentation is the process in which the particles settles down on its own. Microfiltration, ultrafiltration, nano-filtration are defined with respect to the size of grains used in the filters [4].

Pressurized system has been used for the removal of the bacterial and protozoans impurities along with the removal of the dissolved salts in the water. It is used for the treatment of sea water and brackish water (water containing dissolved salts more than the sea water). According to technical suggested measures, the total dissolved salts in the water should be less than 1000mg/l of water, and according to the W.H.O specifications, the TDS should be less than the 250mg/l of water [5].

2. Method of Treatment

In the current work, a hand operated water purification unit is designed, fabricated and tested for output water quality. Four different designs have been tried, one using the commonly available kitchen press, another one based on the pichkari or color sprayer, other two hand operated water purifiers are also designed and developed. The TDS tests were performed for the initial design and the results of these tests have been quite encouraging to go for the final refinement of the design. The optimization of these later developed models has also been attempted. Many components of the equipment such as filter (membrane), cylindrical container, top cover with piston, upper and lower strainers are designed so that a final and efficient product can be developed. Also, the weight and cost of the product are kept in consideration. Fig. 1 shows the drawing of kitchen press press, color sprayer, bottom strainer and cylinder-plunger arrangement used for hand operated water purifier. The top and bottom strainers will remain same in shape for all the designs. The only change for the top and bottom strainers will be in their sizes.

The top and bottom strainers support the filters and perform the primary cleaning by removing the suspended particles and prohibiting these from entering into filters

Fig 1. Kitchen press, color sprayer, bottom strainer and cylinder-plunger arrangement used for hand operated water purifier

Bottom Strainer

IOP Conf. Series: Materials Science and Engineering 1104 (2021) 012024 doi:10.1088/1757-899X/1104/1/012024

This is the main support to the filtering layers. First it supports the filter which is placed in between the top and bottom strainers and second function it performs is the help in the purification of water and it has holes in it of somewhat bigger in size than the holes in top strainer to allow the ease in the flow of water and also to reduce the required pressure to be applied through the rotation of the crank for the flow of water.

Top Strainer

It is there in the equipment to perform the pre-filtration function mainly. It also supports the filter placed in between the top and the bottom strainers but main function is to do pre-filtration. It has some chemical sprayed over it to destroy the dissolved bacteria, pathogens in the water and it has holes of smaller sizes to separate the particulate materials from the water and to prohibit them going into the filter because these particulate matters can choke the filters and it can make the filter to stop doing its work.

First Design Result

NO. of Filters	TDS Values		
	Location1	Location2	Location3
2	680	753	710
3	590	630	625
4	425	515	505
6	450	500	526
8	330	450	375

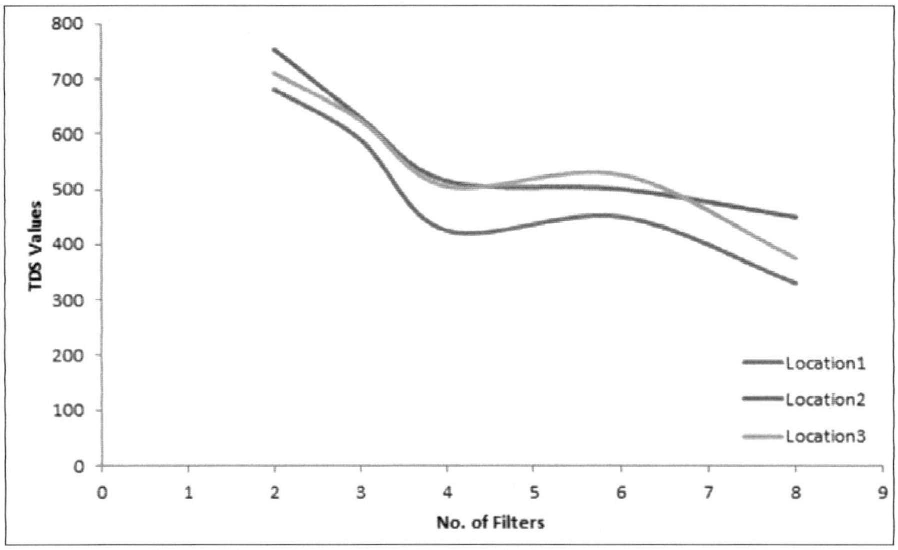

Some defects were found in this design, which were difficult to eliminate. The unevenness in the graph between no. of filters and TDS value is attributed to the following problems:

1. Leakage from the base part of the equipment.

2. Inefficient working or poor working of the equipment.

3. Inefficient pressure generation which is needed for proper working.

Second design Result

NO. of Filters	TDS Values		
	Location1	Location2	Location3
2	650	630	610
3	601	590	530
4	567	515	501
6	475	490	450
8	325	360	340

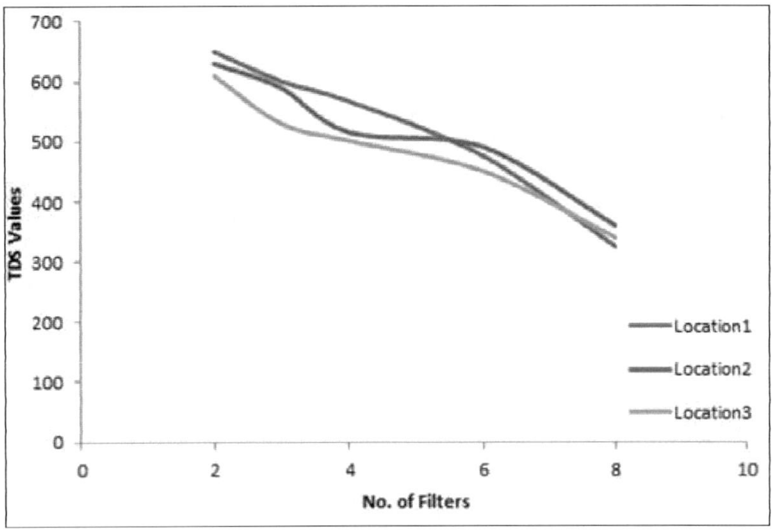

As compared to the initial design, the pressure developed is more, which is resulting in the lower value of final TDS for the same value of the number of filters.As there is still unevenness in the graph. So, there is still some problem of leakage from the basal part of the equipment.

Third Design Result

NO. of Filters	TDS Values		
	Location1	Location2	Location3
2	618	620	625
3	590	595	576
4	445	460	451
6	375	360	368
8	279	267	253

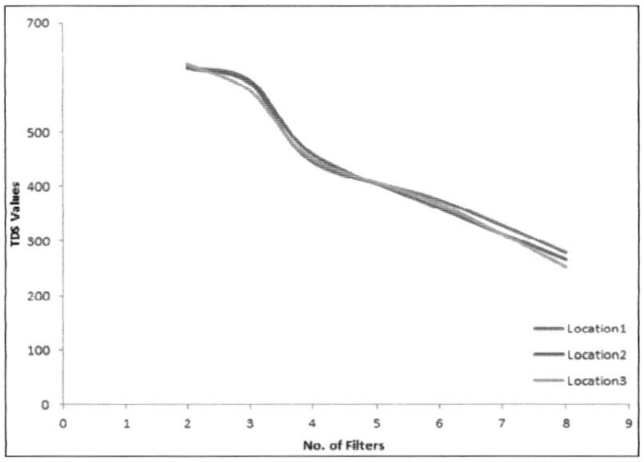

There is a little bit of unevenness in the graph which can be attributed to the leakage occurring from the bottom and top of the equipment. The values of TDS after 2, 3, 4, 6, 8 membranes do not depend upon the initial values of TDS being taken for the purification.

Features of Final Design

1. A detachable input is provided to properly supply water to the cylindrical container without opening and closing the top of the equipment time and again.

2. Cylindrical container is made into two parts and threads are provided so as to allow the ease in cleaning the filters and to optimize the design for efficient working of the equipment.

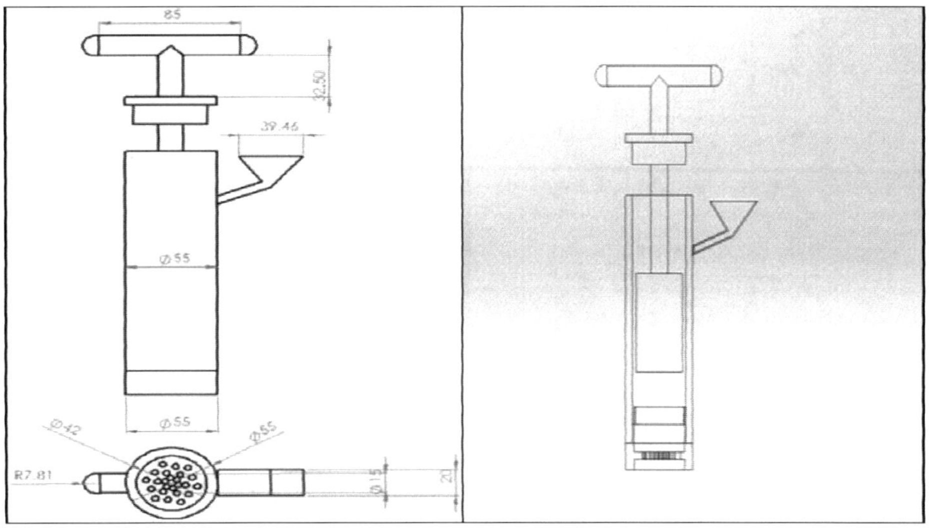

Fig 2. Final improved design for hand operated water purifier

Improved Design Result

NO. of Filters	TDS Values		
	Location1	Location2	Location3
2	621	614	609
3	540	535	528
4	430	428	423
6	312	309	307
8	210	206	203

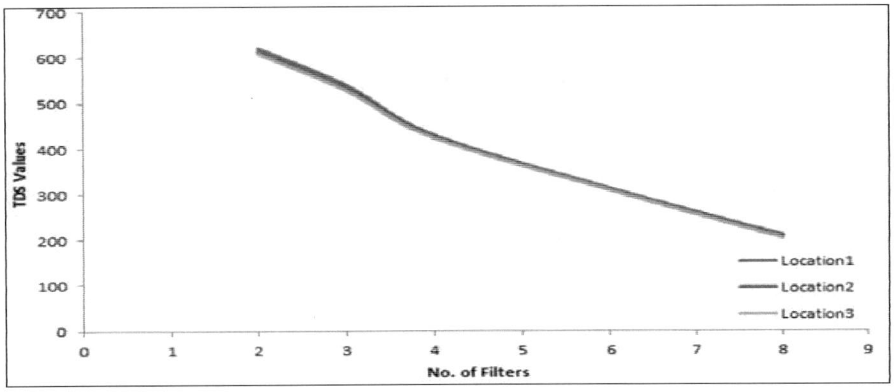

Result and Discussion

The outcome of final improved design over the other three are :

1. The cylindrical container is made into two parts to facilitate the easy cleaning of the filters and to dismantle it when it is not being used, so that it can be easily carried in the bags.

2. It is compact in its design & therefore can be used anywhere during journey, camping, during patrolling.

3. A detachable input is also provided so that the water can be easily filled into the container without removing the piston time and again from the cylindrical container.

4. It is simple in construction and can be operated with ease by anyone.

5.The saturation value of TDS for the existing design is approx. 210 PPM.

Conclusion

1. The values of TDS after 2, 3, 4, 6 , 8 membranes do not depend upon the initial values of TDS being taken for purification .

2. Since all the graphs of the improved designs are almost identical, we can conclude that there is no leakage from the equipment and the final value of TDS is independent of the initial value of TDS being taken for the purification.

References

1. **Shaffaf Ro organization, (estd. in 1995).**

2. **R. Ryan Duport, Talbert N Eisenberg, E Joe Middlebrooks** (Jan 2005). "Research paper on reverse osmosis in treatment of drinking water" WATER QUALRRY SERIES UWRL/Q-82/053

3. A Youtube lecture of **Wasser Entkeimer** on Stripen Sidewinder.

4. A research paper by **Cooperative Research Center for Water Quality and Treatment**, may 1982

5. **Chittaranjan Ray and Ravi Jain** (Jan 2011) "A research paper on drinking water treatment technology-comparative analysis" DOI 10.1007/978-94-007-1104-4_2, © Springer Science + Business Media B.V. 2011

6. **Barbara Daniels & Nancy Mesner**, (June) 2005.A research paper on drinking water treatment system. NR/WQ/2005-24.

IOP Conf. Series: Materials Science and Engineering 1104 (2021) 012025 doi:10.1088/1757-899X/1104/1/012025

Experimental Investigation of WE54 magnesium rare earth metal alloys by ageing treatment

Sachendra[1*], Shailesh Kumar Singh[2], Ujjwal[1], Satyajeet Kumar[1], Kuldeep Singh[3]

[1]Department of Mechanical Engineering, Krishna Engineering College, Ghaziabad, India 201007

[2]Department of Mechanical Engineering, Dronacharya Group of Institutions, Greater Noida, India

[3]Department of Mechanical Engineering, KIET Institute of Engineering & Technology, Ghaziabad, India 201206

Corresponding author : sachendra43@gmail.com

Abstract. Magnesium alloys have potential applications in the automotive, aerospace, aircraft, and electronic industries. Strength of these alloys is due to solid-state precipitates, produced by age-hardening process. The current study deals with the precipitation and hardening behaviour of a commercial quality of magnesium rare earth WE54 alloy to provide a comprehensive report on precipitation and hardening. The Solid Solution treatment of the three samples alloys were performed at a temperature of 525°C and duration of 8 hours. Whereas the ageing treatment was performed at three levels of temperature 170°C,250°C and 300°C corresponding the time duration of 16 hours,4 hours and 4 hours. Mechanical properties of the WE54 rare earth alloy has been analysed along with the microstructure of the age treated specimens. The precipitation sequence and hardening response in the alloy due to ageing and the effects of precipitate shapes on strengthening along with the effect on precipitate behaviour on mechanical properties.

Key words: WE54 alloy, Solid solution, Ageing Treatment, Microstructure, Hardness

1. Introduction

Light weight Magnesium (Mg) alloys have high strength to density ratio. Due to reduction in weight of components made by magnesium alloy, they are used for electronics, automobile, and aerospace applications with great interest. However, magnesium alloys exhibit lower mechanical performance in comparison to aluminium and steel alloys and efforts are being made to enhance their mechanical properties. Tendency for brittleness is found in magnesium and their cast alloys which is due to fracture at twin zones or {0001} basal planes with big grains. But magnesium suddenly shows good deformation behaviour above 225°C, due to the formation of {1011} basal planes, showing that extensive working is possible, above this temperature. Most magnesium alloys in cast and forged character exhibit great processability , machinability, easily machinable and inert gas weldable. It also exhibit good shock absorbing characteristics, due to which life cycle of components made of these material is highly enhanced[1].

The main mechanism for enhancing the mechanical properties is solid-solution hardening or precipitation hardening. Intermetallic phases are formed by magnesium with most alloying elements and the stability of the phase rises with the electronegativity of the alloying element. All rare earth elements (including yttrium) develop eutectic systems of limited solubility with magnesium. Therefore precipitation hardening is quite possible and makes sense. Stable precipitates are formed which raises the corrosion resistance, creep strength, and high temperature strength. Technical alloying elements are yttrium, neodymium, and cerium. Due to the high costs, the usage of these elements are done in high-tech alloys. Zirconium addition leads to tensile strength enhancement without effecting ductility, due to its affinity for oxygen. The developed oxides are structure forming nuclei and enhances grain refining [2].

Precipitation hardening, also known as age hardening, is the most important method of strengthening in magnesium alloys. Precipitation hardening of an alloy occurs with time due to precipitation of a coherent phase from a supersaturated solid solution. During ageing first of all the solute atoms segregate (i.e. cluster) along certain preferential planes of the matrix. Initially, clusters of solute atoms do not possess any definite shape and when these clusters attain definite shapes, they are referred to as zones, more popularly known as Guiner-**Preston zones** or GP zones. Strengthening occurs due to GP-zones and gives rise to first hardness maximum. In Mg-RE alloy, two intermediate metastable phases can precipitate. Intermediate phases are usually partially or sometimes fully coherent. β'' precipitates nucleate and begin to replace GP-I zones, called GP-zones of the second kind (GP-II). Maximum hardness is attained with the formation of these zones as they are more numerous and produce greater distortion than any of the other transition structures. Continued ageing results in the formation of another intermediate phase β and β′ precipitate. This precipitating phase is different from the phase of the alloy system in normal situation [3].

In precipitation-hardened magnesium alloys, strengthening is because of plate-shaped precipitate phases formed on prismatic or basal planes of α-Mg matrix phase[4]. Mg-Y-Nd system have the maximum tensile yield strength and creep resistance, from all its other magnesium alloys developed till now. Conventional age-hardening treatments is responsible for maximum strength of such alloys, as it promotes the formation of metastable *β'* and equilibrium *β* phases as dispersed precipitates, and both of these phases have been reported as forming as plates on $\{1\bar{1}00\}_\alpha$ planes of the magnesium matrix (α) phase. There is a dispersion of irregular, globular particles, in contact with plate-shaped precipitates in the microstructure [5].The performance of the magnesium-rare earth materials at higher temperatures can be a certain by following ways: by matrix of second-phase particles formed at solidification, by precipitation hardening or a mixture of both. The recent developed Mg alloys consist of Y and Zr along with rare earth elements which are accountable for grain refinement [6]. Formation of different kinds of precipitates at different ageing temperature leads to different hardening responses. The hardening precipitation responses during various temperatures ageing was established by XRD patterns. Platelet-like precipitates, globular-like precipitates, densely packed in the matrix, both flake-like and globular precipitates along with their length and size were viewed. The precipitation of β in a range of temperature exhibited the most significant strengthening with hardening effect [7].The most favourable magnesium alloys for commercial purpose developed till now through second phase hardening, in terms of resistance through creep deformation and strength, have been WE54 (Mg-5.2 wt pct Y-2.1 wt pct Nd-2.7 wt pct HRE) and WE-43 alloys based on the Mg-Y-Nd system. It is possible to obtain maximum hardness level in smaller time of ageing by cold work [8]. Based on the above mentioned literature the precipitation and hardening behaviour in the alloy has been studied in the current study due to ageing and the effects of second phase shapes on strengthening along with the effect on precipitate behaviour on mechanical properties.

2. Materials and Methods

The sample prepared for the current study is a magnesium rare earth WE54 alloy, prepared for commercial purpose with approx. composition by weight of 5.1–5.6% Y, 1.5–2.2% Nd, 1.7–2% RE, 0.48% Zr and rest is Mg. The material was collected in extruded condition in form of plate with dimension 150 mm×20 mm. The samples with a diameter of 10 mm and height of 15 mm were machined out of the plate. These samples were used for solution heat treatment followed by aging treatment and the analysis of mechanical properties. During the hardness testing the entire surface were prepared by intermediate grinding on a series of emery paper. Tensile testing specimens were prepared by machining on lath as per the standard. For metallographic examination the heat treated and non heat treated sample were fine polished on a disc polishing machine with polishing alumina powder of size less than 5μm, as grinding media smeared on a blazer cloth. The solution treatment of the samples were carried out in a caliberated box type muffle furnace with an accuracy of temperature of ±5°C under vacuum atmosphere, which was provided by sealing the samples in a quartz glass ampoule with vacuum inside. This treatment was followed by ice water quenching. Further, aging treatment was carried out at different temperature like 170°C, 250°C and 300°C for different possessing time. The heating rate was approximately 5-8°C/s. The heat treatment and ageing treatment operations performed on the samples are coded and listed below.

Table-1 Ageing Treatment of WE54 alloy at different temperature

Code	S S Treatment		Ageing Treatment	
	Temperature in °C	Time in hours	Temperature in °C	Time in hours
A-170	525-530	8	170	16
A-250	525-530	8	250	4
A-300	525-530	8	300	4

3. Results and Discussions

3.1 Microstructural analysis

The present investigation primarily aimed at accessing the solution treatment followed by aging response at different temperature of the magnesium rare earth metal alloy with the help of several characterization techniques. Fig.1 depicts the optical micrograph of all the second phase i.e. $Mg_{12}NdY$ observed in the as received material which was seen through optical microscope, get dissolve during the heat treatment. Yttrium and Neodium, both are powerful solid solution strengtheners and so help in solid solution strengthening of the alloy. The precipitated particle obtained at ageing 170°C and 250°C were too fine and hard to be viewed by optical microscope. However, precipitates obtained after treatment of A-300 were of plate shape and visible by optical microscope.

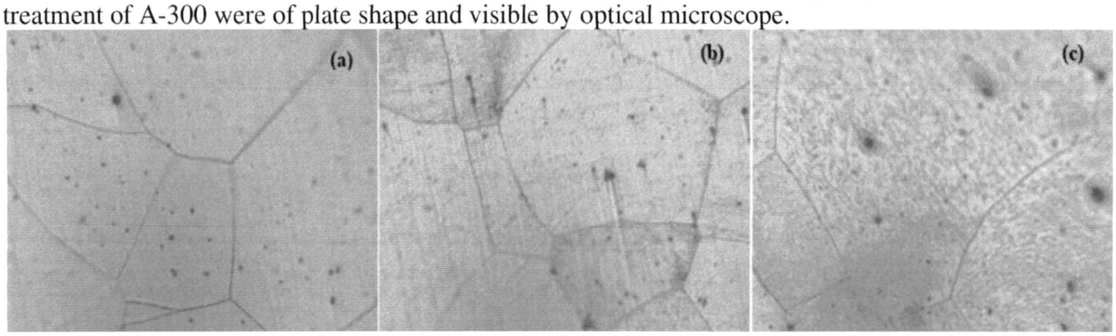

Fig.1 The microstructure of solution treated and aged WE54 alloy at 170°C, 250°C and 300°C

3.2 Hardness of the Mg-RE WE54 alloy

Specimen of Mg-RE WE54 alloy under as-received and heat treatment condition were taken. Solution treatment was followed by ice quench an then aging at different temperatures for a range of time. These alloys were subjected to these conditions to generate the different microstructures having different precipitate phases at different temperature. Efforts have been made to correlate the variation in the hardness values of the generated microstructures to the alloy treated at different temperatures for different time period.

Table-2 Vicker's hardness of WE54 alloy sample under different heat treatment condition

Samples according to heat treatment conditions	**Vickers Hardness (HV)**			
	1st Reading	**2nd Reading**	**3rd Reading**	**Average**
Aged at 170^0C(A170)	103	98.67	101.29	100.98
Aged at 250^0C	88.85	89.96	89.13	89.31
Aged at 300^0C	90.25	84.7	86.16	87.03

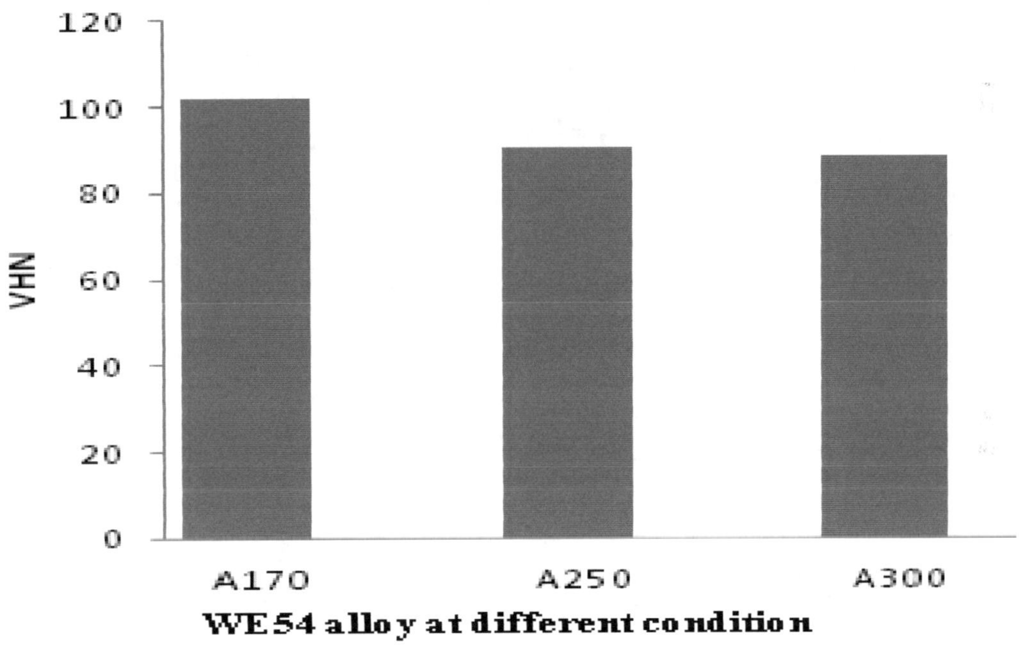

Fig.2 Hardness of the WE54 alloy samples

The maximum hardness was attained at A170 treatment, after that decrease in hardness was found. This means that the alloy have revealed a great response from ageing at 170°C and 250°C but at 300°C, ageing was very less, i.e. Peak hardness increases with decreasing ageing temperature. Also, hardness peak is high at 170^0C, due to lesser infusion of alloying elements.

4. Conclusions

The series of precipitation in Mg-RE WE-54 alloy systems consist of the growth of second stage particle named as β'', β' and β, which were obtained at subsequent aging temperature. All three precipitates form as plates on $\{10\bar{1}0\}_{Mg}$ position and it is hard to differentiate them on the basis of their shape, size and orientation. These prismatic precipitate of plate shape play a predominant role to strengthen the developed alloy. The β' phase, which is body centred orthorhombic, is formed at temperature around 250°C. There is a fine dispersion of these precipitate in the entire phase of the sample and is not possible to view under optical microscopy. Increase in hardness during ageing at 250^{0}C is due to the formation of semi-coherent interface within the mesh. Hetrogeneous nucleation of incoherent precipitates, β occur in grains and on the grain boundaries during aging at 3OO°C. These second phase viewed have face centred cubic structure with their shape and distribution of either plate or spherical, depending on their development temperature.

References

[1] Kainer, K. U. (Ed.). (2003). Magnesium alloys and technology. John Wiley & Sons.
[2] Nie, J. F. (2012). Precipitation and hardening in magnesium alloys. Metallurgical and Materials Transactions A, 43(11), 3891-3939.
[3] Bhargava, A. K., & Sharma, C. P. (2011). Mechanical Behaviour and Testing of Materials; PHI Learning Pvt. Ltd: New Delhi.
[4] Nie, J. F. (2003). Effects of precipitate shape and orientation on dispersion strengthening in magnesium alloys. Scripta Materialia, 48(8), 1009-1015.
[5] Li, Z. C., Zhang, H., Liu, L., & Xu, Y. B. (2004). Growth and morphology of β phase in an Mg–Y–Nd alloy. Materials Letters, 58(24), 3021-3024.
[6] Socjusz-Podosek, M., & Lityńska, L. (2003). Effect of yttrium on structure and mechanical properties of Mg alloys. Materials Chemistry and Physics, 80(2), 472-475.
[7] Xin, R., Li, L., Zeng, K., Song, B., & Liu, Q. (2011). Structural examination of aging precipitation in a Mg—Y—Nd alloy at different temperatures. Materials characterization, 62(5), 535-539.
[8] Nie, J. F. (2002, January). Precipitation and strengthening in selected magnesium alloys. In Magnesium Technology 2002.

IOP Conf. Series: Materials Science and Engineering 1104 (2021) 012026 doi:10.1088/1757-899X/1104/1/012026

A review on exergy analysis of ejector refrigeration system

Sameer Kumar Verma*[1], Satyajeet Kumar[2], Gulshan Sachdeva[1]

[1] National Institute of Technology, Kurukshetra , Haryana ,India
[2] Krishna Engineering College, Ghaziabad, U.P, India
*E-mail: sameerkumarverma1105@gmail.com

Abstract. Owing to the human comfort, the utilization of refrigeration system has increased nowadays. In today's scenario, the air conditioning system has led to the innovative changes in the lifestyle of human beings. The Heat driven ejector refrigeration machine is a favorable alternative to a compression based refrigeration system for the point of view of minimizing energy utilization. As Exergy is directly related to the second law of thermodynamics and the system performance can be only evaluated on the basis of this thermodynamics law. On the basis of efficient performance, it has been concluded that good quality of ejector is more performance oriented rather than other components like condenser, generator and evaporator. This paper presents the idea of ejector refrigeration systems, its exergy analysis and working fluids used. It profoundly investigates the ejector technology and behaviour, exergy analysis, properties of refrigerant and its effect on the ejector refrigeration system.

Keywords: Ejector refrigeration system, Exergy analysis, Exergy destruction, Refrigerants

1.Introduction

The vapour compression refrigeration system (VCRS) face a few extreme difficulties in the coming tenure, the maximum recognized being a limited usage of fluorocarbon refrigerants, spiking the cost of energy and increasing environmental issue. One of the best options is to shift towards the usage of renewable energy. The waste heat can be utilized in order to enhance the system performance characteristics. The ejector does not contain any moving part pump gadget, along with no maintenance cost and lubrication. While talking about ejector system, the working of ejector is same as that of compressor. The ejector system utilizes waste energy or the heat which makes it more engrossing in this energy-conscious span. A lot of research has been done in order to understand the concepts of performance characteristics of ejector system [1-4]. They have suggested that that system performance can be enhanced in terms of running cost, mechanical efficiency and environmental co-relation. The law of entropy states that, analysis of exergy can be used to categorize the position, amount and various sources of exergy destruction. Pridasawas et.al [5] carried out his work on exergy investigation on a ejector system that runs through the utilization of solar energy. He concluded that there is a maximum destruction of exergy occurring in an ejector using butane as a refrigerant. The same conclusion were also drawn by Alexis [6] when used water as working medium. Dahmani et al concluded that maximum exergy destruction takes place in ejector working with R134a (Tetrafluoroethane) as working fluid.

Naturally, the ejector refrigeration system uses water as a working medium (refrigerant). For cooling and refrigeration purpose, steam jet system came in to existence in the mid-1900s [7]. After conducting various experiments, it was found that that the existing system has a very low co-efficient of performance (COP) values. Also, the system was unable to give desired refrigerant effect at 0^0 C.

Consequently, the use of halocarbon refrigerants in ejector refrigeration systems because of more prominent values of COP. Primary requirements for ideal refrigerant is that it has zero ODP and GWP. Chloro-fluoro-carbons(CFCs) such as R12 and R22 etc cause destruction of ozone layer present at the stratosphere due to reaction with chlorine atoms which are present in the refrigerants. The substitution for CFCs are hydro-chloro-fluoro-carbons(HCFCs) which also deplete ozone layer but not as much as CFCs. The replacement for HCFCs are hydro-fluoro-carbons(HFCs).HFCs have ozone depletion potential of zero unlike CFCs and HCFCs. Some examples of HFCs are R134a, R410A, R407C,etc.

2. Ejector Technology

2.1 Ejector Fundamentals

An Ejector is the critical segment in the ER system. The major important function of the ejector is that it can increase the pressure of the refrigerant without using mechanical energy. The utilisation of thermal energy in this makes it even more attractive. It can also reduce environmental problems.

The ejector was first time used by Maurice Leblanc in 1910 as a vapour jet refrigeration cycle. His assembly produced refrigeration effect by utilising waste heat energy. Since steam was used at that time so it is called the steam jet cooling system, later it has been replaced by a system with a mechanical compressor.

An ejector generally consists of four parts-

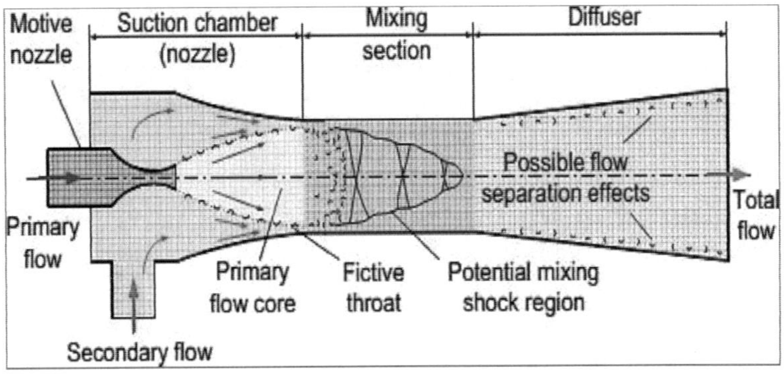

Figure 1: Different sections inside an ejector

Motive Nozzle

The primary fluid enters the motive or primary nozzle which is of convergent-divergent type. In this nozzle velocity increases in the convergent part which leads to a decrease in pressure. At the throat section of the nozzle, Mach number is unity and the velocity becomes sonic. At the divergent part of the nozzle, velocity will accelerate and pressure decreases by obeying the energy balance equation. At the outlet of the nozzle low pressure is created and the flow becomes supersonic.

Suction chamber

As the exit of the nozzle low pressure is created, because of which the pressure difference is caused between the evaporator and the suction chamber. Suction chamber is connected to the evaporator in the ejector. This difference in pressure leads to entry of secondary fluid in the ejector.

Mixing Section

The motive and entrained fluid mixed in the mixing section which is having a constant area section of the ejector. A shock wave is generated at the end of the mixing section due to supersonic flow which causes an abrupt decline in velocity. Also there is a sudden rise in pressure inside the constant area section of the ejector.

Diffuser

In the diffuser section, the blended fluid enters and the kinetic energy of the blended fluid is converted into pressure energy. Therefore the fluid coming out from the diffuser is at an intermediate pressure and velocity is assumed to be zero at the outlet of the diffuser.

2.2 Working principle of the ejector

Figure 2 shows the illustration of an ejector refrigeration system. The main components are generator, ejector, condenser, evaporator, circulation pump and a throttling valve. The primary or motive flow refrigerant in the vapor state enters the generator at high pressure. The low pressure vapor enters through the evaporator. Both motive flow and low pressure vapor are mixed in the mixing chamber of ejector and recovery of pressure take place. After that it enters in to the condenser, where phase changes from vapor to liquid. The phase change will occur when heat is rejected to the environment (Q_{CO}). The fluid coming out from the condenser is splitted in to two different passages. One section enters into the evaporator via expansion device (throttling valve) to produce refrigeration effect (Q_{EV}). The other segment of the fluid is reverted back to the generator where it takes heat and converts into high pressure vapor and finishes the cycle.

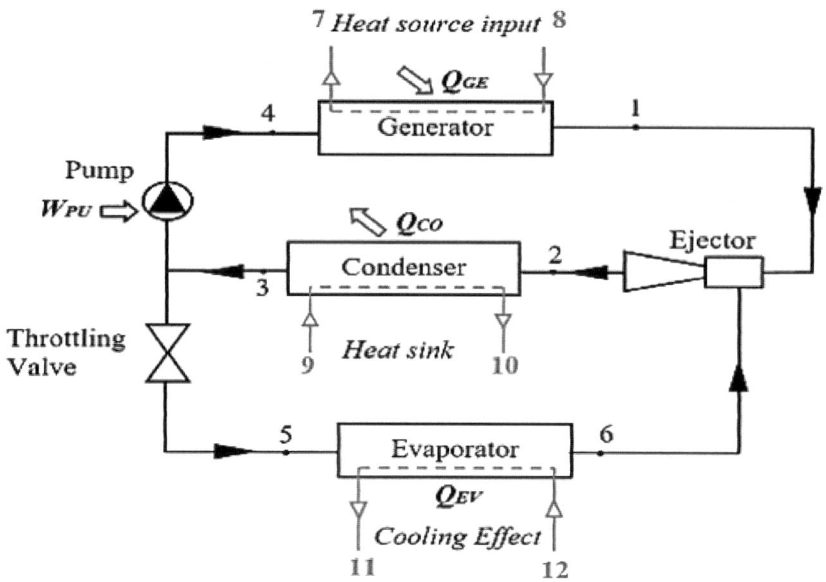

Figure: 2 Ejector refrigeration System

3.1 Conventional Exergetic model

Neglecting the effects of electrical, magnetic, nuclear and surface tension in refrigeration cycle, the total Exergy (E) can be the considered as the combined effect of physical ,chemical, Kinetic and potential exergies (E^{PH}, E^{CH}, E^{KN} and E^{PT}) respectively.

$$E=E^{PH}+E^{CH}+E^{KN}+E^{PT}$$

From above equation, the term E^{CH} , E^{KN} , E^{PT} can be omitted, since there is no chemical reaction occurs in the ejector refrigeration system. Thus exergy E only comprises of physical exergy E^{PH}.

$$E_j=E_j^{PH}=\dot{m}e_j^{PH}=\dot{m}[(h_j-h_o)-T_o(s_j-s_o)]$$

here ,e_j^{PH} can be divided into thermal exergy e^T because of temperature and mechanical exergy

e^M because of pressure.

$$e_j=e_j^{PH}=e_j^T+e_j^M=[(h_j-h_{j,X})-T_0(s_j-s_{j,X})]_{P=const}+[(h_{j,X}-h_{j,0})-T_0(s_{j,X}-s_{j,0})]_{T0=const}$$

The exergy destruction of the k-th component can be expressed as

$$E_{D,k}=E_{F,k}-E_{P,k}$$

The exergy balance for the system can be represented as

$$E_{F,tot}=E_{P,tot}+E_{D,tot}+E_L$$

where E_L represents the loss of exergy in the system. In order to investigate the exergetic performance of the component, both exergy efficiency and exergy destruction is important. The exergy efficiency ratio ($€_k$) is the ratio of exergy of product to the exergy of fuel . Similarly, exergy destruction ratio (y_k) is the ratio of exergy destruction to the total exergy of fuel of overall system.

$$(€_k= E_{P,k}/ E_{F,k},$$

$$y_k= E_{D,k}/ E_{F,tot})$$

IOP Conf. Series: Materials Science and Engineering 1104 (2021) 012026 doi:10.1088/1757-899X/1104/1/012026

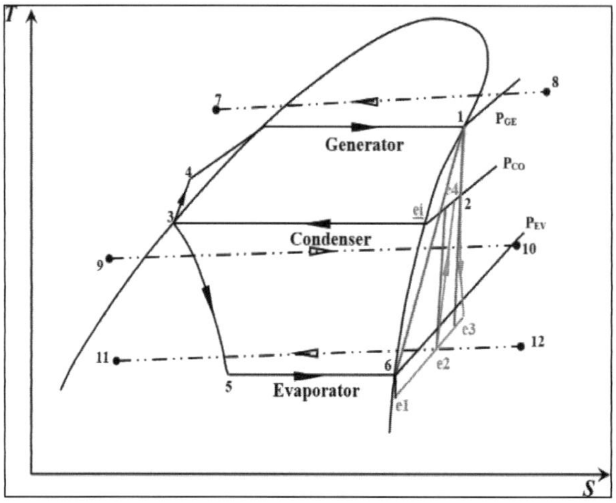

Figure 3: T-s diagram of ejector refrigeration system

3.2 Advanced Exergetic Model

As compared to traditional exergy model, Advance exergy framework is more efficient. According to the advance model, there are two subgroups of total exergy destruction namely endogenous/exogenous and avoidable/unavoidable segments.

3.2.1 Endogenous and Exogenous exergy destruction

Endogenous exergy destruction ($E_{D,k}^{EN}$) is the total amount of exergy that is destroyed in components when the given component is working in real conditions and all other components are assumed to work ideally in the system. Exogenous exergy destruction ($E_{D,k}^{EX}$) is the loss of exergy caused by the remaining components. Therefore, it can be expressed as

$$E_{D,k}=E_{D,k}^{EN}+E_{D,k}^{EX}$$

From this information, it is easy to determine whether the focus of the study should be paid to the k-th component or the leftover component.

Unavoidable and avoidable exergy destruction

The exergy destruction can be further broken into unavoidable and avoidable segments

$$E_{D,k}=E_{D,k}^{UN}+E_{D,k}^{AV}$$

where the unavoidable exergy destruction $E_{D,k}^{UN}$ cannot be reduced due to constraints manufacturing processes, material types etc. The other part of exergy destruction i.e. avoidable exergy destruction $E_{D,k}^{AV}$ is reducable and should be more observed.

Adding the above two approaches

The two splitting approaches can be merged as

$$E_{D,k} = E_{D,k}^{UN,EN} + E_{D,k}^{UN,EX} + E_{D,k}^{AV,EN} + E_{D,k}^{AV,EX}$$

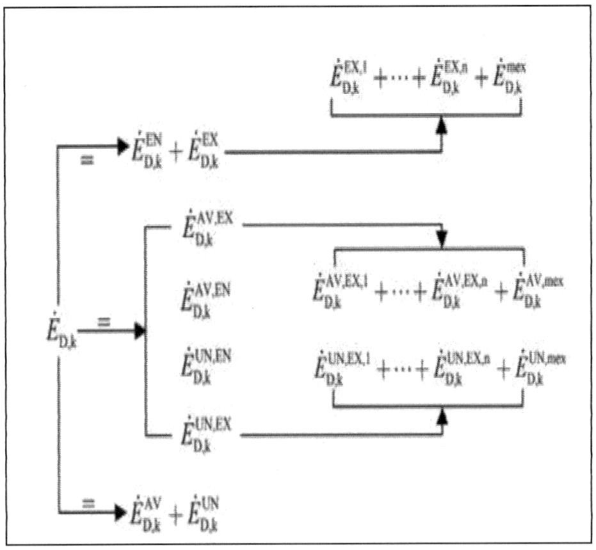

Figure 4: Alternatives for splitting the exergy destruction within the kth component

4. Selection of Working Fluid

The refrigerant is very important in the ejector refrigeration system. A suitable refrigerant can just give efficient system performance, chances of system failure are less and also good for the environment. Additional requirements for selecting working fluids are

- Global warming potential (GWP) and ozone depletion potential (ODP) from the environmental point of view.
- Toxicity, flammability for health point of view.
- Cost and availability of refrigerant.

The essential thermodynamic properties of refrigerant are given below

- The higher critical temperature of the refrigerant is desirable to adjust large variation of generator temperature;
- The specific heat of vapour refrigerant should be as high as possible to limit the degree of superheat whereas that of liquid refrigerant should be low to limit the degree of irreversibility;
- It should be non-toxic and non-flammable.

The working fluids are of three types. These are namely wet, dry and isentropic fluids as shown in Figure 5.

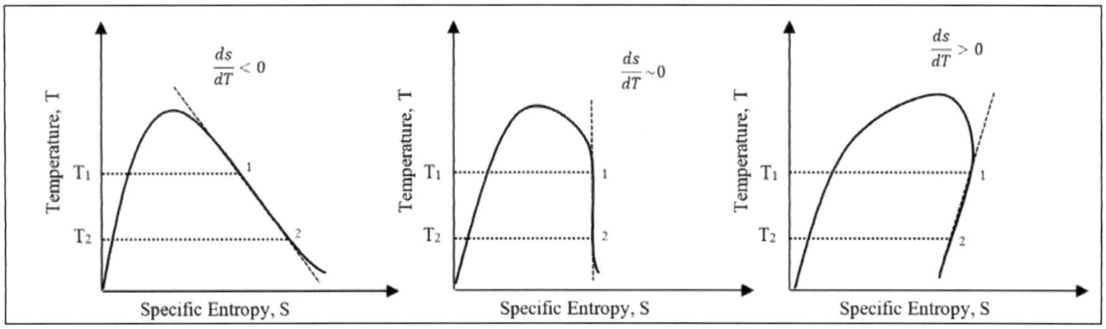

Figure 5: T-S diagrams of three types of refrigerant: (a) wet fluid, (b) isentropic fluid and (c) dry fluid

Figure 5 shows a plot between temperature and entropy of the different working fluid. From figure 5 (a), it can be seen that saturation line for vapour phase is having negative slope. Due to the formation of small vapour droplets at the exit of nozzle, there is a blockage in the effective area ultimately affects the functioning of ejector. In order to avoid such problems, motive flow must be in a superheated condition before entry to the nozzle. During expansion through the nozzle, phase change do not occurs in the dry fluid. It can also be seen that the vapour saturation line is vertical in case of isentropic fluid. Owing to the environmental concern, the refrigerants having zero ozone depletion potential value is recommended. . Generally, the use of ammonia and water is avoided because of high critical pressure of ammonia and also it is toxic in nature while for water its COP is low. According to Jianyong et al generally "R134a, R152a, R245fa, R290, R600 and R600a are used as working fluid in the ejector refrigeration system. From thermodynamic characteristics, there is no working fluid which is perfectly suitable for the ER system, each one has its own advantages and disadvantages.

IOP Conf. Series: Materials Science and Engineering 1104 (2021) 012026 doi:10.1088/1757-899X/1104/1/012026

5. Conclusion

Ejector refrigeration system is an advantageous technology for producing a desired refrigerating effect by utilising waste or low-grade heat with various working fluids. In the present paper ejector technology, exergetic models, properties of refrigerant and its impact on the ejector refrigeration system are discussed in depth. Two exergetic models are discussed i.e., conventional exergetic model and advanced exergetic model. In the conventional model fuel-product concept is discussed. In the advanced exergetic model, exergy destruction can be classified into four parts avoidable, unavoidable, endogenous and exogenous it is more accurate as compared to the conventional model. The ejector allows the use of several refrigerants but only those refrigerants are advantageous which have zero ODP and zero GWP which prevent from environment-related problems and also suitable for safety point of view along with this the refrigerants have a positive slope on T-s diagram for the proper functioning of the ejector.

References

[1] W.Pridasawas, P.Lundqvist, A year-round dynamic simulation of a solar-driven ejector refrigeration system with iso-butane as a refrigerant, Int.J. Refrig.30 (2007) 840-850.

[2] V.M.Nguyen, S.B.Riffat, P.S.Doherty, Development of a solar-powered passive ejector cooling system, Appl. Therm. Eng. 21 (2001) 157-168.

[3] S.A.Sherif, D.Y.Goswami, G.D.Mathur, S.V.Iyer, B.S.Davanagere, S.Natarajan, F.Colacino, A feasibility study of steam-jet refrigeration, Int. J. Energy Res.22 (1998) 1323-1336.

[4] M.Dennis, K.Garzoli, Use of variable geometry ejector with cold store to achieve high solar fraction for solar cooling, Int. J. Refrig. 34 (2011) 1626-1632.

[5] Pridasawas W, Lundquist P. An exergy analysis of a solar-driven ejector refrigeration system . Sol Energy 2004; 76:369-79.

[6] Alexis GK. Exergy analysis of ejector-refrigeration cycle using water as working fluid. Int J energy Res 2005;29:95-105.

[7] ASHRAE. Steam-jet refrigeration equipoment. In:1983 Equipment Handbook, Ch. 13. ASHRAE, Atlanta, GA, U.S.A., 1983.p. 13.1-13.6.

[8] Varga, S., Lebre, P.S., Oliveira, A.C., 2013a. Readdressing working fluid selection with a view to designing a variable geometry ejector. Int. J. Low Carbon Tech. 0, 1-11.

[9] Jianyong Chen, Hans Havtun, Bjorn Palm, Screening of working fluids for the ejector refrigeration system.

[10] Jianyong Chen, Hans Havtun, Bjorn Palm Conventional and advanced exergy analysis of an ejector refrigeration system.

Experimental Investigation of the AISI H-13 die steel by Wire-Cut EDM

Ujjwal [1*], Sachendra [1], Shailesh Kumar Singh [2]
1 Department of Mechanical Engineering, Krishna Engineering College, Ghaziabad, India 201007
2 Department of Mechanical Engineering, Dronacharya Group of Institutions, Greater Noida, India
Corresponding author- ujjwalchauhan@gmail.com

Abstract---
The current study is done to provide a comprehensive report on the machining of AISI H-13 Die Steel with bare and coated wire electrodes. The study is carried out for the pulse duration-on time and pulse duration-off time process parameters effect on both electrodes and surface roughness is taken as response parameter of material along with the best viable combination of process parameter for machining.
Key words: AISI H-13 Die Steel, Wire-cut EDM, Surface Roughness.

1. Introduction

In engineering world, new techniques are being used to ease the machining process. Especially non-conventional energy sources like light, sound, electrical, mechanical, chemical, electrons and ions are new attractive concepts. With the new continues industrial and technological growth, recent developments to machine harder and difficult materials, which have wide application in various fields like nuclear engineering, aerospace and other industries. Still the processing time and surface texture with smooth surface are the major concerns for operators [1].

With daily challenges of machining materials with high strength & temperature resistant and difficult-to-machine, suggestion comes out to be to use EDM process. EDM machining is used to remove material even for small batches [2]. The moment of electrode is controlled by numerically control system. The smooth instant spark occurs sufficiently, it exhibits a pressure between workpiece and electrode which results in of a very high temperature and at such high pressure and temperature (8000 C-12000 C) that some metals get melted and evaporated at the same time. The investigation of Conductive zirconias containing 22 to 44 vol% NbC or Tic was done through wire-cut EDM. They evaluate the surface roughness and machining rate by varying process parameters like pulse duration and duty factors and share the feasible conditions for the process [3]. The various process parameters pulse time, open circuit voltage, wire speed and dielectric fluid pressure was experimentally investigated on wire-cut EDM process on AISI 4140 steel having 10mm thickness with 0.25 brass wire electrode to see the effect on response parameters which are surface texture and cutting speed. His findings are that increasing pulse time, wire speed, open circuit voltage and dielectric fluid pressure increases the surface roughness and cutting rate [4]. The investigation was also done by wire-cut EDM for machining $Al2O3p/6061Al$ composite by taking different levels of variables like pulse duration to identify the effects on response parameters. All experiments were conducted on a FANUC W1 CNC EDM machine with 0.25mm brass wire electrode. As well as, they see the impact of wire electrode while machining the $Al2O3p/6061Al$ composite because of its repeatedly breaking, which emphases on the work suggesting locations of possible wire breakage and the reason for same during machining [5].

2. Materials and Methods

FANUC ROBOCUT Alpha 1-iE CNC Wire-Cut EDM was used to carry out the experiments. AISI H-13 die steel is being used as workpiece material for present investigation. AISI H-13 is special hot-worked chromium tool die-steel with good hardness and toughness properties. Tool steels are used to construct the die components subject to wear. The nominal composition is shows the composition of the AISI H-13 Die Steel with a significant contribution of carbon (0.356 %), Silicon (0.885%), Chromium (5.109%), molybdenum (1.220%) and rest is Fe.

Table 1 : **Composition of Work piece Material in percentage**

Fe	C	Si	Mn	P	S	Cr	Mo	V
90.750	0.356	0.885	0.439	0.015	0.016	5.109	1.220	0.802

The roughly cuboid plate is being machined on milling machine so that it becomes a complete cuboid with 210mm x 70mm x 31mm plate. After that, it goes on grinding machine where it is been grind to remove all inaccuracies in the workpiece so that it become ready for machining on Wire-cut Edm machine. The machine has a mechanism of a wire feed through the workpiece via a hole which is already drilled into the workpiece to allow the pass through to cut. The workpiece is being machined by two different wires i.e. hard brass wire (uncoated wire) and zinc coated brass wire (coated wire) in this experiment. Zinc coated brass wire consists of a layer (approx .6 micron) zinc coating over a core brass alloy.

In the present work, the Taguchi's method has been used to plan the experiments. L9 orthogonal array is being used for experiment conduction. Signal to noise ratio (S/N) is used for analysis of responses. The loss function is linked with the quality matrix of the system. The loss associated with the process of product can be minimized by maximizing the S/N ratio. The (S/N) ratio in terms smaller the better (SB) is calculated. The process parameters like pulse duration -on time and pulse duration-off time for the present investigation is being taken is shown in table with levels.

Table 2 : Machining parameters and their levels

Process Parameter	Symbol	Units	Level 1	Level 2	Level 3
Pulse-on Time	Ton	μs	6	8	10
Pulse-off Time	Toff	μs	18	20	22

The Taylor-Hobson Surtronic3+ surface roughness tester is used to measure surface roughness.

3. Results and Discussion

The surface roughness was noted down with two no. of trials with same process parameters under same conditions to avoid any bias in investigation. The work piece with their raw data and S/N ratio for each work piece are given in table for each trial.

IOP Conf. Series: Materials Science and Engineering 1104 (2021) 012027 doi:10.1088/1757-899X/1104/1/012027

Exp. No.	Surface Roughness with Zinc coated wire electrode			Surface Roughness with brass wire electrode		
	Surface Roughness (µm)		S/N Ratio	Surface Roughness (µm)		S/N Ratio
	R1	R2		R1	R2	
1	2.32	2.35	-7.36592	2.54	2.58	-8.16506
2	2.51	2.49	-7.95887	2.71	2.69	-8.62733
3	2.59	2.61	-8.29953	2.79	2.83	-8.97435
4	3.10	3.05	-9.75719	3.19	3.21	-10.1030
5	2.88	2.90	-9.21801	3.10	3.12	-9.85525
6	2.74	2.72	-8.72331	2.91	3.07	-9.51653
7	3.25	3.22	-10.1976	3.59	3.54	-11.0414
8	3.12	3.09	-9.84133	3.32	3.35	-10.4620
9	3.19	3.21	-10.1030	3.41	3.43	-10.6806

Effect of process parameter on the surface roughness with brass electrode and coated electrode is being shown below in figure (1) and figure (2).

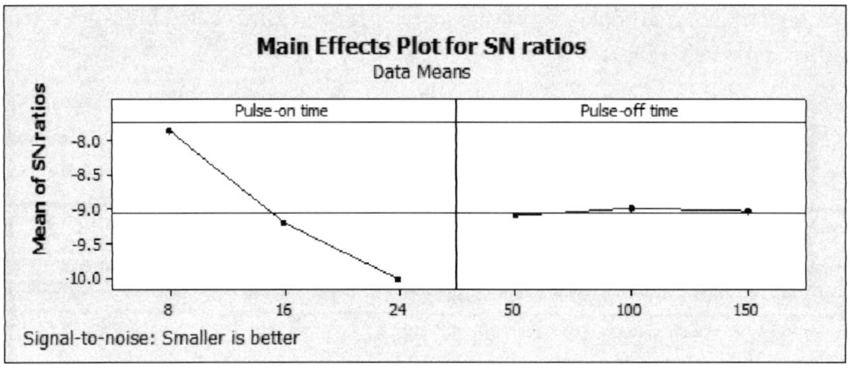

Figure 1 : Effect of process parameter on surface roughness with zinc coated wire (S/N ratio)

Figure (1) shows that the first value of Pulse duration-on time (Ton_1) and the second value of Pulse duration-off time ($Toff_2$) provide the minimum value of surface roughness with zinc coated electrode. The same extent of the parameters (Ton_1 and $Toff_2$) are the feasible optimum levels for least surface roughness with zinc electrode in Wire-cut EDM process are achieved by raw data analysis.

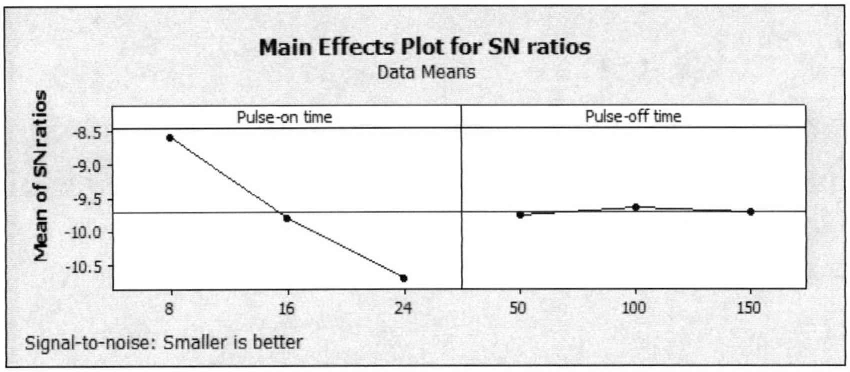

Figure 2 : Effect of process parameter on surface roughness with brass wire (S/N ratio)

Figure (2) shows that the third value of Pulse duration-on time (Ton_1) and the first value of Pulse duration-off time ($Toff_2$) provide the low value of surface roughness with brass electrode. The same extent of the parameters (Ton_1 and $Toff_2$) are the feasible optimum levels for the minimum surface roughness with brass wire electrode in Wire-cut EDM machining are resulted by raw data analysis.

4. Conclusions

The surface roughness values for AISI H-13 die steel after machining with coated electrode is less than AISI H-13 die steel machined with brass electrode. The lesser surface roughness is desirable. Zinc-coated wire electrode leads to better surface texture compare to brass electrode. The brass wire with zinc coating facilitates a higher tensile strength for wire electrode. Wire electrodes having tensile strength on higher side shows good thermal resistance property in high temperature and maintain straightness even with vibration and mechanical tension in wire electrode. As well as, the uniform layer of zinc on core provides good discharge characteristics. Electrodes which have properties like discharge characteristics and higher tensile strength can be achieved through finer discharge which can be created with zinc coated wire. Low surface roughness is being achieved in case of coated electrode machining surface as compare to brass wire electrode, overall quality of machining surface get improved after machining with coated electrode. It will further help to achieve better responses.

References

[1] Lazarenko, B. R., & Lazarenko, N. (1943). About the inversion of metal erosion and methods to fight ravage of electric contacts. WEI-Institute, Moscow in Russian.

[2] Hasçalýk, A., & Çaydaş, U. (2004). Experimental study of wire electrical discharge machining of AISI D5 tool steel. Journal of Materials Processing Technology, 148(3), 362-367.

[3] Matsuo, T., & Oshima, E. (1992). Investigation on the optimum carbide content and machining condition for wire EDM of zirconia ceramics. CIRP annals, 41(1), 231-234.

[4] Tosun, N. (2003). The effect of the cutting parameters on performance of WEDM. KSME international journal, 17(6), 816-824.

[5] Yan, B. H., Tsai, H. C., Huang, F. Y., & Lee, L. C. (2005). Examination of wire electrical discharge machining of Al2O3p/6061Al composites. International Journal of Machine Tools and Manufacture, 45(3), 251-259.

IOP Conf. Series: Materials Science and Engineering 1104 (2021) 012028 doi:10.1088/1757-899X/1104/1/012028

Assessment of Environmental Performances of Small and Medium Scale Indian Industries in the context of Green Supply Chain Management (GSCM)

Sanjeev Kumar[1*], Somnath Chattopadhyaya[2] and Sandeep Tiwari[1]
[1]Krishna Engineering College, Ghaziabad, India. Email: sanjeevkg9@gmail.com, sandeeptiwari1970@gmail.com
[2]IIT (ISM), Dhanbad, India. Email: somuismu@gmail.com

Abstract:
GSCM is extremely important for recent manufacturing systems. In an environmentally awaken atmosphere GSCM has an indispensible role to play to reduce the carbon foot print of different manufacturing operations. In the Indian context, the small and medium scale enterprises are unaware of environment friendly supply chain operations. Therefore there is an urgent need to develop a scale for implementing GSCM practices in those manufacturing domain. The objective of this paper is to investigate in formulating a widely acceptable scale for evaluation of green supply chain (GSC) system practices among those manufacturers. On the basis of the data collected from 119 Indian enterprises, measurement models of GSCM practices were evaluated. The assessments are done though the questionnaire from Small and Medium size enterprises (SMEs). Role of environmental performance in the green supply chain system for those enterprises is critically analyzed in this work. Conclusive strategies of this attempt will enable those manufacturers to perform effectively in dealing with GSCM issues.

Key words: Green Supply Chain Management, Survey, Environmental performance.

1. Objective:

Principal objective of the study is to do a comparative analysis of environmental performance between small and medium scale Indian industries in the viewpoint of GSC system. This critical comparison will empower the India SMEs to do the needful.

2. Literature review:

There are only a few attempts to critically analyze the environmental performance in GSCM. Various approaches to GSCM practice have been identified by various researches; yet they are very brief and have limited scope. One of them is by Quinghu Zhu et al (2008) [16] where the researchers conceptualize GSCM practices implementation.

A supply chain environmental performance assessment for telecommunication sectors was developed by Horvath, 1999 [8]. The assessment was based on a life-cycle assessment.. Relationships between Green Supply Chain Management practices implementation and performance with a focus on the moderating effects of quality and lean practices were investigated by Zhu and Sarkis (2004) [23]. A number of GSCM practices implemented by Chinese enterprises to improve their performance were described by Zhu et al. (2005) [24].

IOP Conf. Series: Materials Science and Engineering 1104 (2021) 012028 doi:10.1088/1757-899X/1104/1/012028

3. The green supply chain:

Small and medium scale enterprises generate substantial value addition in the manufacturing sector, but their performance for implementing environmental regulations is not at all acceptable. In most of the cases they are the main culprit for polluting the environment. Due to this fact the implementation of GSCM practices is absolutely crucial for environmental perspective. Therefore, there is a burning need to formulate a scale to evaluate the environmental performance of the SMEs. GSC involves three approaches namely: environment, strategy and logistics. Figure 1 suggested by. (Breno Torres Santiago Nunes, Sergio Marques Junior, Rubens Eugenio Barreto Ramos, 2004)[1]. In context to Gilbert, 2001[5], green supply chain initiative provide competitive advantages such as low costs. In addition to low cost, Gilbert (2001) [5] stated that GSC also opens up new markets for organization.

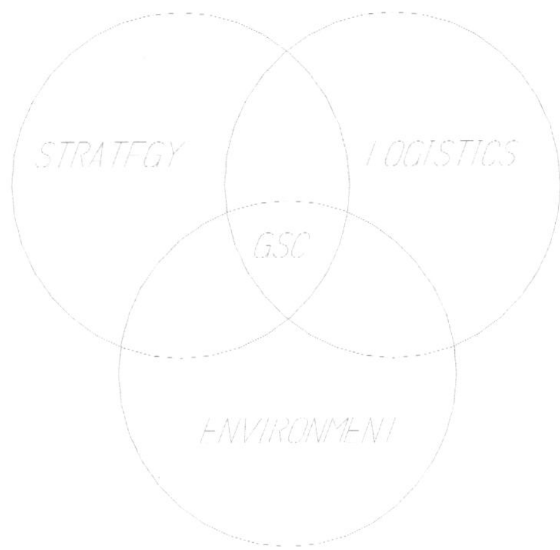

Fig. 1 Approach of green supply chain

With the inception of the concept of GSC, manufacturers have started stressing upon the greener products design. For changing product design and for improving the recycling of the products, the manufacturer has to work in close proximity with recycled parts suppliers.

4. Research methodology:

Best Practices questionnaire have been developed. These questionnaire consist 10 underlying statements. Countries like China and Spanish industries are also critically investigated to set a benchmark of performance of Indian SMEs in the domain of implementing GSCM practices. To analyze the scores of Indian SMEs in different modules are being develop to comprehensively express its real performance.

The target respondents for this survey were requested to indicate through a five-point Linkert scale (Excellent/Completely agree-5 , Very good/ Rather agree-4, Good/ Partly agree-3, Average/ Rarely agree-2, Below average/Completely disagree-1,), the extent to which it perceived its enterprises implementing each of the dimensions of environmental performance and green supply chain management practices. The number of Indian SMEs is very large and analyzing the performance of each of them is quite cumbersome task. So some representative industries are selected for a gross performance trend in terms of environment friendly manufacturing. The 119 SMEs are sufficient to comprehend the overall performance towards GSCM.

For each of the 10 statements item analysis was conducted between the highest and lowest group through a "**mean score**", "**t-test**" and **"p" test**.

IOP Conf. Series: Materials Science and Engineering 1104 (2021) 012028 doi:10.1088/1757-899X/1104/1/012028

5. Result and comparative analysis:

5.1 Comparative analysis of "Environmental Performance":

Air emission control is important for reducing environmental pollution and subsequently implementing green supply chain management. Similarly, minimization of solid waste is crucial for green operations. Furthermore, recycling and recovery reduces environmental hazards and ensures cost reduction. Additionally, waste water treatment is also essential as per the rules and regulations of the ministry of environment. Table- 1 and Graph- 1 explains about the effectiveness and draws comparative analysis between Environmental Performance factors and 10 dimensions underlying for Indian small and medium sized enterprises. On the whole 119 industries have been considered (59 from medium scale industry and 60 from small scale industry).

Table: 1 Comparative analsysis of mean score of Environmental Performance

Sl. No.	Dimension	SMALL SCALE Mean (S.D)	MEDIUM SCALE Mean (S.D)	MEAN DIFF.	RANK	t	p
I	Minimization of air emission	2.266 (0.756)	2.745 (0.842)	0.479	8	3.262	0.001
Ii	Minimization of solid waste	2.400 (0.741)	2.762 (0.727)	0.362	3	2.695	0.008
Iii	Improve production procedure/method for reducing waste/scrap	2.183 (0.624)	2.517 (0.731)	0.334	2	2.664	0.009
Iv	Recovery through sale of scrap and used/rejected material	1.600 (0.785)	1.881 (1.068)	0.281	1	1.635	0.105
V	Recovery through sale of excess capital equipment	1.483 (0.650)	1.915 (1.193)	0.432	6	2.457	0.015
Vi	Recovery through sale of old/obsolete equipment	1.474 (0.727)	1.898 (1.213)	0.424	5	2.300	0.023
Vii	Recycling of waste water	1.466 (0.675)	1.949 (1.195)	0.483	9	2.717	0.008
Viii	Optimization of man power resources in production process	2.483 (0.911)	2.898 (0.687)	0.415	4	2.801	0.006
Ix	Reduction of consumption for hazardous materials	2.050 (0.768)	2.711 (0.910)	0.661	10	4.281	0.000
x	Reduction in frequency of environmental accidents	1.800 (0.859)	2.271 (0.961)	0.471	7	2.816	0.006

ICOTRIME 2020 IOP Publishing
IOP Conf. Series: Materials Science and Engineering 1104 (2021) 012028 doi:10.1088/1757-899X/1104/1/012028

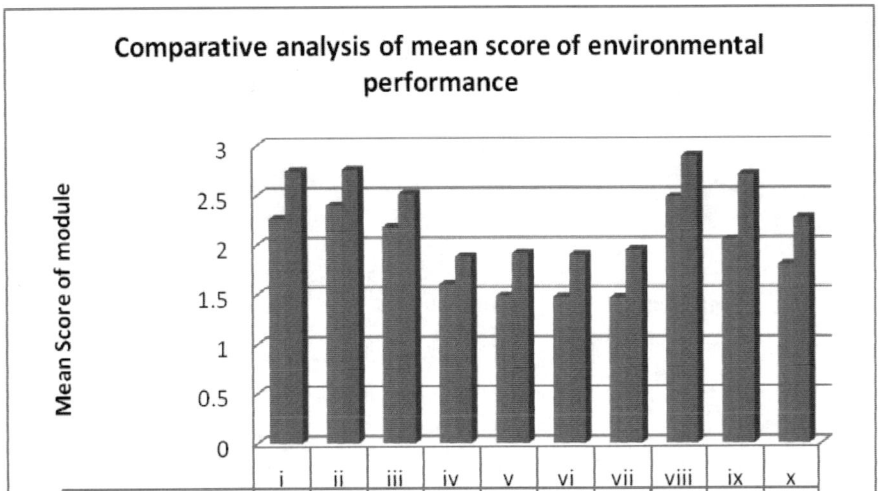

	i	ii	iii	iv	v	vi	vii	viii	ix	x
■ SMALL SCALE	2.26	2.4	2.18	1.6	1.48	1.47	1.46	2.48	2.05	1.8
■ MEDIUM SCALE	2.74	2.76	2.51	1.88	1.91	1.89	1.94	2.89	2.71	2.27

Graph: 1- Comparative analysis of mean score of Environmental Performance

6. Comparative result analysis of different country:

It is very much necessary for analyzing the comparative assessment of these Indian SMEs with respect to some foreign counter parts. Air emission control, waste water management, recycling and recovering are the important modules of activities in all form of manufacturing plants. The comparative assessment of these modules is to be carried out to comprehend the gravity of the situation. The scores related to the performance of Chinese and Spanish industries are collected from the existing literature. United nations have set some pollution and emission norms for all the countries in the world. So a comparative assessment scale is very much important for analyzing the pollution related performance of these SMEs of the different parts of the world and bare minimum standard has to be implemented of every country without paying heed to their geographical locations, financial status and the overall discipline of the countries. The researchers have worked on various modules of GSCM practices implementation in different countries, for different time periods, for different classes of industries. The mean scores (percentage arrived with linkert scale adopted by researchers) of the some modules studied are tabulated below:

Table 2: Analysis of comparative result of different country for GSC.

Modules/Equivalent modules	Maximum recommended score by ISO 14001 /Certification body for implementing an environmental management system	Javier et al, 2008[9](Operations management practices linked to adoption of ISO 14001: An empirical analysis of Spanish manufacturers [Spain]	Qinghua et al, 2008,[15] Firm-level correlates of emergent green supply chain management practices in the Chinese context [China]	A Comparative Analysis of Environmental Performance between Small and Medium Scale Indian Industries for Implementation of Green Supply Chain System **[India]**
Linkert Scale	**in %**	**6 (in %)**	**5 (in %)**	**5 (in %)**
Minimization of air emission	100	4.98 (83.0)	---	2.505 (50.1)
Minimization of solid waste	100	5.34 (89.0)	---	2.581 (51.6)
Improve production procedure/method for reducing waste/scrap	100	4.37 (72.8)	3.670 (73.4)	2.350 (47.0)
Recovery through sale of scrap and used/rejected material	100	3.98 (66.3)	3.480 (69.6)	1.740 (34.8)
Recovery through sale of excess capital equipment	100	---	3.340 (66.8)	1.699 (33.9)
Recovery through sale of old/obsolete equipment	100	4.01 (66.8)	3.430 (68.6)	1.686 (33.7)
Recycling of waste water	100	3.81 (63.5)	3.350 (67.0)	1.707 (34.1)
Optimization of man power resources in production process	100	3.99 (66.5)	----	2.690 (53.8)
Reduction of consumption for hazardous materials	100	4.26 (71.0)	3.570 (71.4)	2.380 (47.6)
Reduction in frequency of environmental accidents	100	4.58 (76.3)	3.650 (73.0)	2.035 (40.7)

ICOTRIME 2020 IOP Publishing
IOP Conf. Series: Materials Science and Engineering 1104 (2021) 012028 doi:10.1088/1757-899X/1104/1/012028

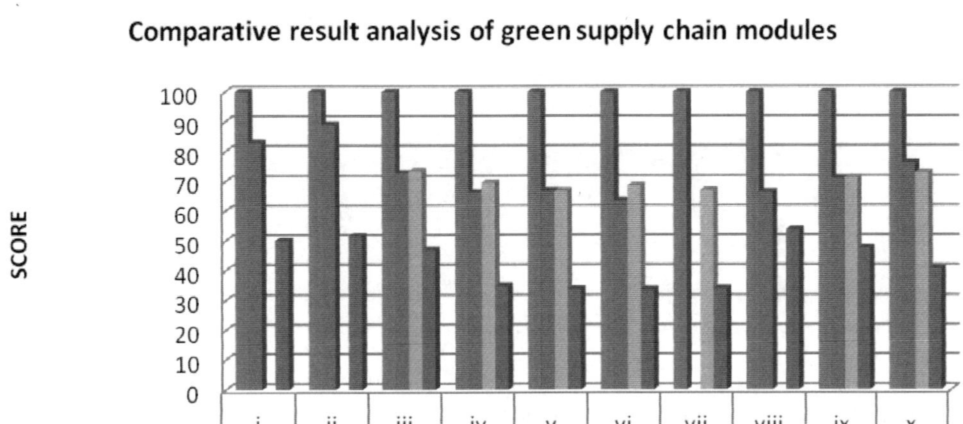

	i	ii	iii	iv	v	vi	vii	viii	ix	x
Highest Standard	100	100	100	100	100	100	100	100	100	100
Spain	83	89	72.8	66.3	66.8	63.5		66.5	71	76.3
China			73.4	69.6	66.8	68.6	67		71.4	73
India	50.1	51.6	47	34.8	33.9	33.7	34.1	53.8	47.6	40.7

Graph 2: Comparative result analysis of different country for green supply chain module

It is observed that in above foreign industries, the impact of GSCM practices and implementation is higher. It is also observed that for all the modules, Indian industries score significantly less compared to the industries of the other countries. In India, the awareness regarding the implementation of GSCM practices is far from satisfactory. It is suggested that more emphasis should be given for GSCM practices for effective recognition and satisfactory implementations. The management of Indian industries, especially small and medium sized industries, should be more focused to the above modules for better implementation of GSCM practices.

7. CONCLUSION

A detailed study about the performance of the Indian SMEs is carried out and the scores of the Indian industries are compared with Spanish and Chinese counterparts. It is revealed from the comparison that most of the Indian enterprises are operating at a very low level in terms of implementing GSCM. In most of the modules, the scores of Indian enterprises are the least. It is evident from this analysis that the awareness regarding GSCM is exceptionally poor for the manufacturing operations in India. Therefore, a uphill task is ahead of the Indian SMEs to come to the international level. Pollutions due to the manufacturing processes mainly in the urban and suburban areas are mostly done by Indian SMEs. In different sectors, the SMEs has to incorporate the awareness programs related to the green procurement, green processing, green waste management at various levels. These lower level performances of the Indian SMEs have to be audited on regular basis under a common platform for alleviating them from the bottom most level. This paper identifies the pragmatic situations of Indian SMEs and their extreme level of non-performance towards GSCM implementation.

REFERENCES:

1. Breno Torres Santiago Nunes, Sergio Marques Junior, Rubens Eugenio Barreto Ramos, (2004). A theoretical approach for green supply chain, Federal University Do Rio Grande Do Norte Industrial Engineering Program.

2. Dianjun Qin , Xiaojing Guo (2011). 'Research on implementation strategies of green supply chain management'. Applied Mechanics and Materials, 84-85, pp.757.

3. G. Kannan, P. Sasikumar and K. Devika (2010). 'A genetic algorithm approach for solving a closed loop supply chain model: A case of battery recycling'. Applied Mathematical Modelling, 34 (3), pp.655.

4. Gilbert, S. (2000) Greening supply chain: Enhancing competitiveness through green productivity: Tokyo: Asian Productivity Organization.

5. Gilbert, S. (2001) Greening Supply Chain: Enhancing Competitiveness through Green Productivity, Tapei, Taiwan, p 1.

6. Guillen-Gosalbez, G., Grossmann, I.E. (2009). Optimal design and planning of sustainable chemical supply chains under uncertainty. AIChE Journal 55(1), p 99.

7. Henriques, I. and P. Sadorsky (1999). The Relationship Between Environmental Commitment and Managerial Perceptions of Stakeholder Importance. Academy of Management Journal, 42 (1), p 87.

8. Horvath, A. (1999). Supply chain environmental assessment of the telecommunications sectors. Intemational Symposium on Electronics & the Environment, San Francisco, Califomia, p 146.

9. Javier Gonzalez – Benito, Oscar Gonzalez – Benito,(2008). Operations management practices linked to the adoption of ISO 14001: An empirical analysis of Spanish manufacturers. Int. J. Production Economics, 113, p 60.

10. K.C.Shang, C.S.Lu, S.Li (2010). 'A taxonomy of green supply chain management capability among electronic related manufacturing firms in Taiwan'. Journal of environmental management, 91(5), pp.1218.

11. Kaushik Sridhar (2011). 'The nature of environmental problems and environmental degradation'. International Journal of Business Excellence, 4(2), pp.202.

12. Ninlawan C., Seksan P., Tossapol K., and Pilada W. (2010). The Implementation of Green Supply Chain Management Practices in Electronics Industry. Proceedings of the International Conference of Engineers and Computer Scientists,Hong Kong, vol.III, March 17-19.

13. Prahinski, C., Kocabasoglu, C. (2006). Empirical research opportunities in reverse supply chain. Omega 34 (6), p 519.

14. Q. Lu, W. Li, B. Sundarakani, S. Cai, R. De Souza1, M. Go (2008). Green Supply Chain: How Does It Affect Current Supply Chain Practice? Proceedings of the 2008 IEEE IEEM, p 1128.

15. Qinghua Zhu, Joseph Sarkis, James J. Cordeiro, Kee-Hung Lai (2008). Firm-level correlates of green supply chain management practices in the Chinese context, Special Issue on Logistics: New perspectives and challenges, 36 (4), p 577.

16. Qinghua Zhu, Joseph Sarkis, Kee-hung Lai (2008). Confirmation of a measurement model for green supply chain management practices implementation. Int. J. Production Economics 111, p 261.

17. Ramudhin A., Chaabane, A. (2010). Carbon market sensitive sustainable supply chain network design. International Journal of Management Science and Engineering Management, 5 (1), p 30.

18. Sarkis, J. (2003). A strategic decision making framework for green supply chain management. Journal of Cleaner Production 11 (4), p 397.

19. Sheu, J. B.; Chou, Y. H.; Hu, C. C. (2005). An integrated logistics operational model for green-supply chain management. Trans. Res., 41 (4), p 287.

20. Snir, E.M.(2001). Liability as a catalyst for product stewardship. Production and Operations Management 10 (2) p 190.

IOP Conf. Series: Materials Science and Engineering 1104 (2021) 012028 doi:10.1088/1757-899X/1104/1/012028

21. V.R.Pramod and D.K.Rawat (2010). 'System modeling of telecom service sector supply chain: a SAP-LAP analysis'. International Journal of Business Excellence, 3 (1), pp.38.

22. Yasutaka Kainuma, Nobuhiko Tawara (2006). A multiple attribute utility theory approach to lean and green supply chain management. Int. J. Production Economics 101, p 99.

23. Zhu, Q., Sarkis, J. (2004).Relationships between operational practices and performance among early adopters of green supply chain management practices in Chinese manufacturing enterprises. Journal of Operations Management 22 (3), p 265.

24. Zhu, O.; Sarkis, J.; Geng, Y. (2005). 'Green supply chain management in china: pressures, practices and performance. International Journal of Operations and Production Management, 25 (5), pp.449.

25. Zsidisin, G.A., Siferd, S.P. (2001). Environmental purchasing: A framework for theory development. European Journal of Purchasing and Supply Management 7 (1), p 61.

Study of Effect of Mach Number on Coefficient of drag of Sounding Rocket by Computational Method

Ashutosh Kumar[1*], Rakesh Kumar Shukla[2], Kushagra S. Gusain[3], Aakash Sadar[4]
Department of Mechanical Engineering, Krishna Engineering College, Ghaziabad India

Abstract— This research paper conducts a study of aerodynamic drag coefficient of a sounding rocket named Rohini300 MK-II. There are various factor which plays important role in rocket stability and aerodynamic drag coefficient is one of them. Zero-lift drag coefficient, C_D (i.e. drag coefficient at attack angle of zero degree) is evaluated and its behavior is observed with the different mach numbers. This paper defines the development of a CFD model of the RH 300 MK-II in the modeling software, then importing the 3D model into CFD software and then the creation of CFD domain, meshing of the domain of CFD, application of various boundary conditions and then obtaining plots and results for coefficient of drag. And Finally Result is cocluded to get its Zero Drag coefficient.

Keywords – ANSYS Fluent 19, Drag Coefficient, Mach Number, RH 300 MK-II Sounding Rocket

1. INTRODUCTION

A. RH 300 MK-II Sounding Rocket

Indian Space Research Organisation (ISRO) developed Rohini – Sounding Rocket Series for meteorological and atmospheric study. RH 300 MK-II is a sounding rocket of Rohini sounding rocket series which is used for aerodynamic study. It is a single stage, fin stabilized sounding rocket which weighs nearly 500kg with a diameter of 305mm and overall length of 5.6m. A maximum thrust of 65kN is generated for about 20 seconds by the solid propellant boosters motor which powers the sounding rocket.The spin rockets function for 300 milliseconds and are then ejected out. The commands for all the activities after launch are provided by the vehicle's centralized sequencer-encoder.

Programmable S-band transmitter transmits the data collected by the various systems of the sounding rocket to the tracking station.

B. Computational Fluid Dynamics (CFD)

It is a branch of fluid mechanics that applies numerical methods and algorithms to solve problems that involve fluid flows is called CFD. Computer Software (like Ansys Fluent) are used to perform the calculations and analysisof liquid or gases. Initial validations of these softwares are performed using a wind tunnel with the final validation coming in full-scale testing, e.g. flight tests.

2. METHODOLOGY

A. Modeling of the RH-300 MK-II Sounding Rocket

The modeling of the RH 300 MK-II, is used in this paper. It is applied by using Solid Works model 2013. The External flow is solved and analyzed by using this solid work model. Some sketch commands are used in particular sounding rocket modeling are line, circle, ellipse, spline etc. The commands used are extrude, revolve, extrude cut,

IOP Conf. Series: Materials Science and Engineering 1104 (2021) 012029 doi:10.1088/1757-899X/1104/1/012029

mirror, etc. While solving and modeling of this sounding rocket, proper care should be taken because of any improper way of modeling leads overlapping of the geometry.

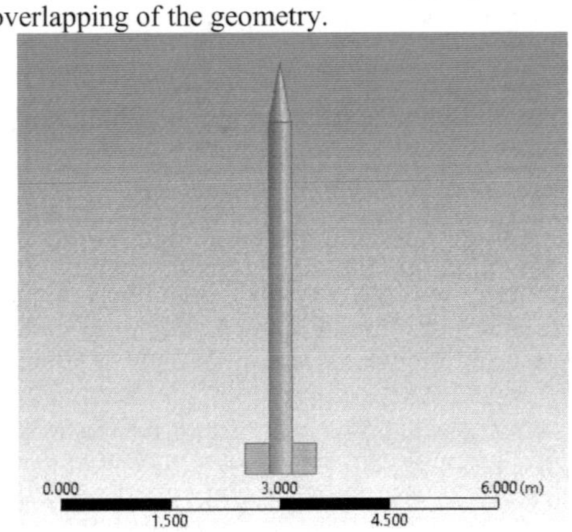

This enclosure around the sounding rocket is made in ANSYS Workbench 19.

B. Meshing

The meshing of the sounding rocket and atmosphere created around it, is done using ANSYS Fluent 19. The continuum for this as ANSYS Workbench is imported, then the continuum is sub divided into different categories like inlet, outlet, wall and rocket body. So the required meshing conditions are used and the continuum is meshed. We used Cut Cell meshing technique with inflation near the sounding rocket body.

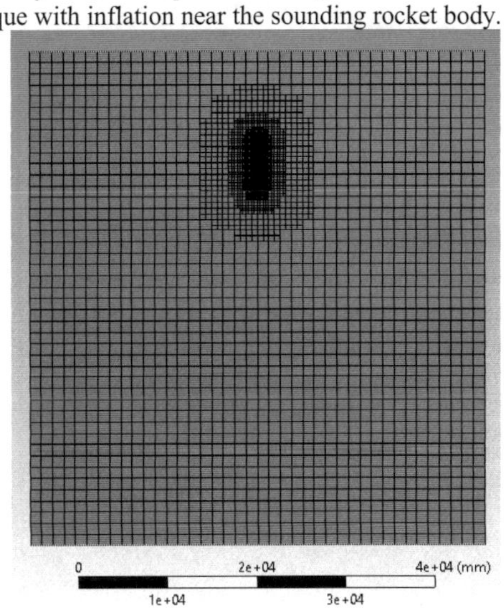

The continuum is given a mesh size of 15mm and the rocket volume is given a fine mesh size of 5mm since it is a complex geometry. The difference between the element sizes can be seen in fig 5.

After the meshing of the continuum it is then exported to ANSYS Fluent Setup 19, where in the flow analysis over the sounding rocket is done.

C. Setup of the Continuum

The simulation on the domain is done in ANSYS Fluent Setup 19. Initially the meshing of the domain is checked and once the software approves it, the models, materials and boundary conditions are set. The meshing of the sounding rocket and atmosphere created around it, is done using ANSYS Fluent 19. The continuum for this as ANSYS Workbench is imported, then the continuum is sub divided into different categories like inlet, outlet, wall and rocket body.

The set up conditions are 1.Model 2. Working Material 3. Boundary Conditions 4. Solutions.

Generally SST model is applied in simulation and SST is shear-stress transport which is used for the specific purpose.
The working Material is Air which is used for this purpose of analysis.
The boundary conditions are applied for calculating Mach number or velocity at inlet and outlet of the system by using the CFD analysis method. We analyzed RH 300MK-II in the speed range of 0 to 900 m/s. Therefore, the boundary condition at inlet for the continuum is given as a range of 0 m/s to 900 m/s.

Solution are selected for the analysis purpose and relaxation factor for momentum and pressure to get the desired kinetic energy turbulent dissipation rate and turbulent velocity which is at are set as 0.5 and for the turbulent kinetic energy, turbulent dissipation rate and turbulent viscosity is set to 0.75, 0.75 and 1 respectively.

3. RESULTS AND DISCUSSIONS

The expected graphical results are set to drag coefficient. The graphs shown below shows the value of coefficient of drag on the rocket at different Mach Numbers. Solution are selected for the analysis purpose and relaxation factor for momentum and pressure to get the desired kinetic energy turbulent dissipation rate and turbulent velocity.

A. Graphs
1. Mach 0.5

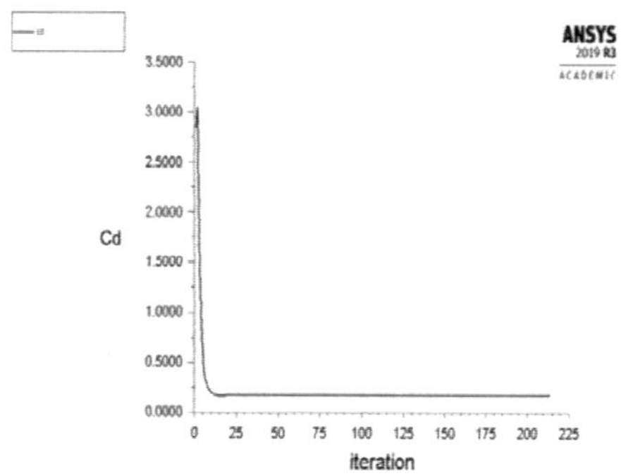

It can be seen that the value of Coefficient Drag for Mach No. 0.5 has converged on the value 0.4 at 220 iterations
2. Mach 1

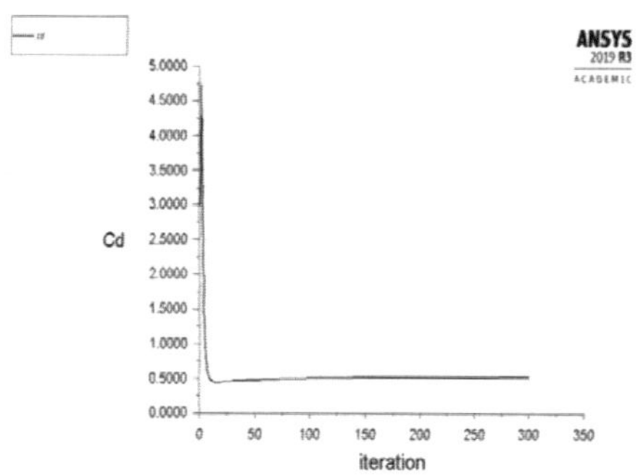

IOP Conf. Series: Materials Science and Engineering 1104 (2021) 012029 doi:10.1088/1757-899X/1104/1/012029

It can be seen that the value of Coefficient Drag for Mach No. 1 has converged on the value 0.5 at 300 iterations.

3. Mach 1.5

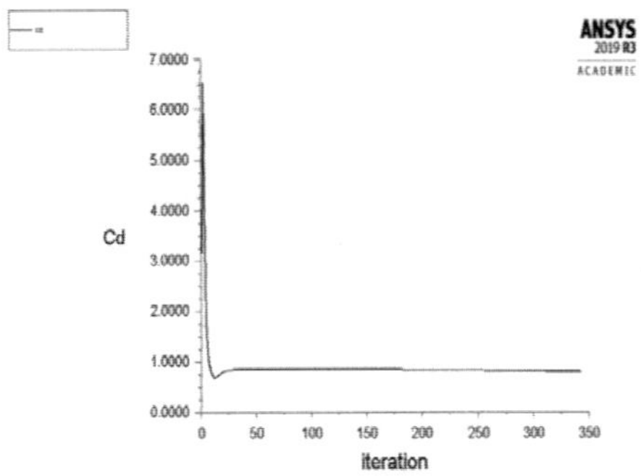

It can be seen that the value of Coefficient Drag for Mach No. 1.5 has converged on the value 0.75 at 350 iterations.

4. Mach 2

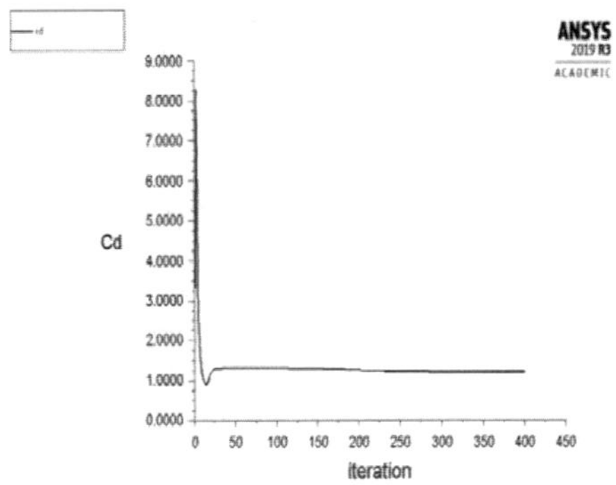

It can be seen that the value of Coefficient Drag for Mach No. 2 has converged on the value 1.2 at 400 iterations.

5. Mach 2.62

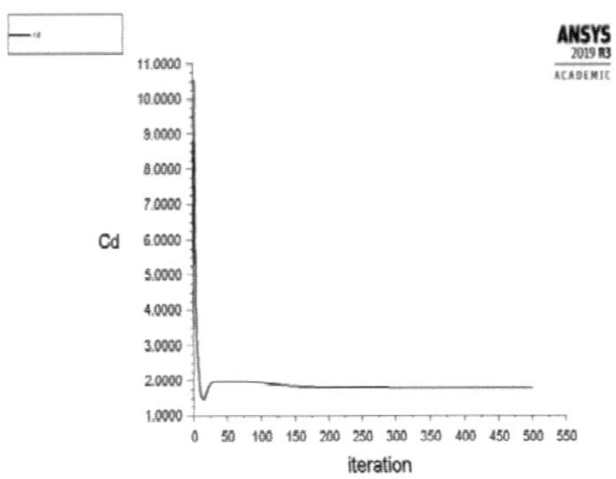

It can be seen that the value of Coefficient Drag for Mach No. 2.62 has converged on the value 1.75 at 500 iterations.

4. CONCLUSION

After the simulation, the obtained results (i.e. coefficient of drags) are plotted against mach numbers. The plotted graph gives the clearer picture of the aerodynamic drag coefficient.

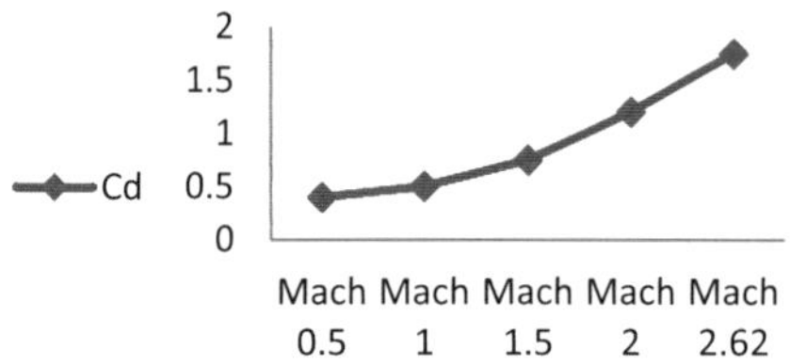

As we can see in the figure 8.1, the zero lift drag coefficient at Mach 0.5, Mach 1, Mach 1.5, Mach 2 and Mach 2.62 is 0.4, 0.5, 0.75, 1.2 and 1.75 respectively. With the increase of velocity the drag coefficient increases exponentially.

The zero lift drag coefficient obtained from the CFD analysis lie within satisfactory range. So, it can be concluded that the sounding rocket RH 300 Mk-II is aerodynamically (in terms of drag) sound. It is important for getting desired coefficient of performance to set the value. Mach Number is deciding parameter for this set up. The velocity is also important for its performance measurement and decision analysis. Hence, The zero drag is used to get for CFD analysis lie within satisfactory range. So, it can be concluded that the sounding rocket RH 300 Mk-II is aerodynamically (in terms of drag) sound.

5. FUTURE SCOPE

Scope of future work is very wide as there are numbers of project work possible. One of thepossible future project works can be the comparison of CFD simulations with the current practical analysis. It can give us a clear picture how the CFD analysis is close to practical analysis.

Another potential future project work could be related different nose designs. As we know that coefficient of drag is majorly dependent on the design of the sounding rocket. So, with change in design of nose of rocket, we can simulate the results in Ansys Fluent and compare them with each other at different mach numbers. This could give us a more feasible conclusion on the design aspect in terms of aerodynamics.

REFERENCES

(1).AlgimantasFedaravičius, SigitasKilikevičius, ArvydasSurvila, Investigation on theaerodynamic characteristics of a rocket-target for the system "Stinger, JVEInternational Ltd. Journal of Vibroengineering, Dec 2015, Vol. 17, Issue 8. ISSN1392-8716.

(2). Aerodynamics Modeling of Sounding Rockets by KristofferHammargren, 2018.

(3). Olalekan, A.H. Attah, O.C. Bisola, O. Jimoh, H. Atinuke, Ojo. (2014). FlightTrajectory Simulation of a Single Stage Sounding Rocket. International Journal of
Scientific & Engineering Research, Volume 5, Issue 11, November 2014. ISSN 2229-5518.

(4). Fedaravičius A., Kilikevičius S., SurvilaA. Optimization of the rocket's nose andnozzle design parameters in respect to its aerodynamic characteristics. Journal ofVibroengineering, Vol. 14, Issue 3, 2012, p. 1390-1398.

(5). Menter F. R., Kuntz M., Langtry R. Ten years of industrial experience with the SSTturbulence model. Turbulence, Heat and Mass Transfer 4, Begell House, 2003, p.625-632.

Applications of solar Photovoltaic's to power stand-alone area and street lighting

Om Prakash*[*1], Satyajeet Kumar[1], Dr. Pramod Kumar[2]

[1]Krishna Engineering College, Ghaziabad, U.P, India
[2]Dr. B. R. Ambedkar national institute of technology, Jalandhar, Punjab, India
[*]*Email: omprakash007mech@gmail.com*

Abstract. This paper is an introduction to the topics of photovoltaic solar energy conversion and with the help of solar illuminations; street lightings are employed in the surroundings of NIT JALANDHAR. The paper presents the Energy problem in the world: Growing consumption of energy. Solar radiation is described as a main source of renewable energy as well as possible applications of this energy by means of thermal conversion and photovoltaic conversion. The functionality of stand alone solar system is completely different than hybrid kind of solar system. Photovoltaic effect in the p-n junction and physical processes in the photovoltaic structures of solar cells are presented. The performance of each solar panel has been co-related through current and voltage (I-V) curve. The panel size and battery required to operate a better functionality is the key idea of solar photovoltaic system design and its smooth operation. The required number of batteries should be inline such a way that it does fulfil the requirement of power output.

Keywords: *Renewable* Energy, Photovoltaic, *Solar radiation, p-n junction, solar cells.*

1. Introduction

Photovoltaic conversion of solar energy involves a wide range of knowledge, energy problems properties of solar radiations reaching the earth surface's, photovoltaic effect in the p-n junction properties of semiconductors which are important in photovoltaic conversion, technology in solar cells, construction of solar photovoltaic panel modules and arrays, control system of photovoltaic power station as well as ecological and economic problems of photovoltaic installations.

2. Basic theory of solar cell

Solar cells are of two types as per the concept of physics and these are N-Type and P-Type. When light falls on the semiconductor device it form generally two type of charges. i.e +ve and –ve chaege which is called Holes.

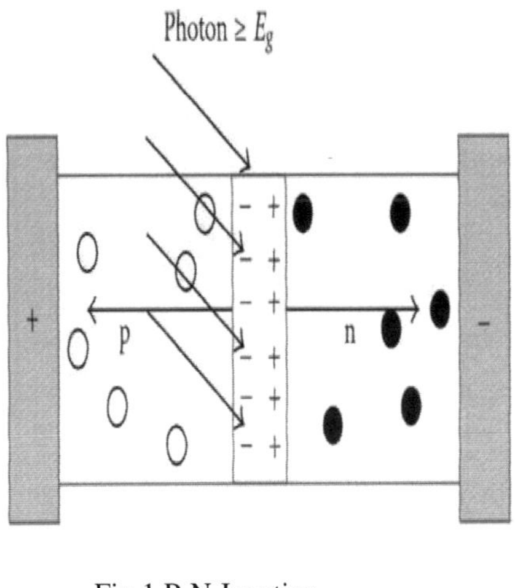

Fig.1 P-N Junction

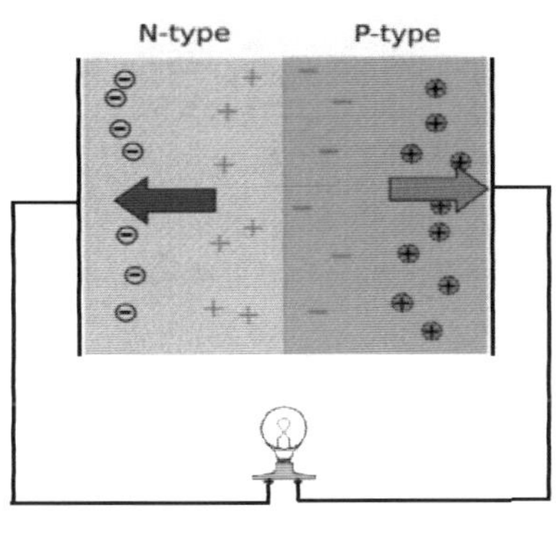

Fig.2 Current flow direction

2.1 Solar Cell Modelling

Modeling of solar cell can be expressed by many ways in software packages like MATLAB & P- SPICE etc. and there are many methods to represent a model as like Mathematical block modeling, Embedded MATLAB Programming and Physical block modeling. Here physical block of solar cells are used for the modeling of PV module.

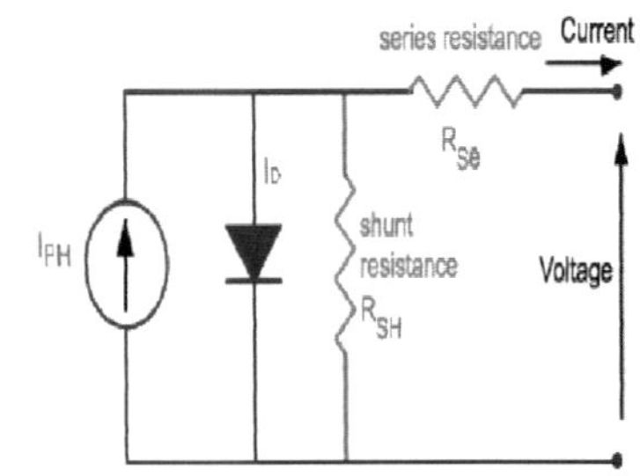

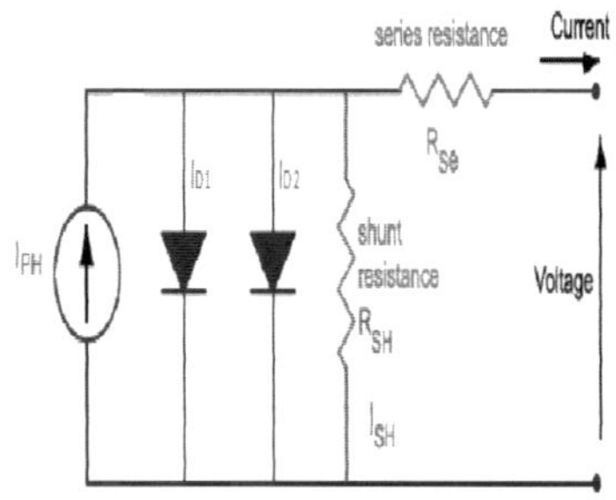

2.1 Characteristic Curve Module of solar panel working

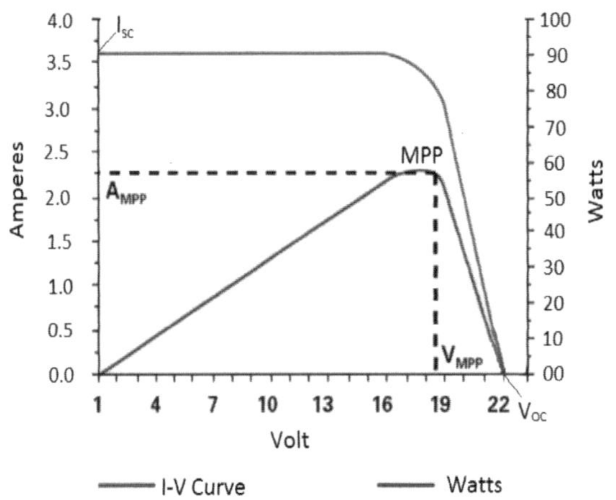

Fig.5 Standard I-V Characteristic of a Solar Panel

2.2 Photovoltaic conversion efficiency

The main outcome of any photovoltaic cell is its efficiency and which is defined by a formulation given below which contains the ratio of peak power to the value of output of solar power.

$$\eta = \frac{Vmp.Imp}{I\left(\frac{KW}{m^2}\right).A(m^2)}$$

From the above given expression Vmp denotes the voltage when power is at its peak. I denotes solar intensity when considering unit meter square as area. A is denoting the area of exposed surface when solar radiation falls upon. The condition on which the pv system will give maximum efficiency depends upon the environmental conditions of power point tracking. If we anyhow manage to track maximum

IOP Conf. Series: Materials Science and Engineering 1104 (2021) 012030 doi:10.1088/1757-899X/1104/1/012030

radiation on the surface exposed we will get more and more radiation to convert into solar power and then store in the batteries for future usages.

2.3 The solar irradiation in Jalandhar, Punjab, India

Geographical information of Experimental set up. Latitude -31.3260152, Longitude- 75.5761829 Figure below shows the solar irradiance data of Jalandhar and optimum tilt of solar panels on monthly basis. Source: http://www.synergyenviron.com.

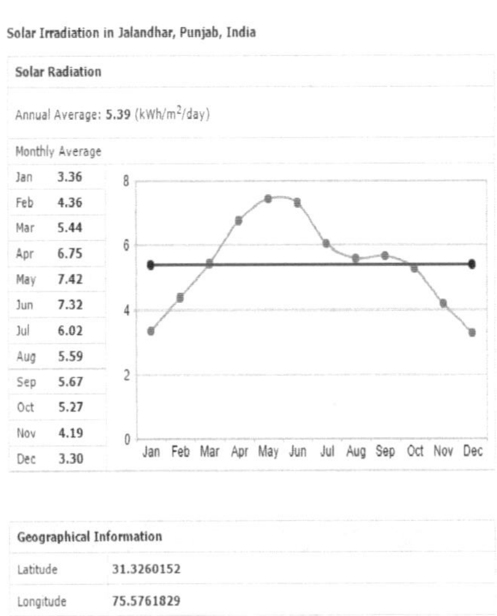

Fig. 6 The solar irradiance data for Jalandhar

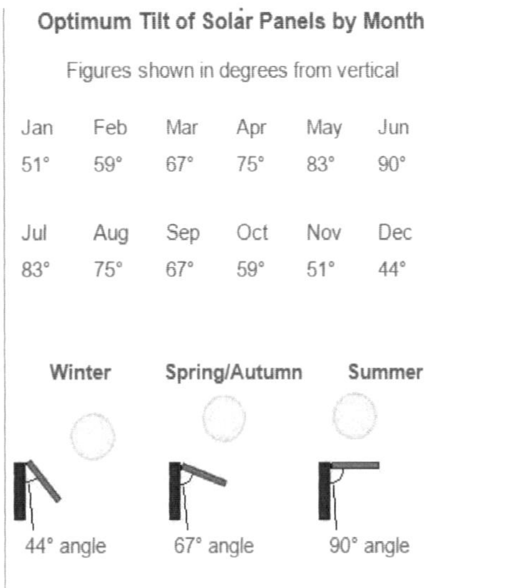

Fig.7 Optimum angle of solar panels tilt at Jalandhar

3 Experimental specification & installation of Solar panel

Both figure and table shows the solar panel & specification of the panel installed at NIT Jalandhar.

NIT Jal

(Pmax)	120W
(Vmp)	17.70V
(Impp)	6.78A
(Voc)	21.60V
(Isc)	7.76A
Maximum voltage of the system	1000 V
Tolerance	±5 %

:l

4. Results and Discussions

In the present work the analysis of solar panel based on CFL lights have been done. Various parameters such as Battery sizing, number of modules and wire sizing have been calculated and results of which are presented in this paper.

(i) Load estimation under study

Load	Watts	H/day	Number	Watt-Hr.
CFL	11	10	20	2200

(ii) Battery Sizing

Days/Week – 7

$$\text{DC Wh/day} = \frac{\text{Watts} \times \text{daily use(hours)} \times \text{days per week}}{7}$$

$$= \frac{22 \times 10 \times 7}{7}$$

=220Wh/day

(iii) Daily energy supplied to inverter

Good inverters provide efficiency as high as 97%.Let us consider the inverter efficiency as 93%, then the energy supplied by the battery to inverter input should be

$$= \frac{2200}{0.93} = 2365.59 \text{ Wh}$$

(iv) Deciding the system voltage

Here the conversion of DC Voltage to AC voltage will play an important role because when it come across inverter the voltage will become AC. But before that the voltage will remain DC. The typical terminal voltage of a sample PV System is taken as 12V. And for the design and its fabrication we also employed a 12 voltage battery.

(v) Determination of battery capacity for a given load

We generally consider the Batteries DoD in the range between 60% to 80%. Let us suppose a scenario where the Batteries are of 12V and power output is 75 AmpH capacities which having the DoD value as 70%. It reflects that out of 100 AmpH batteries capacities we have only 75*0.7=52.5 AmpH is of usable.

The capacity of required charge $= \frac{2365.59}{12}$ Ah =197.13A. Therefore the number of battery required

$$= \frac{197.13}{75 \times 0.7} \text{Ah}$$

=3.75, i.e., It means that there will be need of 4 batteries of having 75 Ah capacities.

We need to round off this number; thus we need 4 batteries to supply the required charge capacity. These 4 batteries should be connected in parallel. But 75 Ah batteries are of 12V only, so the battery bank needs to supply the charge at 12 V (system voltage). Therefore, in order to get 12V, four 12V batteries should be connected in parallel to get 12V terminal voltage and the capacity will get added.

(vi) Daily energy generated by PV panels:

Let us consider that battery efficiency is 85%. Therefore, the energy supplied at the input of battery terminal should be:

$$= \frac{2365.59}{0.85} \text{Wh}$$

=2784.0Wh

Suppose we have a circuit of the efficiency 90%. In this scenario the circuit controller input should be 3093.3 Wh. So what could be the amount of energy generation out of it is the value which has been found as 3093.38 Wh from the calculations. And this energy generation is on daily basis.

(vii) Solar radiation, capacity and number of panels:

Total Ah generated by the PV panels should be $= \frac{3093.38}{12}$Ah =257.78 Ah

The values which we take as reference for the solar input radiation is 1000 Watt per metre square. Generally as per the geographical conditions india witness the maximum solar radiation for 5 to 7 hours at peak time. And the value of corresponding output is 5000 Wh/mquare.

(viii)The amount of total current in Amp which should a solar panel produce is given by the expression :

$$=\frac{257.78\text{ Ah}}{6\text{h}} = 42.96\text{ A}$$

Therefore number of modules will be decided on the basis of panel rating of $W_p$and I_p. This is 120Wp and 6.78A correspondingly. Number of modules $=\frac{42.96}{6.78}=6.33\approx 7$. So we need 7 module of 120W_p to provide us 42.96 A current.

(ix) Wire sizing

PV source circuit

I_{SC} per module =7.76 A, Number of modules=7, I_{SC} =7.76 A

Considering Environmental Effect Isc = 7.76×1.25 =9.7 A

By 80% rule: Isc=9.7×1.25 = 12.125 A

Wire size = 10 mm^2.

(x) Calculation

There are two bulbs running through each solar panel each of 11 watts. In most of the days it works for 10 hours per day so energy produced by the 10 panels.

=22× 10× 10 Wh =2200 Wh

≅2.2KWh.

So 10 solar panels running for 10 hrs. Each day produces 2.2 KWh energy.

The unit cost of Electricity nowadays as per the Govt. is 6 Rupees/unit. So savings of electricity Bill each day is

=2.2 × 6 = 13 Rupees/day = 4818 Rupees/year Savings.

So if solar panels are replaced in the place of Electricity supply it can save enormous amount of money.

5. The Performance characteristics of solar panels on I-V curve

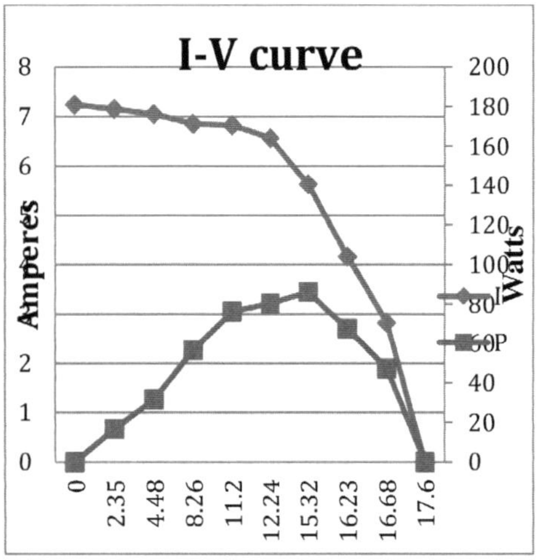

Fig. 9 I-V Curve of Solar Panel At Civil Deptt. No.2

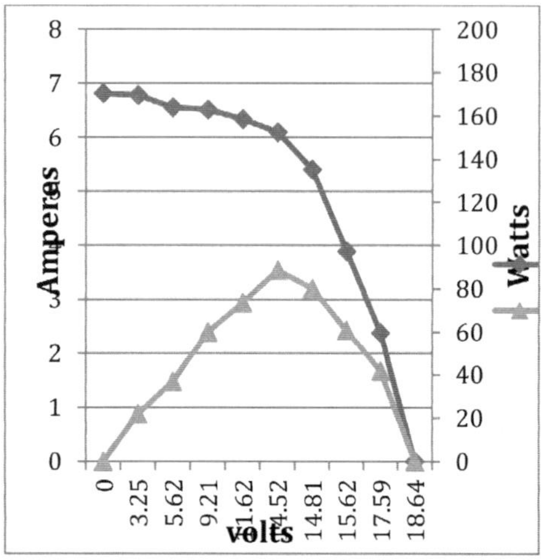

Fig.10 I-V Curve of Solar Panel at Gate

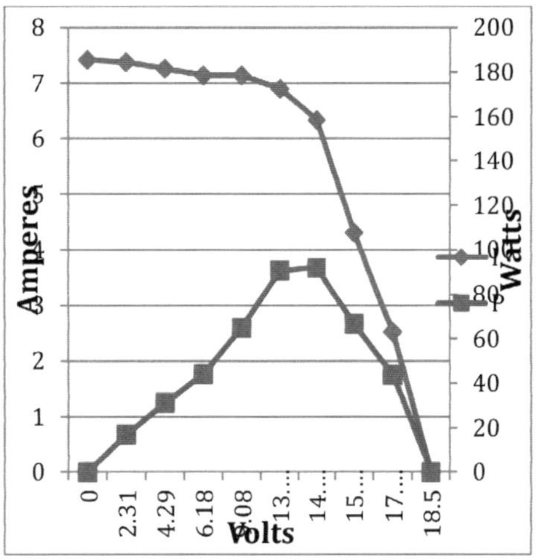

Fig.11 I-V Curve of Solar Panel at Hostel No. 1 main gate 1

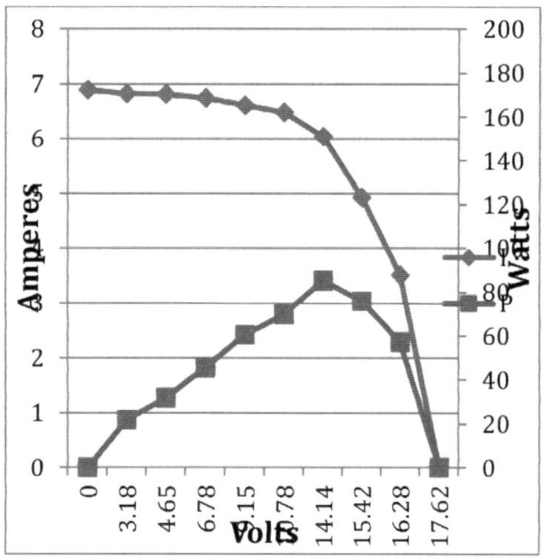

Fig.12 I-V Curve of Solar Panel at

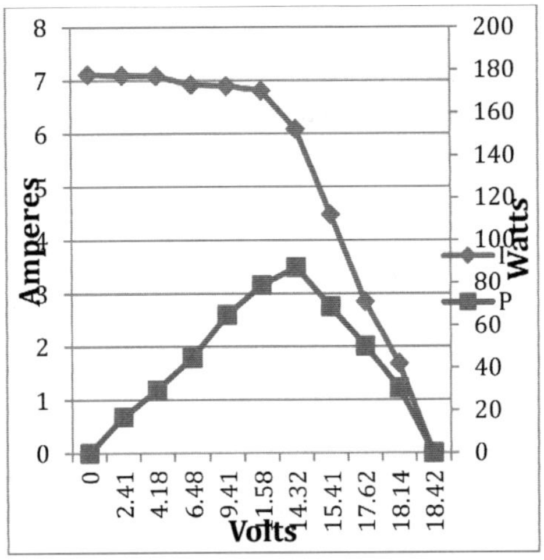

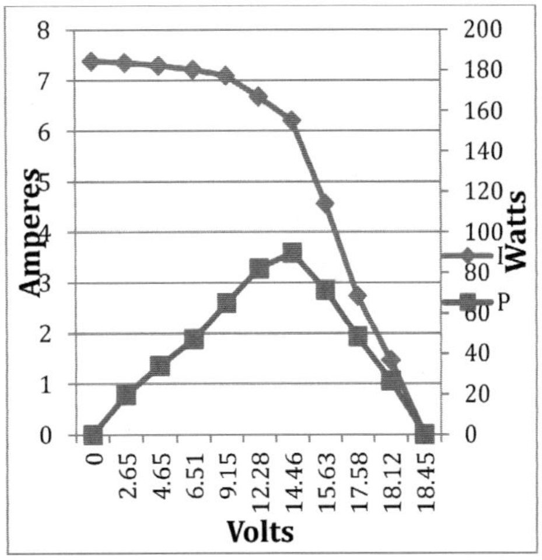

Fig.13 I-V Curve of Solar Panel at director's house Fig.14 I-V Curve of Solar Panel at guest house

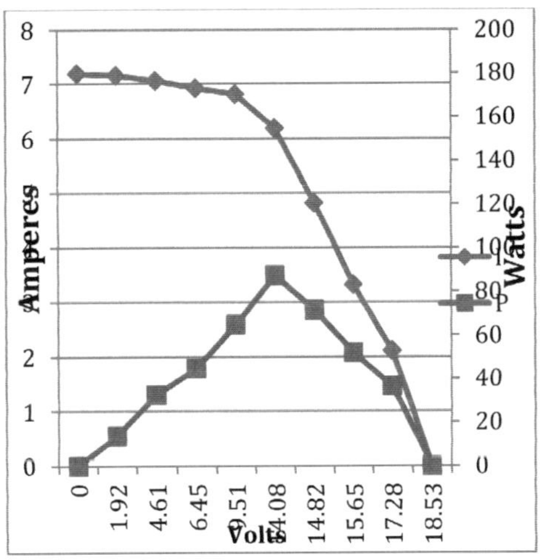

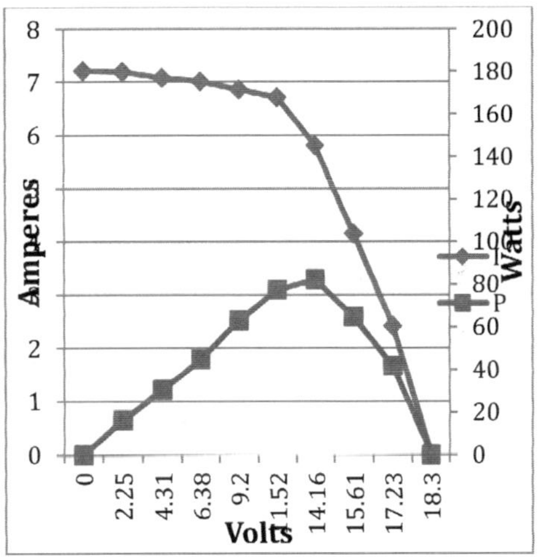

Fig.15 I-V Curve of the Solar Panel at library gate 2 Fig.16 I-V Curve of the Solar Panel at main

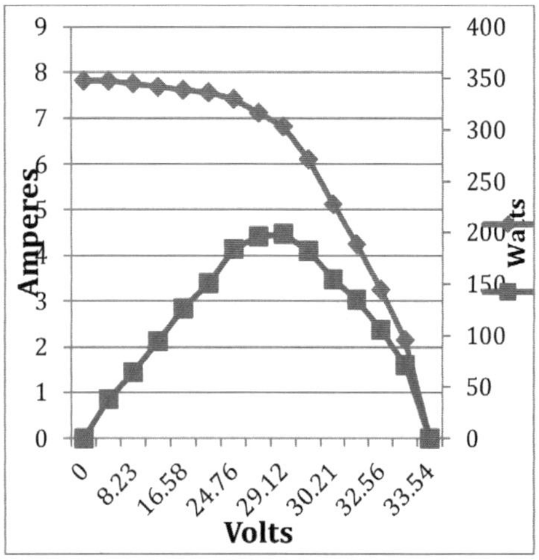

Fig.17 I-V Curve of the Solar Panel at mechanical Department

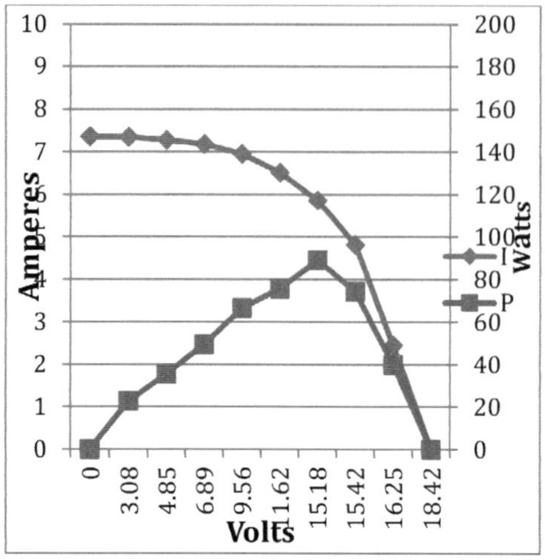

Fig.18 I-V Curve of the Solar Panel at mega girl's hostel

6. Conclusion

The analysis done for NIT Jalandhar solar panels installed it is concluded that if solar panels of Off-Grid type are mounted with Inverted and is connected with Grid and power is supply, it can save enormous amount of electricity Bill. If all the solar panels in NIT, Jalandhar are in working condition it can save Rupees 4818 Every year. The solar panels in Mechanical department and at hostel No.1 power output is the best among all as they produce around 86 % and 77 % Efficiencies. And solar panel at Main Gate 2 has the least efficiencies of 69 %.

References

[1]. A.A.Ghoneim (2006) "Design optimization of photovoltaic powered water pumping systems" International Journal on Energy Conversion and Management Vol.47, pp.1449— 1463.

[2] A.Hamidat, B.Benyoucef(2009) "Systematic procedures for sizing photovoltaic pumping system, using water tank storage" International Journal on Energy Policy. Vol.37, pp. 1489-1501.

[3] Andrea Feltrin, Tomomi Meguro, Elisabeth Van Assche, Takashi Suezaki, Mitsuru Ichikawa, Takashi Kuchiyama, Daisuke Adachi, OsamuInaki, KuntaYoshikawa, Gensuke Koizumi, HisashiUzu, Hiroaki Ueda, Toshihiko Uto, Takahisa Fujimoto, Toru Irie, Hironori Hayakawa, Naoaki Nakanishi, Masashi Yoshimi, Kenji Yamamoto."Advanced light trapping designs for high efficiency thin film silicon solar cells". International Journal on Solar Energy Materials and Solar Cells. Vol. 119(2013), pp. 219-227.

[4] C. Becker, D.Amkreutz, T.Sontheimer, V.Preidel, D.Lockau, J.Haschke, L. Jogschies, C.Klimm, J.J.Merkel, P.Plocica, S.Steffens, B.Rech.(2013) " Polycrystalline silicon thin film solar cells: Status and perspectives". International Journal on Solar Energy Materials and Solar Cells.Vol. 119 (2013), pp.112-123.

[5] F.Kaule, W.Wang, S.Schoenfelder "Modelling and testing the mechanical strength of solar cells" International Journal on Solar Energy Materials & Solar Cells Vol.120 (2014), pp.441-447.

[6] GhaniaAzzouzia, WahibaTazibt (2012) "Improving silicon solar cell efficiency by using the impurity photovoltaic effect" International workshop on Modification and Analysis of Materials for Future Energy Sources.Vol. 41, pp.40-49.

[7] GhaniaAzzouzi, Mohamed Chegaar (2011). "Impurity photo voltaic effect in silicon solar cell doped with sulphur: A numerical simulation" International Journal on Physics.Vol. 41, pp.40-49.

[8] Hyun-junSoh, JeonghoonYoo, Daekeun Kim(2012) "Optimal design of the light absorbing layer in thin film silicon solar cells". International Journal on Solar Energy.Vol.86, pp.2095-2105.

[9] International Energy Agency, "Methodology Guidelines on Life Cycle Assessment of Photovoltaic Electricity", IEA PVPS Task 12, Subtask 20, LCA Report IEA-PVPS T12-01:2009 October 2009.
[10] Hall James and Hall Jeffrey, "Evaluating the Accuracy of Solar Radiation Data Sources", Solar Data Warehouse, February 2010.
[11] C.R. Osterwald, A. Anderberg, S. Rummel, and L. Ottoson, "Degradation Analysis of Weathered Crystalline-Silicon PV Modules", 29th IEEE PV Specialists Conference, New Orleans, Louisiana, May 20-24, 2002.

Comparative Study of Electrical Discharge Machining, Electrical Discharge Drilling and Electrical Discharge Diamond Drilling of Nimonic 75

Rajeev Kumar[1], Somnath Chattopadhyaya[2], Sanjeev Kumar[3], G.K. Singh[4], U.K. Vates[5]

[1,2]Department of Mechanical Engineering, IIT(ISM) Dhanbad, Jharkhand-826004, India
[3]Department of Mechanical Engineering, Krishna Engineering College, Ghaziabad, UP-201007, India
[4]Federal Technical & Vocational and Educational Training Institute, Addis Abbaba, Ethopia
[5]Department of Mechanical Engineering, Amity University, Noida, UP-2001313, India

Email:[1]errajeev79@gmail.com,[2]somuismu@gmail.com,[3]sanjeevkg9@gmail.com, [4]gksinghits@yahoo.com, [5]ukvates@amity.edu

Abstract. Due to wide range of applicability of hard and difficult-to-machine materials like super-alloys (Ti-alloys, Inconel, and Haste-alloy etc), advanced ceramics, composite materials, and high technical characteristics of above-mentioned materials, suitable advanced machining processes are conducted. EDM is one of the most popular advanced machining processes. But the EDM process has low metal removal rate (MRR) so EDD is another machining process as it has relatively higher MRR as compared to EDM. Here electric discharge diamond drilling (EDDD), a hybrid-EDM process comes into the picture in order to improve the machining performance characteristics. The present research aimed to compare the process performance in electrical discharge cutting, drilling and diamond abrasive drilling. The effects of process parameters such as peak current (I), pulse-on time (Ton), and pulse-off time (Toff) were studied on metal removal rate (MRR) and tool wear rate (TWR) in electrical discharge diamond drilling of Nimonic 75.

1. Introduction

Advanced engineering materials such as super-alloys, titanium alloys, sialon, advance composites, high strength temperature resistant materials (HSTR) have emerged recently, which added pace in aircraft industries and related technology. These materials possess properties like high hardness, fatigue strength, toughness, corrosion resistance, creep resistance and poor thermal diffusivity etc. However, these properties impose challenges to traditional machining processes in terms of production of high surface roughness, frequent cutting tool failure, inaccurate geometrical dimension and production of complex features. To meet these challenges, unconventional machining processes (UMPs) have come into existence. Theses UMPs are electrical discharge machining process (EDM), electric discharge drilling (EDD) electrochemical machining (ECM), jet machining process (JMPs), ultrasonic machining processes (USM) and so on. Recently, a new drift has been introduced to merge the features of different conventional and unconventional machining processes. Such machining processes are called as hybrid machining processes (HMPs). HMPs are developed to take advantage of each of the constituent machining processes and diminish the disadvantages of each constituent process. It has been observed that, hybrid machining process improves the machining performances, increases the capabilities of the constituent processes, and widen the area of application of the constituent processes.

In the EDM process, the constant tool is utilized for making a blind hole and EDM drilling, a rotating tool electrode is used for making a blind hole.

The metal bonded diamond abrasive tool is used for making the blind hole is known as electric discharge diamond drilling (EDDD). Some author makes an approach in these directions such as Shu et al. [1] drilled holes in mild steel HPM50 and WC P20 materials.

It was found that the Ra value of the surface produced by EDDD is improved. They used a tool electrode made of fine copper-bonded SiC abrasive in EDDD process. Effect of pulse current, duty factor, grain size and wheel speed were estimated on surface roughness and wear rate based on an experimental investigation in EDDD. Surface roughness and wear rate of HSS has been predicted and optimized using an artificial neural network [2]. EDDG is also reported on Al-SiC metal matrix composite using an artificial neural network. The effect of influencing independent variables like duty factors, Ton, abrasive wheel speed (AWS) and current on SS and metal removal rate (MRR)have been analyzed during EDDG. Responses were optimized dup to the adequate level such as a higher value of AWS and lower value of Ton, duty factor (DF) and pulse current, [3]. The effects of Ton, Toff and dielectric flux density were also examined over the temperature distribution during the EDM process [4]. Surface integrity of D2 steels was investigated in EDDSG process by using scanning electron microscopy (SEM) test. An EDDSG experiment reviles that by increasing peak current, MRR increases. If the rotational speed is increasing, it leads to increase the mechanical abrasion [5]. Optimization of SR and MRR is already reported on HSS under EDDSG having few important inputs [6]. Experimental investigations had been done of EDDG on Al–10 wt.% IC Composite [7]. The workpiece has been cut with the help of EDDCG using metal bonded diamond assisted grinding wheel. Due to the influence of dielectric, work surface and the diamond assisted wheel is separated. Abrasion takes place due to sparking and influence of diamond grains having inter-electrode gap-width (gw) which is less than protrusion height (P_h) [8]. It is evident that EDDCG (discharge diamond cutoff grinding) process is applied for material removal on titanium alloy (Ti-Al-Mo-V). MRR and SR (surface roughness) are optimized for the same using response surface experimental design [9]. EDDCG process was also applied on metal matrix nano-composite to estimate the effects of rotational speed, gap current, Ton and DF on surface roughness (Ra) and material removal rate (MRR)[10]. Theoretical studies of WEDM critical parameters and their effect on Monel alloy have been evaluated for the SR and MRR. In the present study, SR and MRR were selected as response parameters for the independent variables such as air inlet pressure, pulse off time, water flow rate, pulse-on time and wire feed, were taken as input parameters [11]. To estimate the effects of individual independent parameters on SR and MRR, one factor at a time approach is being adapted during EDDFSG (electrical discharge diamond face surface grinding) hybrid process on metal matrix composite (MMC).Recast layer formation on the machined surface is also evaluated by using Scanning Electron Microscopy (SEM). This process was optimized for higher MRR with good surface finish and hard recast layer [12]. A magnetic aided electric discharge machining of Aluminium-based MMC was performed in order to improve its process performance. Outcome reviles that an improvement of approx. 12% in MRR and reduction in recast layer thickness at high spark and magnetic energy [13].

The electric discharge diamond grinding (EDDG) has been used in different modes of operation like surface grinding and cut-off grinding for machining of difficult-to-cut materials and found an improved response of material removal rate [14-15]. A. Okade et al. [16] Proposed a method for curved hole EDM drilling performance improvement using a suspended ball electrode by workpiece vibration. The results have shown that this method really improved curve hole EDM drilling by workpiece vibration. Sagar U. Sapkal et al [17] studied optimization of micro EDM drilling of Ti6Al4V by using copper tungsten tool electrode and applied response surface methodology (RSM) based on the central composite design (CCD). Pulse-on time (Ton), discharge voltage (V_d), capacitance (C) and electrode rotation speed (RPM) were considered as input process parameters whereas material removal rate (MRR), side gap width, and taper ratio as response parameters. The results show that for MRR, discharge voltage, capacitance and electrode rotation speed influenced the most. Similarly, side gap width was found influenced by Ton, C, RPM. In contrast, the taper ratio found influenced by Ton, C. Grzegorz et al. [18] studied the influence of the length of the tool electrode on the performance of micro-EDM. They derived characteristics of micro-drilling in stainless steel samples depending on the length of the tool electrode. MRR, tool wear ratio, machining accuracy were the main concerned taken by authors. M Klinev et al. [19] performed EDM drilling on non-conductive materials in deionized water. During the experiment, they applied an assisting conductive layer to the surface of the workpiece to start the discharge process. It was found that ZrO2 was able to machine easily in deionized water. By this method, a hole up to 1.5 mm depth in ceramic material can be drilled in the presence of deionized water. Raju et al. [20] studied the multi-performance operation of micro-EDM drilling of Inconel 600 alloy using grey relational analysis (GRA). The EDM feed rate, voltage (V), capacitance (C), pulse-on time (Ton), pulse-off time (Toff) were considered as input parameters whereas MRR, taper angle, overcut, diametric variance at entry and exit of micro-hole were considered as performance measures. The results show that capacitance (C) influenced the most to taper angle followed by voltage and feed rate.

From the literature review, we found that few researchers are striving to improve the machinability of such advanced materials using unconventional and hybrid machining technique. The present study aimed to create a blind hole in Nimonic alloy using different machining processes such as EDDD, EDD and EDM. It is evident that EDDD is one of the hybrids techniques which combination of diamond drilling (DD) and EDM, the EDDD process as shown in Fig.1. In EDDD process, materials remove takes place from the surface by simultaneous influence of diamond grains, continuous-discrete electrical sparks resulting mechanical abrasion and electrical sparks erosion, respectively. A metal bonded diamond abrasive rotating tool electrode has been implemented to get adequate metal removal through EDDD process in Nimonic alloy (alloy of nickel-chromium having good corrosion and heat resistance). EDD process contents rotating cylindrical drills inbuilt with EDM set up to produce a hole in the workpiece material. Through the EDM process tool with zero speed is utilized for making the hole. In this research, authors performed the comparative study among these processes EDDD, EDD and EDM based on one factor at a time (OFAT) approach. The most important critical drilling parameters like peak current, pulse-on time and pulse-off time are considered as input factors. The effect of process parameters are analyzed on material removal rate (MRR) and tool wear rate (TWR) to compare the process performances through various ways. The results show that the value of MRR is higher in EDDD as compared to EDD and EDM whereas TWR is lowest in case of EDDD, compared to EDD and EDM respectively.

2. Materials and methods

From the literature review, we found that few researchers are striving to improve the machinability of such advanced materials using unconventional and hybrid machining technique. The present study aimed to create a blind hole in Nimonic alloy using different machining processes such as EDDD, EDD and EDM. It can be said that the EDDD is one of the hybrid techniques which combine the diamond drilling (DD) and EDM, the EDDD process as shown in Fig.1. In EDDD process, materials removal takes place from the surface by simultaneous influence of diamond grains, continuous-discrete electrical sparks resulting mechanical abrasion and electrical sparks erosion, respectively. A metal bonded diamond abrasive rotating tool electrode has been implemented to get adequate metal removal through EDDD process in Nimonic alloy. EDD process contents rotating cylindrical drills inbuilt with EDM set up to produce a hole in the workpiece material. Through the EDM process tool with zero speed is utilized for making the hole. In this research, authors performed the comparative study among these processes EDDD, EDD and EDM based on one factor at a time (OFAT) approach. The most important critical drilling parameters like peak current, pulse-on time and pulse-off time are considered as input factors. The effect of process parameters are analyzed on material removal rate (MRR) and tool wear rate (TWR) to compare the process performances through various ways. The results show that the value of MRR is higher in EDDD as compared to EDD and EDM whereas TWR is lowest in case of EDDD, compared to EDD and EDM respectively.

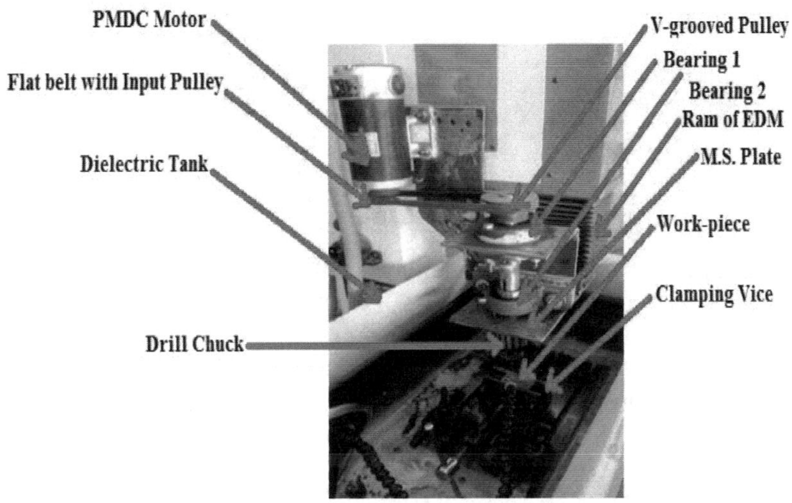

Figure 1. Electrical discharge diamond drilling (EDDD) set up

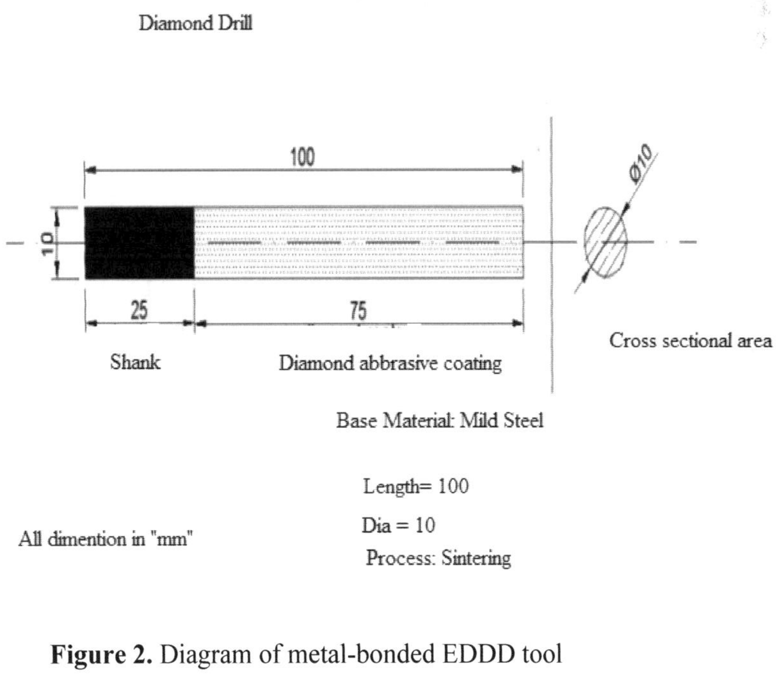

Figure 2. Diagram of metal-bonded EDDD tool

IOP Conf. Series: Materials Science and Engineering 1104 (2021) 012031 doi:10.1088/1757-899X/1104/1/012031

3. Results and discussion

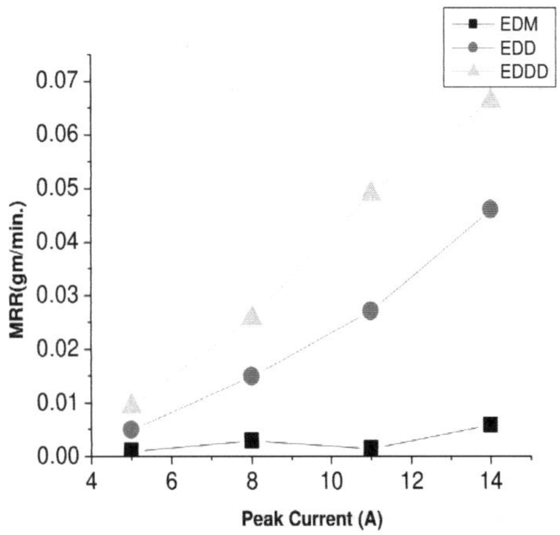

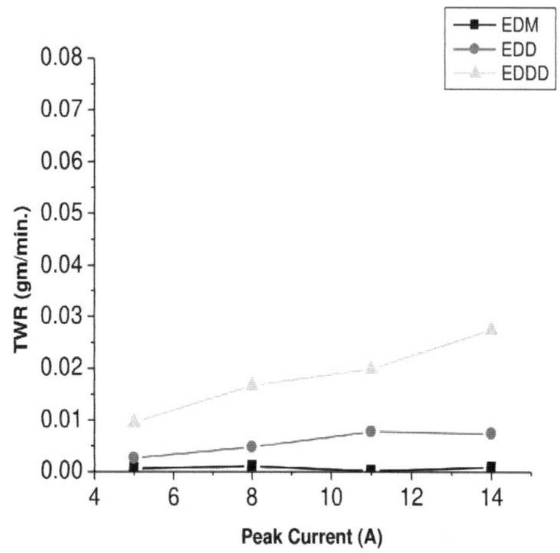

Figure 3. Effect of peak current (I) on MRR **Figure 4.** Effect of peak current (I) on TWR

The present graph of Fig. 3 is representing that comparatively high material removal takes place in EDDD with compare to EDD and EDM process. It is evident that mechanical abrasion type metal removal is also an important consideration along with thermo-physical MRR due to which more amount of metal removal reported from workpiece in EDDD than others. On the other hand, TWR is also reported slightly higher than EDD and EDM, which is not favourable for the industry as Fig. 4. Instead of using the die-sinking EDM, WEDM may be set up used in EDDD which TWR does not matter at all.

It is also very clear from Fig. 3 to Fig. 6 that MRR and TWR both are increased by increasing peak current and Ton during every process but its values are comparatively high for EDDD due to mechanical abrasion and thermo-physical means of material depletion from work and tool materials.

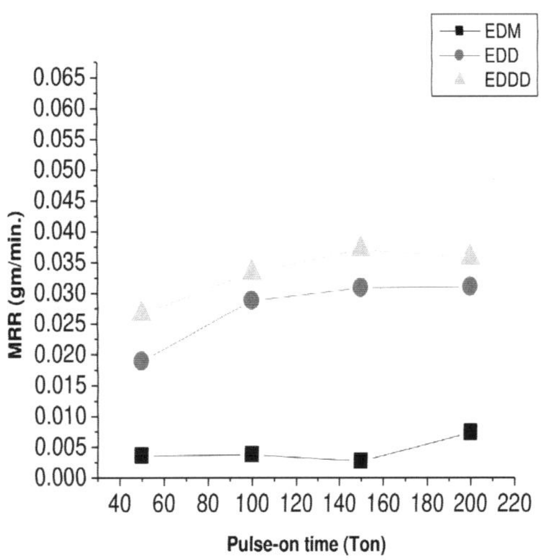

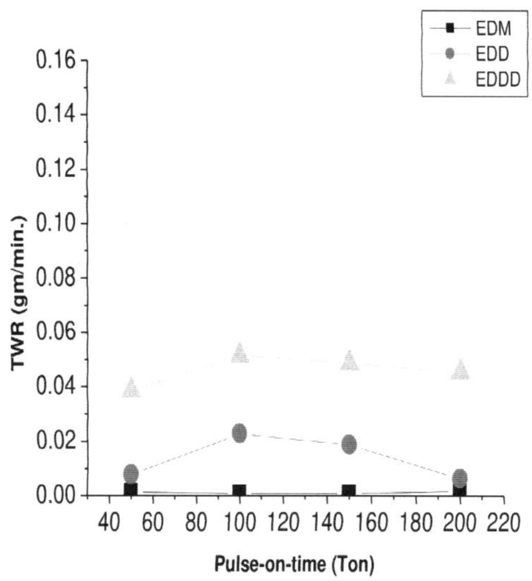

Figure 5. Effect of pulse-on time (T_{on}) on MRR **Figure 6.** Effect of pulse-on time (T_{on}) on TWR

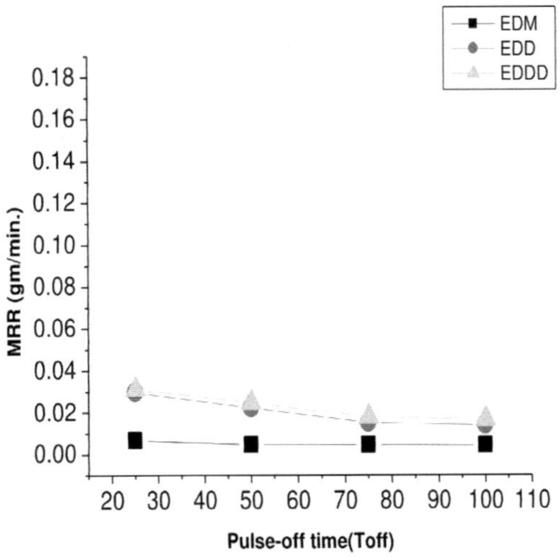

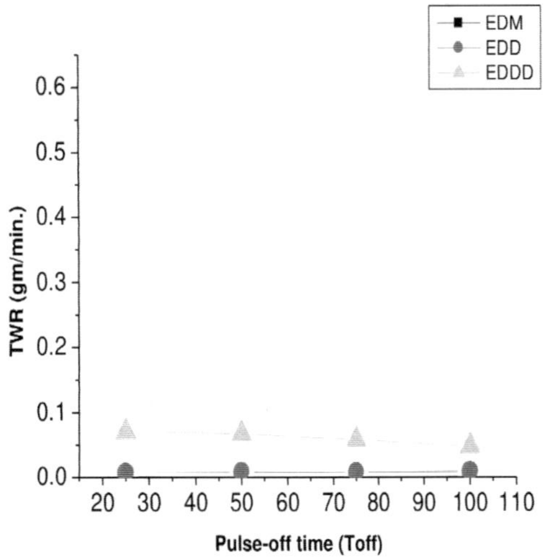

Figure7. Effect of pulse-off time (T_{off}) on MRR **Figure 8.** Effect of pulse-off time (T_{off}) on TWR

From Fig.7 & Fig.8, MRR and TWR found decreases with an increase in pulse off time (Toff) and increases with increasing pulse-on time (Ton). The optimal value of Ton and Toff may be selected for these particular applications.

As per the comparative study of electrical discharge cutting, drilling and diamond drilling process on Nimonics 75, it is evident that EDDD is most adaptive hybrid drilling technique. EDDD exhibits comparatively very high MRR and TWR at high peak current, pulse on time and high drill speed. Responses may be further optimized to improve the surface integrity of machined components as well as the process performances.

3.1. Scanning electron microscopy (SEM) analysis

The micrographs of the surface of three selected samples have been taken using scanning electron microscope (SEM) and which are shown in Fig.9(a),(b) & (c) respectively. In order to make a difference in surface morphology of all these three samples which are machined by EDM, EDD and EDDD. From Fig. 9 (a), it can be seen that more amount of recast layers in the sample has deposited on the machined surface.

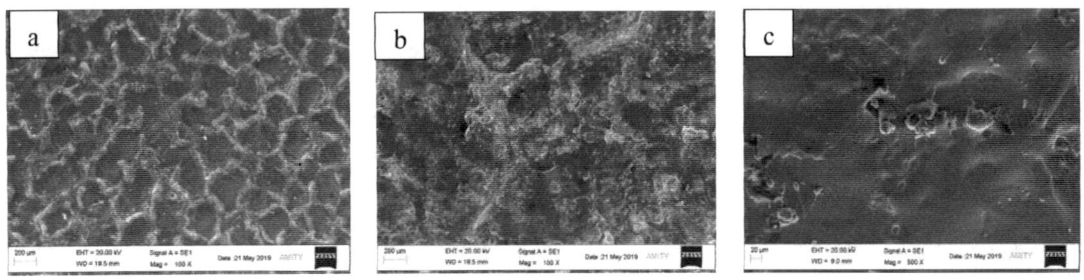

Figure.9. SEM Images of surface for sample machined with a) EDM at I= 8A, Ton=150μs, Toff=75μs and DS=0 RPM b) EDD at I= 8A, Ton=150μs,Toff=75μs and DS=825 RPM c) EDDD at I= 8A, Ton=150μs, Toff=75μs and DS=825 RPM

Table 1: Comparison of response parameters among EDM, EDD and EDDD

Response parameters	EDM		EDD		EDDD	
	Min	Max	Min	Max	Min	Max
MRR (gm/min.)	0.0014	0.0074	0.0015	0.046	0.0081	0.662
TWR (gm/min.)	0.0013	0.0016	0.0048	0.031	0.0455	0.600

4. Conclusion

Based on the above experimental results, the following conclusion is drawn:
i) The performance of EDDD process has found quite better than the EDD and EDM. EDDD process found to be the best process because of simultaneously used electrical erosion and mechanical abrasion resulting in high MRR. But in the case of the EDD process, electrical erosion with rotating tool electrode helps to increase MRR than EDM. As in EDM, only electrical erosion takes place so MRR found lowest.
ii) In EDD, higher the peak current, higher the MRR as it removes softens material easily but less than EDDD because both electrical erosion and mechanical abrasion take place simultaneously causing higher MRR.
iii) In EDM, only electrical erosion takes place but no abrasion and rotary action so causing the least amount of MRR found.
iv) Higher pulse-on time (Ton), both MRR and TWR are higher as longer duration of on-time melts both electrode materials more.
v) Higher the pulse-off time (Toff), lower the MRR and TWR because of lesser available time to be on (T_{on}).
SEM analysis was also confirmed that maximum MRR found in the case of EDDD.
Based on the evidence available from the experiment conducted above, the author suggested that EDDD is the best among all the mentioned processes.

References

[1] Shu K. M, Shih HR and Tu GC 2006 Electrical discharge abrasive drilling of hard materials using a metal matrix composite electrode *Int J Adv Manuf Tech.***29** 678-687
[2] Kumar S. & Choudhary S.K. 2007 Prediction of wear and surface roughness in electro-discharge diamond grinding Journal of Materials Processing Technology. **191** 206–209
[3] Sunder S. and Yadava V. 2012 Development, experimental investigation and neural network modelling of surface-electrical discharge diamond grinding of Al-SiC metal matrix composite. *International Journal of Abrasive Technology* **5(3)** 223–244
[4] Balaji P.S. and Yadava V. 2013 Three dimensional thermal finite element simulation of electro-discharge diamond surface grinding *Simulation Modelling Practice and Theory* **35** 97–11
[5] Choudhary R., Kumar H. and Singh S. 2013 Machining Performance and Surface Integrity of AISI D2 Die Steel Machined Using Electrical Discharge Surface Grinding Process. *Journal of Materials Engineering and Performance* **22 (12)** 3655-3665
[6] Sunder S. and Yadava V. 2014 Modeling and optimization of material removal rate and surface roughness in surface-electrical discharge diamond grinding process International *Journal of Industrial and Systems Engineering* **17(2)** 133–151
[7] Agrawal S.S and Yadava V. 2015 Development and Experimental Study of Surface-Electrical Discharge Diamond Grinding of Al–10wt% SiC Composite *Journal of The Institution of Engineers (India,* **97(1)** 1–9
[8] Yadav S.K.S. and Yadava V. 2011 Neural network modeling and multi-objective optimization of electrical discharge diamond cut-off grinding (EDDCG) *International Journal of Abrasive Technology* **4(4)** 346–362
[9] Yadav S.K.S. and Yadava V. 2013 Experimental Investigation of Electrical Discharge Diamond Cut-Off Grinding of Ti Alloy. *Materials and Manufacturing Processes* **28** 557–561

[10] Yadav S.K.S. and Yadava V. 2013 Experimental Investigations to Study Electrical Discharge Diamond Cut-off Grinding (EDDCG) Machinability of Cemented Carbide *Materials and Manufacturing Processes* **28(10)** 1077-1081
[11] Arun Kumar N.E. and Suresh Babu A: 2018 Influence of input parameters on near-dry WEDM of Monel alloy. *Materials and Manufacturing Processes* **33(1)** 85-92
[12] Yadav Ram Singar and Yadava Vinod. 2018 Experimental investigations on EDDFSG of hybrid MMC *Materials and Manufacturing Processes* **32(2)** 135-144
[13] Bains Preetkanwal Singh, Sidhu Sarabjeet Singh and Payal H.S. 2016 Study of magnetic field-assisted ED machining of metal matrix composites Materials and Manufacturing processes, **31(14)** 1889-1894
[14] Agarawal SS and Yadava V. 2013 Modeling and prediction of materials removal rate and surface roughness in surface electrical discharge diamond grinding process of metal matrix composites *Mater Manuf Process* **28** 381– 389
[15] Yadav SKS and Yadava V. 2013 Experimental investigation of electrical discharge diamond cut- off grinding of Ti-alloy. *Mater Manuf Process* **28** 557–561
[16] A. Okade, A. Yamaguchi and K.Ota. 2017 Improvement of curved hole EDM drilling performance using a suspended ball electrode by workpiece vibration *CIRP Annals-Manufacturing Technology* 66,189-192
[17] Sagar U. Sapkal and Pravin S. Jagtap. 2018: Optimisation of Micro EDM Drilling Process Parameters for Titanium Alloy by Rotating Electrode. Procedia Manufacturing **20** 119–126
[18] Grzegorz Skrabalak 2018 Influence of electrode tool length on the micro EDM drilling performance *Procedia CIRP* **68** 594 – 598
[19] Kliuev, M., Maradia, U 2016 Boccadoro, M., Perez, R., Stirnimann, J., & Wegener, K.: Experimental study of EDM-drilling and shaping of SiSiC and SiC *Procedia CIRP* **42(1)** 191-196
[20] Raju B. Bhosle and S. B. Sharma. 2017 Multi-performance optimization of micro-EDM drilling process of Inconel 600 alloy *Materials Today: Proceedings* **4** 1988–1997

IOP Conf. Series: Materials Science and Engineering 1104 (2021) 012032 doi:10.1088/1757-899X/1104/1/012032

Performance analysis of ERS using R134a - An experimental investigation (Part 1) – Ejector design and performance analysis

Vikas Kumar[1*], Sandeep Tiwari[1], Gulshan Sachdeva[2], Vaibhav Jain[3], Vishal Mittal[1], Paras Sharma[1], Vipin Rajput[1], & Homenidhi Bhardwaj[1]

[1*]Mechanical Engineering Department, Krishna Engineering College, Ghaziabad (India)
Corresponding author Email: vikas.burman21@gmail.com.
[2]Mechanical Engineering Department, National Institute of Technology, Kurukshetra (India)
Email: gulshansachdeva@nitkkr.ac.in
[3]Mechanical Engineering Department, MAIT, New Delhi (India)
Email: vaibhavursathi@gmail.com
Abstract. The present study proposed a mathematical model for calculating the entrainment ratio and area-ratio for the single-phase ejector working double choking conditions. The model contains gas dynamic equations and is used to calculate performance of ejector for R134a and R141b refrigerants. The model is also validated with the experimental results of the previous researcher and found results are in good coherence with average deviation of -0.287% and 2.48% for area-ratio and entrainment ratio respectively.

Keyword: Mathematical model, ejector, performance.

1. Introduction

Refrigeration systems using these days are either consumes large amount of electric energy (i.e., in case vapor compression refrigeration system) or very bulky (i.e., in case of vapor absorption refrigeration system). The demand of refrigeration system working by consuming heat energy is very high and researchers are focusing their attention to design and develop heat assisted refrigeration systems. The ejector refrigeration system (ERS) is an attractive system to meet the current requirements. ERS consumes low temperature heat which makes it suitable to operate on solar energy or industrial waste heat. Another advantage of the ERS is its simplicity in design, low initial investment and light in weight. The schematic diagram of ERS is shown in figure 1. The liquid refrigerant condensed by the condenser is pumped by a liquid pump to the generator where the liquid refrigerant gets converted into the high-pressure vapor refrigerant by absorbing latent heat. This high-pressure vapor refrigerant is used as motive fluid by the ejector to suck the low-pressure vapor refrigerant from the evaporator and keeps the evaporator at low pressure. This low-pressure refrigerant is also known as secondary refrigerant. Both primary and secondary refrigerant mixes with each and moves together to the condenser at an intermediate pressure where the mixed refrigerant is condensed by releasing its latent heat in the condenser.

The main and principal component of the system is ejector. An ejector can be defined as a compressing unit which utilizes the fluid energy of the high pressure, high temperature incoming fluid also known as primary fluid. When this motive fluid passes through the convergent-divergent primary nozzle, it expands and accelerate to a very high velocity normally above speed of sound and thus creates a very low-pressure environment in the suction chamber which is responsible for the generation of a suction effect. This suction force entrains and draws the low-pressure vapor refrigerants from the evaporator. While passing through the constant area section of the ejector, these both fluids mix with each other and when this mixed fluid moves through the diffuser section then the pressure is again raised on the compensation of the loss of kinetic energy.

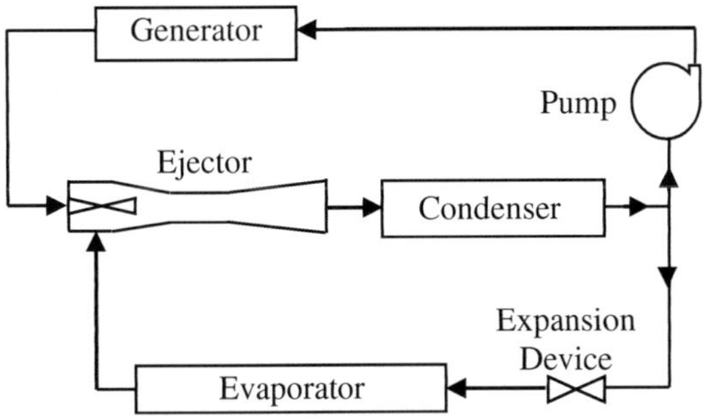

Figure 1 Schematic diagram of ERS.

A large amount of literature is available on the performance analysis of the ERS. But most of the available studies is theoretical and a few researchers has analyzed the ERS for various ejector geometry and various operating conditions. Huang et al. [1] has developed a mathematical model using R141b for analyzing of the single-phase ejector using gas dynamics equations. They have studied the behavior of the ejector but not analyzed the complete system in detail. They use eleven different ejector geometries for validating the results of the developed mathematical model. Chou et al. [2] had also developed a mathematical model to study the ejector's behavior but instead of focusing on performance analysis, they have analyzed the choking condition's influence on the various section of the ejector by varying the condenser pressure. They use three different refrigerants i.e., R113, R141b and steam in their analysis. Kumar and Sachdeva [3] proposed a novel 1-D mathematical model for predicting the and fining the geometry of a single-phase ejector. In addition to the gas dynamics equation, they also use some new concepts in their models i.e., Prandtl's mixing length, Kelvin-Helmholtz instability, Prandtl-Meyer expansion wave and Baroclinic effect. By considering real gas equations Tyagi and Murty [4] had developed a mathematical model using R11 and R113. The generator temperature is kept low i.e., in between 70-90°C. While keeping the condenser temperature constant at 30°C and evaporator temperature at 0°C, they found 0.43 and 0.38 COP with R11 and R113 respectively.

Boumaraf and Lallemand [5] formed a mathematical model to simulate the performance of ERS, which separately included the efficiencies of each and every component. They used R141b and R600a as the working fluids in their simulation for optimized and off operating conditions. They established their results with the previously available experimental and theoretical 1-D models [1]. COP of the system was found to increase with the decrease in source temperature but it decreased the critical condenser pressure of the system. Zhu and Li [6] studied and proposed a novel model for calculating performance of an ejector for both dry and wet vapor as working fluid at the critical conditions. Instead of utilizing simple thermodynamic relations, they used real velocity approach and integrated the velocity function for finding mass flow rates of primary and secondary fluid. Assuming the mixing layer in between the primary and secondary fluid in chocking conditions, all the equations were derived accordingly.

Elakhdar et al. [7] developed a one-dimensional mathematical model to understand the frictional losses between the primary and secondary fluid flow during mixing and its effect on the performance. They reported the error difference of 4.27% in the entrainment ratio obtained from their model and the actual experimental results of Huang at el. [1]. The concept of Fanno flow to introduce frictional

IOP Conf. Series: Materials Science and Engineering 1104 (2021) 012032 doi:10.1088/1757-899X/1104/1/012032

compressible flow in the mixing chamber of the ejector was introduced in the mathematical model by Satheesh and Ooi [8]. Moreover, the average heat capacity ratio was calculated at different sections in the ejector rather than assuming a constant heat capacity ratio like other researchers. An average error of only 4% in the entrainment ratio was projected with that of the experimental results of Huang et al. [1].

2. Mathematical model

A one-dimensional mathematical model has been developed and proposed to evaluate performance of single-phase ejector for known dimensions and operating conditions using gas dynamics compressible flow equations.

Assumptions: Following assumptions have been made for the analysis:

- Primary/motive nozzle is working without shock wave and flow exit is above Mach one.
- To increase the model efficiency appropriate coefficient are selected on account of friction and mixing losses.
- The heat transfer rate in between the ejector wall and the working fluid is assumed to be zero i.e., an adiabatic flow.
- An assumption is taken into considered in the 1-D model that both fluids start mixing with each other at hypothetical throat.
- Both fluid mixes with each other in the constant area section.
- The secondary fluid is assumed to be choked at inlet face of the constant area section.
- The velocity of the flow while exiting from the diffuser is assumed to be zero.
- The refrigerant inside the ejector is assumed to be dry vapors.

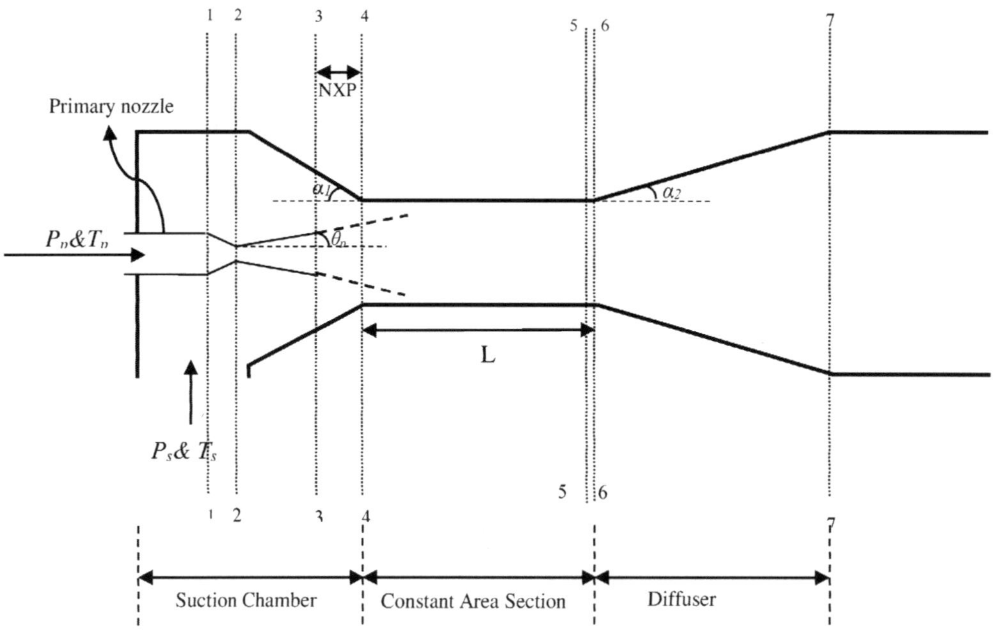

Figure 2 Single phase Ejector geometry with sections

2.1 Governing gas dynamic equations

The refrigeration systems are normally designed for a specific application that contains a certain evaporator temperature. Consequently, the ejector must be designed for desired refrigeration effect and evaporator temperature. Furthermore, for obtaining desired cooling effect and evaporator temperature, the entrained secondary fluid mass and the suction pressure must be maintained constant by the designed ejector. Here in this case, it is assumed that the mass flow rate of primary fluid provided by the generator is fixed at a specified saturated generator pressure and has some velocity at input to the primary nozzle of the ejector. Hence, this velocity can be calculated by the formula below:

$$\dot{m}_p = \rho_{p\_1} \times A_{p\_1} \times V_{p\_1} \quad (1)$$

where $\dot{m}$ and ρ are the primary fluid mass flow rate and the density respectively. The generator pressure (P_{gn}) and temperature (T_{gn}) is used to calculate the stagnation temperature (T_{p_1}) and pressure ($P_{p,1}$) of the primary fluid at inlet of the primary nozzle isentropic relations [9, 10]:

$$T_{p\_1} = T_{gn} + \frac{V_{p\_1}^2}{2 \times C_{p\_gn}} \quad (2)$$

$$\frac{P_{p\_1}}{P_{gn}} = \left(\frac{T_{p\_1}}{T_{gn}}\right)^{\left(\frac{\gamma_{gn}}{\gamma_{gn}-1}\right)} \quad (3)$$

The mass flow rate of the primary/motive fluid calculated using isentropic gas dynamics relation [10, 11]:

$$\dot{m}_p = A_{p\_2} \times P_{p\_1} \times \left\{\frac{\gamma_{p\_1}}{\dot{R} \times T_{p\_1}} \times \left(\frac{2}{\gamma_{p\_1}+1}\right)^{\left(\frac{\gamma_{p\_1}+1}{\gamma_{p\_1}-1}\right)}\right\}^{\frac{1}{2}} \times (\eta_p)^{\frac{1}{2}} \quad (4)$$

Frictional losses produced into the primary nozzle is taken to consideration increase the model efficiency i.e., isentropic efficiency of the primary nozzle (η_p). It is assumed that the hypothetical throat of the secondary fluid exists at the inlet section of the CAS at section 4-4. The pressure of the secondary fluid ($P_{s,4}$) is evaluated by the relation as under at section 4-4 [9, 10]:

$$\frac{P_{s\_4}}{P_{ep}} = \left(\frac{2}{\gamma_{ep}+1}\right)^{\left(\frac{\gamma_{ep}}{\gamma_{ep}-1}\right)} \quad (5)$$

The assumption of equal pressure of both primary and secondary fluid in the present model is considered [11].

$$P_{s\_3} = P_{p\_3} \quad (6)$$

It is also found numerically that difference in the velocities of the both i.e., primary and secondary fluid is very large. So, both fluids may not mix with each other until the difference in their velocities may not reduce considerably.

At section 3-3 i.e., at exit plane of the primary nozzle the Mach number (M_{p_3}) of the primary/motive fluid has been calculated by the following relation [9, 10]:

$$\frac{P_{p\_1}}{P_{p\_3}} = \left\{1 + \left[\left(\frac{\gamma_{p\_1}-1}{2}\right) \times \left(M_{p\_3}\right)^2\right]\right\}^{\left(\frac{\gamma_{p\_1}}{\gamma_{p\_1}-1}\right)} \quad (7)$$

The area of cross-sectional of the motive nozzle at its exit (A_{p_3}) is evaluated as below [10, 11]:

$$\frac{A_{p\_3}}{A_{p\_2}} = \frac{1}{M_{p\_3}} \times \left[\left(\frac{2}{\gamma_{p\_2}+1}\right) \times \left\{1 + \left[\left(\frac{\gamma_{p\_2}-1}{2}\right) \times \left(M_{p\_3}\right)^2\right]\right\}\right]^{\frac{1}{2}\left(\frac{\gamma_{p\_2}+1}{\gamma_{p\_2}-1}\right)} \quad (8)$$

The "*Prandtl-Meyer expansion wave*" has been used for evaluating divergence angle of the motive nozzle as follows [12]:

$$\theta_p = \{\nu(M_{p\_3}) - \nu(M_{p\_2})\}/2 \quad (9)$$

Here θ_p is the nozzle divergence angle and $\nu(M_{p,2})$ and $\nu(M_{p,3})$ are the Prandtl-Meyer functions [12].

$$\nu(M_{p\_3}) = \left(\frac{\gamma_{p\_3}+1}{\gamma_{p\_3}-1}\right)^{\frac{1}{2}} \times \tan^{-1}\left\{\left(\frac{\gamma_{p\_3}-1}{\gamma_{p\_3}+1}\right)^{\frac{1}{2}} \times \left({M_{p\_3}}^2 - 1\right)^{\frac{1}{2}}\right\} - \tan^{-1}\left({M_{p\_3}}^2 - 1\right)^{\frac{1}{2}} \quad (10)$$

$$\nu(M_{p\_2}) = \left(\frac{\gamma_{p\_2}+1}{\gamma_{p\_2}-1}\right)^{\frac{1}{2}} \times \tan^{-1}\left\{\left(\frac{\gamma_{p\_2}-1}{\gamma_{p\_2}+1}\right)^{\frac{1}{2}} \times \left({M_{p\_2}}^2 - 1\right)^{\frac{1}{2}}\right\} - \tan^{-1}\left({M_{p\_2}}^2 - 1\right)^{\frac{1}{2}} \quad (11)$$

The Mach number (M_{p_4}) of the motive flow is evaluated by [9, 10, 11]:

$$\frac{P_{p\_4}}{P_{p\_3}} = \left[\frac{1+\left((\gamma_{p\_3}-1)/2\right){M_{p\_3}}^2}{1+\left((\gamma_{p\_3}-1)/2\right){M_{p\_4}}^2}\right]^{\frac{\gamma_{p\_3}}{\gamma_{p\_3}-1}} \quad (12)$$

At hypothetic throat, both fluids are assumed to be at same pressure [1].

$$P_{p\_4} = P_{s\_4} \quad (13)$$

Cross-sectional area at section 4-4 for the primary fluid can be calculated by relation

$$\frac{A_{p\_4}}{A_{p\_3}} = \left[\frac{\phi_p/M_{p\_4}}{1/M_{p\_3}}\right] \times \left[\frac{\left(2/(\gamma_{p\_3}+1)\right)\left(\left(1+\left((\gamma_{p\_3}-1)/2\right)\right){M_{p\_4}}^2\right)}{\left(2/(\gamma_{p\_3}+1)\right)\left(\left(1+\left((\gamma_{p\_3}-1)/2\right)\right){M_{p\_3}}^2\right)}\right]^{\left(\frac{\gamma_{p\_3}+1}{2(\gamma_{p\_3}-1)}\right)} \quad (14)$$

Secondary fluid mass flow rate can be calculated as:

$$\omega \times \dot{m}_p = \dot{m}_s = P_{ep} \times A_{s\_4} \times \left[\left(\frac{\gamma_{ep}}{\dot{R} \times T_{ep}}\right) \times \left(\frac{2}{\gamma_{ep}+1}\right)^{\left(\frac{\gamma_{ep}+1}{\gamma_{ep}-1}\right)}\right]^{\frac{1}{2}} \left(\eta_s^{\frac{1}{2}}\right) \quad (15)$$

The total area of cross-section at section 4-4 (A_4) is the sum of the area of the primary and secondary fluid:

$$A_4 = A_{p\_4} + A_{s\_4} \quad (16)$$

The secondary fluid flow's area (A_{s_3}) is calculated by relations:

$$\frac{P_{s\_4}}{P_{s\_3}} = \frac{\left(1 + \frac{\gamma_{s\_3}-1}{\gamma_{s\_3}} M_{s\_3}^2\right)^{\frac{\gamma_{s\_3}}{\gamma_{s\_3}-1}}}{\left(1+\frac{\gamma_{s\_3}-1}{2}\right)^{\frac{\gamma_{s\_3}}{\gamma_{s\_3}-1}}} \quad (17)$$

$$\frac{A_{s\_3}}{A_{s\_4}} = \frac{1}{M_{s\_3}}\left[\frac{1 + \frac{\gamma_{s\_3}-1}{2} M_{s\_3}^2}{\frac{\gamma_{s\_3}+1}{2}}\right]^{\frac{\gamma_{s\_3}+1}{2(\gamma_{s\_3}-1)}} \quad (18)$$

Both fluids mix with each other in between section 4-4 to 5-5. Momentum and energy balance equations is used to calculate velocity and the temperature [1]:

$$\varphi_m\left(\dot{m}_p V_{p\_4} + \dot{m}_s V_{s\_4}\right) = \left(\dot{m}_p + \dot{m}_s\right) V_5 \quad (19)$$

$$\dot{m}_p\left(Cp_{p\_4} T_{p\_4} + \frac{V_{p\_4}}{2}\right) + \dot{m}_s\left(Cp_{s\_4} T_{s\_4} + \frac{V_{s\_4}}{2}\right) = \left(\dot{m}_p + \dot{m}_s\right)\left(Cp_5 T_5 + \frac{V_5}{2}\right) \quad (20)$$

A normal shock is assumed at section 6-6 i.e., at the inlet face of the diffuser. And by using the following relations, pressure and temperature after the diffuser can be calculated:

$$M_6 = \frac{{M_5}^2 + \left(\frac{2}{\gamma_5 - 1}\right)}{\left(\frac{2{M_5}^2 \gamma_6}{\gamma_6 - 1}\right) - 1} \quad (21)$$

$$\frac{P_6}{P_5} = \frac{M_5\sqrt{1+M_5{}^2(\gamma_5-1)/2}}{M_6\sqrt{1+M_6{}^2(\gamma_5-1)/2}} \tag{22}$$

$$\frac{T_6}{T_5} = \left[1 + \left(\frac{2\gamma_5}{\gamma_5+1}\right)\left(M_5{}^2 - 1\right)\right]\left[\frac{2+(\gamma_5-1)M_5{}^2}{(\gamma_5+1)M_5{}^2}\right] \tag{23}$$

The flow is assumed to be stagnant at the exit of the diffuser and the flow properties is calculated by the following relations:

$$\frac{P_7}{P_6} = \left(1 + \frac{\gamma_6-1}{2}M_6^2\right)^{\frac{\gamma_6}{\gamma_6-1}} \tag{24}$$

$$\frac{A_7}{A_{4,final}} = exp\left\{\frac{(P_7-P_6)(1-M_6{}^2)}{\rho_6 M_6{}^2\gamma_6 R T_6}\right\} \tag{25}$$

3. Results and numerical procedure

The temperature (T_p), pressure (P_p) and the mass flow rate ($\dot{m}_p$) of the primary fluid at the inlet of primary nozzle; pressure (P_s), temperature (T_s) of the secondary fluid in the evaporator and critical condenser pressure (P_c) are the inputs to the mathematical model. The numerical procedure has been carried out in Engineer Equation Solver (EES) [13].

The model is designed to calculate the entrainment ratio for the desired input conditions. Additionally, it also calculates the area-ratio of the single-phase ejector. The present model is used to calculate the entrainment ratio and the area-ratio for R141b and R134a refrigerants.

3.1 Model Validation

The results of the proposed mathematical model are first validated by comparing the model's results with the results obtained by Huang et al. [1]. Figure 3 shows the deviation of the results of the area-ratio obtained by the present model while comparing with Huang et al. [1]. The average deviation of -0.285% is found in area-ratio while comparing with the experimental results of the Huang et a. [1]. Similarly, figure 4 shows the comparison for the entrainment ratio between the present model and the experimental results of the Huang et al. [1]. Graphs show that the results obtained by the proposed model ae in good coherence with experimental results of Huang et al. [1].

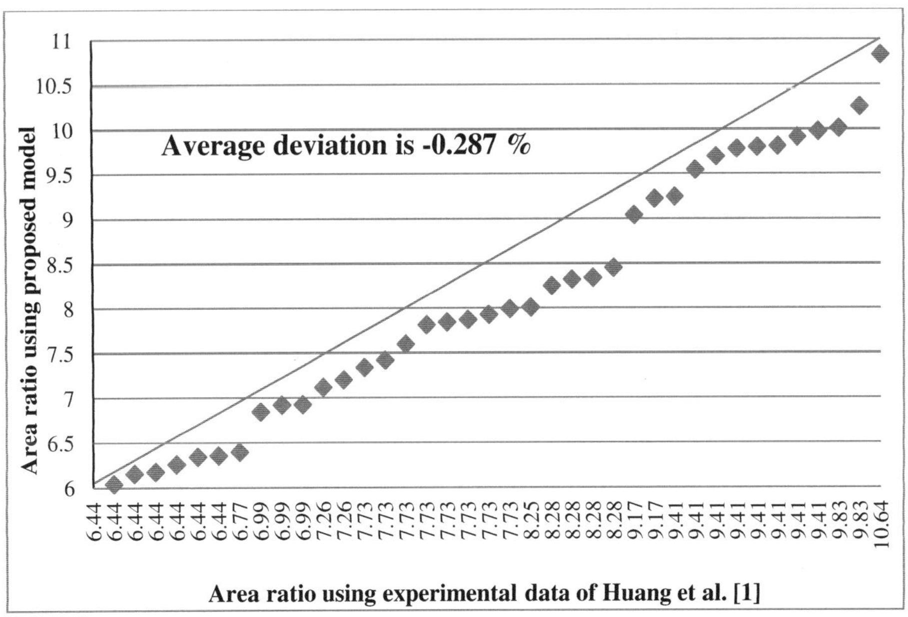

Figure 3 Area ratio's comparison between results obtained by proposed model and that of Huang et al. [1]

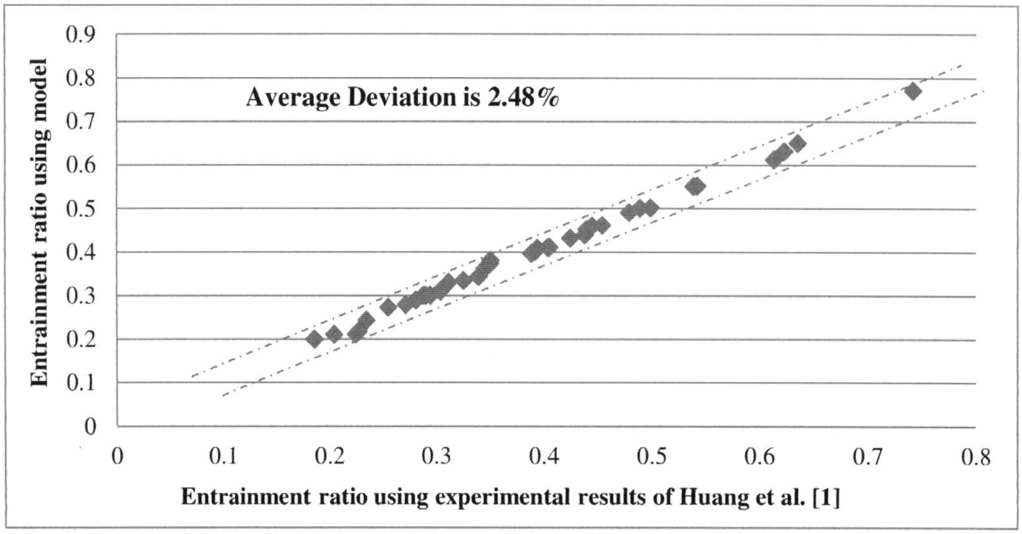

Figure 4 Entrainment ratio's comparison between the results obtained using proposed model and that of Huang et al. [1]

Table 1 Geometries of ejector used in the experiments

Sr. No	Ejector	Primary nozzle geometry			Secondary nozzle geometry			Area Ratio = (A_4/A_{p2})
		Inlet Area (A_{p1}) (mm^2)	Throat Area (A_{p2}) (mm^2)	Exit Area (A_{p3}) (mm^2)	CAS Area (A_4) (mm^2)	CAS's Length (L) (mm)	(NXP) (mm)	
1	I	31.7	3.14	10.3	31.7	80	2.54	10.08
2	II	31.7	3.14	10.3	31.7	120	2.54	10.08
3	III	31.7	4.91	15.9	31.7	80	2.54	6.45
4	IV	31.7	4.91	15.9	31.7	120	2.54	6.45
5	V	31.7	7.07	20.6	67.2	80	2.54	9.51
6	VI	31.7	7.07	20.6	67.2	120	2.54	9.51

Table 1 shows the ejector geometry calculated by the proposed model for R134a refrigerant. The input parameters for the model are throat area of the primary nozzle, temperature and pressure of generator, condenser and evaporator. These dimensions are calculated for 84°C generator temperature, 30°C condenser temperature and 10°C evaporator temperature.

4. Conclusion

The paper proposed a mathematical model of single-phase ejector to evaluate the performance analysis and a few ejector geometries. The results obtained by the model for R141b refrigerant is validated by comparing them with the experimental results of Huang et al. [1] and the results found are in good coherence. The proposed model is also used to calculate the performance and geometry of the single-phase ejector using R134a refrigerant. All dimensions and performance parameters are discussed in the next research paper of the sequence.

Acknowledgement
This research is funded under the "**Visvesvaraya Research Promotion Scheme**" funded by Dr. A. P. J. Abdul Kalam Technical University, Lucknow, India.

References

[1] Huang BJ, Chang JM, Wang CP and Petrenko VA, A 1-D analysis of ejector performance, International Journal of Refrigeration, vol. 22, pp. 354-364, 1999.
[2] Chou SK, Yang PR, Yap C, Maximum mass flow ratio due to secondary flow choking in an ejector refrigeration system, International journal of refrigeration, vol. 24, pp. 486-499, 2001.
[3] Kumar V and Sachdeva G, 1-D model for finding geometry of a single phase ejector, Energy, vol. 165, vol. 75-92, 2018.
[4] Tyagi KP and Murty KN, Ejector-compression systems for cooling utilizing low grade waste heat, Heat recovery systems, vol. 5, No. 6, pp. 545-550, 1985.
[5] Boumaraf L and Lallemand A, Modeling of an ejector refrigeration system operating in dimensioning and off-dimensioning conditions with the working fluids R142b and R600a, Applied thermal engineering, vol.29, pp. 265-274, 2009.
[6] Yinhai Zhu and Yanzhong Li, Novel ejector model for performance evaluation on both dry and wet vapors ejector, International journal of refrigeration, vol. 32, pp. 21-31, 2009.
[7] Elakhdar M, Nehdi E, Kairouani L and Tounsi N, Simulation of an ejector used in refrigeration systems, International journal of refrigeration, vol. 34, pp. 1657-1667, 2011.
[8] N Satheesh Kumar and Kim Tiow Ooi, One dimensional model of an ejector with special attention to Fanno flow within the mixing chamber, Applied thermal engineering, vol. 65, pp. 226-235, 2014.
[9] Imrie BW, Compressible Fluid Flow, Butterworth & Co (Publishers) Ltd, Bristol, London, 1973.
[10] Genick BM, Fundamentals of Compressible Fluid Mechanics, Boston, MA, USA, 2002.
[11] Pjush K Kundu and Ira M Cohen, Fluid Mechanics, second edition, Chapter – 16, Academic Press, California, USA, 1990.
[12] Anderson JD, Modern Compressible Flow, Second edition, McGraw Hill Publishing Co, 1990.
[13] Engineer Equation Solver, version 2015, Academic Professional V10.097 National Institute Technology, Kurukshetra, India.

IOP Conf. Series: Materials Science and Engineering 1104 (2021) 012033 doi:10.1088/1757-899X/1104/1/012033

Performance analysis of ERS using R134a - An experimental investigation (Part 2) – System design

Vikas Kumar[1*], Sandeep Tiwari[1], Gulshan Sachdeva[2] and Vaibhav Jain[3], Devanshu[1], Paras Raghav[1], Ankush Nagar[1] & Uttam[1]

[1]Mechanical Engineering Department, Krishna Engineering College, Ghaziabad (India)
Corresponding author email: vikas.burman21@gmail.com.
[2]Mechanical Engineering Department, National Institute of Technology, Kurukshetra (India)
E-mail: gulshan4you@gmail.com
3 Mechanical Engineering Department, MAIT, New Delhi (India)
Email: vaibhavursathi@gmail.com

Abstract. The paper presents a mathematical model based on the 1-D gas dynamic equations used to calculate the design and dimensions of the ERS. In the model, the design of the heat exchangers i.e., generator, condenser and evaporator are calculated for the specified ejector geometry. Additionally, the fabrication methodology is also discussed in detail.

Keywords: Ejector, system design, heat exchanger.

1. Introduction

This paper is the continuation part of the pervious paper in which the single-phase ejector is analyzed for the performance analysis and geometry by using 1-D mathematical model. This paper presents a mathematical model for designing an ejector refrigeration system (ERS) which can be further used for the experimental fabrication. It comprises the design of the condenser, generator and evaporator of the ERS. A very few researchers have presented the complete mathematical model for designing ERS in their research paper.

ERS is an attractive technology to herness the low-grade heat energy. There are number of various sources from which the low-grade heat energy can be easily obtained like thermal energy from solar radiations, industrial waste-heat, bio-gasifier, geothermal etc. There are a few important advantages of ERS i.e., simple design, low maintenance and operational cost. On the other hand, coefficient of performance of these systems is low and they are very sensitive to the atmospheric conditions. Apart from mathematical and CFD models, a few researchers have experimentally analyzed the system behavior and performance. To increase the performance of ERS, they focused on the geometrical and operational parameters.

Eames et al. [2] analyzed a single-phase double choked ERS using steam as a working fluid. The experiments were conducted for generator temperature between 120 to 140°C, evaporator temperature 5 to 10°C and condenser pressure from 34 to 60mBar; an average difference of 17% in COP was reported with that of the numerical findings. Aphornratana and Eames [3] conducted experiments using steam as a refrigerant, but with a movable primary nozzle. COP of the system was found to decrease and the critical condenser pressure to increase, when the nozzle exit position was moved away from the face of the constant area section and vice-versa. The generator conditions were maintained constant. Chunnanond and Aphornratana [4] experimentally investigated the effect of expansion angle, effective position and the shock wave position in an ejector refrigeration system while working with steam. The expansion angle is the angle at which the primary fluid fans out from the primary nozzle. The effective position is the position at which the secondary fluid gets choked and an imaginary nozzle is formed. The shock wave position was tracked by employing 12 pressure transducers. They found that the nozzle exit position controlled the position of effective area. When the primary nozzle was moved away from the mixing chamber, it reduced the primary fluid flow expansion inside the constant area section which in turn increased the available area for the secondary fluid mass flow rate. The increased secondary mass flow rate improved the entrainment ratio of the ejector, but it reduced the critical condenser pressure. The experimental results also revealed that increase in expansion angle decreased the entrainment ratio and thus shifted the shock wave position towards the primary nozzle. Rungtrakoon et al. [5] investigated the effect of eight different primary nozzles at the evaporator temperature of 7.5°C while the generator temperature was varied from 110°C to 150°C. Increase in primary nozzle throat diameter decreased the entrainment ratio of the ejector, but it could now be operated at higher condenser pressure.

2. Mathematic model

A mathematical model based on 1-D gas dynamic equations is used to calculate the dimensions of the various heat exchangers used in the ERS experimental setup. The complete program for finding the design of the ERS is given below:

Program for finding tube lengths of heat exchangers in Ejector Refrigeration System (ERS)

```
"Known information"
T_g=357[k]; T_evp=285[k]; d_t=0.0019[m]; f_p=0.9; f_m=0.8; n_p=0.99; n_s = 0.65; d_p_1=0.00635[m];
d_3=0.0042[m]; T_c_s=303[k];
P_c_s=Pressure(R134a,T=T_c_s,x=0.9);
P_g=Pressure(R134a,T=T_g_sat,x=1);
P_evp=Pressure(R134a,T=T_evp_sat,x=1);
T_evp_sat=T_evp-2;
T_g_sat=T_g-4;
"Area calculation"
A_3=(pi#*(d_3^2))/4;
A_p_1=(pi#*(d_p_1^2))/4;
A_t=(pi#*(d_t^2))/4;

"primary mass flow rate"
m_p = ((P_g*A_t)/(sqrt(T_g)))*(sqrt((Gamma/R)*((2/(Gamma+1))^((Gamma+1)/(Gamma-1)))))*(sqrt(n_p));
Gamma = cp_g/cv_g;
cv_g=Cv(R134a, T=T_g, x=1);
cp_g=Cp(R134a, T=T_g, x=1);
R=(UGC/MW)*1000; UGC = 8.314[J/gmol.K];
MW=MolarMass(R134a);
rho_p=Density(R134a,T=T_g,x=1);

"Mach number & Pressure"
(A_p_1/A_t)^2=(1/((M_p_1^2))*(((2/(Gamma+1))*(1+(((Gamma-1)/2)*(M_p_1^2))))^((Gamma+1)/(Gamma-
1))));
(P_g/P_p_1)=((1+((Gamma-1)/2)*(M_p_1^2))^(Gamma/(Gamma-1)));

"Secondary fluid pressure at y-y"
(P_evp/P_s_y)=((1+((Gamma-1)/2))^(Gamma/(Gamma-1)));
(P_p_y/P_p_1)=((((1+(((Gamma-1)/2)*(M_p_1^2)))/(1+(((Gamma-1)/2)*(M_p_y^2))))^((Gamma/(Gamma-
1)))));
```

```
(A_p_y/A_p_1)=((f_p/M_p_y)*(((2/(Gamma+1))*(1+(((Gamma-
1)/2)*(M_p_y^2))))^((Gamma+1)/(2*(Gamma-1)))))/((1/M_p_1)*(((2/(Gamma+1))*(1+(((Gamma-
1)/2)*(M_p_1^2))))^((Gamma+1)/(2*(Gamma-1)))));
P_p_y=P_s_y;
P_p_y=P_m;
A_3=A_p_y+A_s_y;

"Mass flow rate of secondary fluid"
m_s=((P_evp*A_s_y)/(sqrt(T_evp)))*(sqrt((Gamma/R)*((2/(Gamma+1))^((Gamma+1)/(Gamma-
1)))))*(sqrt(n_s));

"Temperature of both fluids"
T_g/T_p_y=(1+(((Gamma-1)/2*(M_p_y^2))));
T_evp/T_s_y=(1+((Gamma-1)/2));

"MIxed flow at section m-m"
f_m*((m_p*v_p_y)+(m_s*v_s_y))=(m_p+m_s)*v_m;
(m_p*((cp_g*T_p_y)+((v_p_y^2)/2)))+(m_s*((cp_g*T_s_y)+((v_s_y^2)/2)))=(m_p+m_s)*((cp_g*T_m)+((v_m
^2)/2));
v_p_y=M_p_y*c_p_y;
v_s_y=c_s_y;
c_p_y=sqrt(Gamma*R*T_p_y);
c_s_y=sqrt(Gamma*R*T_s_y);
M_m=v_m/c_m;
c_m=sqrt(sqrt((Gamma*R*T_m)^2));

"MIxed flow across shock"
(P_3/P_m)=(1+(((2*Gamma)/(Gamma+1))*((M_m^2)-1)));
(M_3)^2=((1+(((Gamma-1)/2)*(M_m)^2)))/((Gamma*(M_m^2))-((Gamma-1)/2));
P_c/P_3=((1+(((Gamma-1)/2)*((M_3)^2)))^(Gamma/(Gamma-1)));
(T_m)/(T_3)=(1+(((2*Gamma)/(Gamma+1))*((M_m^2)-1)))*((2+((Gamma-
1)*(M_m^2)))/((Gamma+1)*(M_m^2)));
T_c=Temperature(R134a,P=P_c,x=0);

"Heat exchangers design"
h_bar_boiler=Flow_Boiling_avg(Fluid$, T_g_sat, G, d_p_1, x_in, x_out, q_boiler, 'Horizontal');
Fluid$='R134a';
G_boiler=m_p/A_p_1;
x_in=0;
x_out=1;
h_boiler_1=Enthalpy(R134a,T=T_g_sat,x=0);
h_boiler_2=Enthalpy(R134a,T=T_g_sat,x=1);
h_fg_boiler=h_boiler_2-h_boiler_1;
q_boiler=m_p*h_fg_boiler;
k=k_('Copper', T_g_sat);
Call External_Flow_Cylinder(Fluid$, T_inf, T_surface,  P_water, u_inf, D_outer_pipe: F_d\L, h_water, C_d,
Nusselt, Re);
T_inf=90+273;
T_surface=84+273;
P_water=101325[Pa];
u_inf=0.7[m/sec];
D_outer_pipe=0.00635[m];
q_boiler=(1/((1/(h_bar_boiler*(Pi*d_3*L_boiler)))+((ln(d_p_1/d_3))/(2*Pi*k*L_boiler))+(1/(h_water*Pi*d_p_
1*L_boiler))))*LMTD_boiler;
LMTD_boiler=6[K];
q_boiler=U_overall_boiler*Pi*d_p_1*L_final_boiler*LMTD_boiler;
```

```
h_evp_1=Enthalpy(R134a,T=T_evp_sat,x=0);
h_evp_2=Enthalpy(R134a,T=T_evp_sat,x=1);
h_fg_evp=h_evp_2-h_evp_1;
q_evp=m_s*h_fg_evp;
d_evp=0.009525[m];
q_evp=U_overall_evp*Pi*d_evp*L_final_evp*LMTD_evp;
LMTD_evp=4[K];
h_bar_evp=Flow_evp_avg(Fluid$, T_evp_sat, G_evp, d_p_1, x_in, x_out, q_evp, 'Horizontal');
G_evp=m_s/A_p_1;
Call External_Flow_Cylinder(Fluid$, T_inf_evp, T_surface_evp,  P_water, u_inf, D_outer_pipe: F_d\L,
h_water)
T_inf_evp=283;
T_surface_evp=292;
q_evp=(1/((1/(h_bar_evp*(Pi*d_3*L_evp)))+((ln(d_p_1/d_3))/(2*Pi*k*L_evp))+(1/(h_water*Pi*d_p_1*L_evp)
)))*LMTD_evp;
h_cond_1=Enthalpy(R134a,T=T_c_s,x=0);
h_cond_2=Enthalpy(R134a,T=T_c_s,x=1);
h_fg_cond=h_cond_2-h_cond_1;
m_t=m_s+m_p;
q_cond=m_t*h_fg_evp;
q_cond=U_overall_cond*Pi*d_p_1*L_final_cond*LMTD_cond;
LMTD_cond=5[K];
h_bar_cond=Flow_cond_avg(Fluid$, T_cond_sat, G_cond, d_p_1, x_in, x_out, q_cond, 'Horizontal');
Call External_Flow_Cylinder(Fluid$, T_inf_cond, T_surface_cond,  P_water, u_inf, D_outer_pipe: F_d\L,
h_water,
T_inf=308;
T_surface=303;
q_cond=(1/((1/(h_bar_cond*(Pi*d_3*L_cond)))+((ln(d_p_1/d_3))/(2*Pi*k*L_cond))+(1/(h_water*Pi*d_p_1*L
_cond))))*LMTD_cond;
"{End of Program}"
```

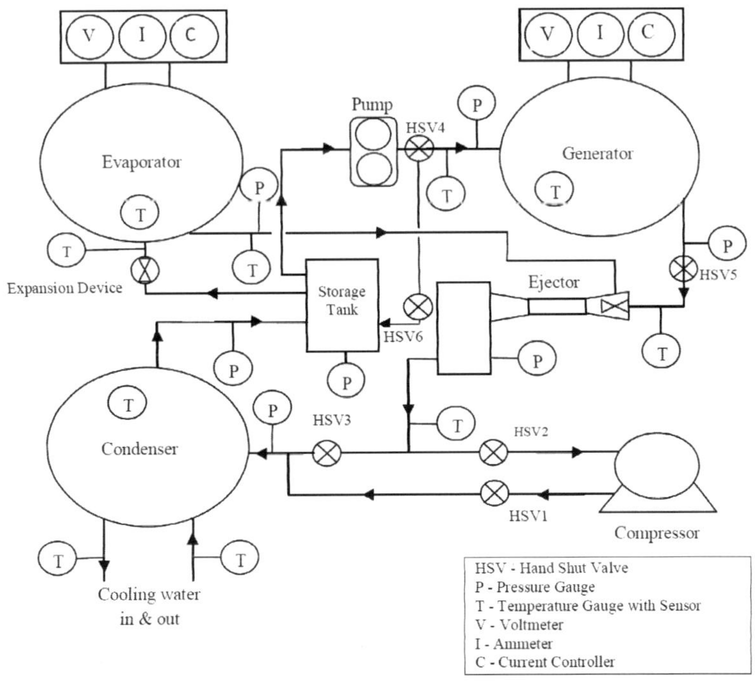

Figure 4.3 Schematic diagram of ERS [6]

3. Experimental test bench

An experimental test rig has been fabricated using R-134a refrigerant. The actual photograph of the set-up is shown in figure 4.2. The experimental setup consists of four main components: generator, condenser, evaporator, and ejector. The test facility is having various measuring devices i.e., , temperature sensors, pressure gauges, ammeter, voltmeter, indicator, etc.

3.1 Heat exchangers

The heat exchangers i.e., evaporator, condenser and generator are designed in tube in shell type heat exchangers. A 6.35mm diameter Copper tube in coiled shape has been used for the fabrication of the condenser and the generator, but the evaporator has 9.52mm diameter copper tube. The length of the copper tube in the evaporator, generator, and the condenser is 15.24, 24.384 and 30.48meters respectively. These copper tubes are spiraled and then placed in three insulated containers/shells. The copper tube has been used for the refrigerant, and shell for the external fluid in all the heat exchangers. A 4.5kW electric heater with a current controller has been placed in the generator to provide heat to the external fluid, which is 40% concentrated glycol solution. A mechanical recirculating pump is used as a stirrer fitted in the generator. It recirculates the glycol solution to make uniform temperature distribution among the container. To minimize the heat interaction between the heat exchangers and the surrounding, 9mm thick industrial insulation is used to insulate all the connecting pipes used in the experimental test facility. The shell of the evaporator has a known quantity of water, which was continuously cooled by the refrigerant flowing in the coils of the evaporator. An electric resistance heater of 3kW capacity with a current controller has been used to provide the heating effect, same as the cooling effect produced in the evaporator. This way the temperature of water in the evaporator shell does not change and so the energy consumed by the heater of the evaporator is reasonably assumed to be equal to the cooling effect produced by the ERS. The uniform temperature of the water present in the evaporator is maintained by a mechanical stirrer. To remove the latent heat of the refrigerant in the condenser fresh water at a constant temperature is circulated. And after gaining the heat from the condenser this water is drained. The liquid refrigerant inlet in the evaporator and the generator has been provided at the bottom of the coil, but in the condenser; as the refrigerant is in vapor phase at its inlet, it was provided at the top of the coil.

3.2 Ejector

The ejector is made up of four different parts i.e., primary nozzle, constant area section, suction chamber and diffuser. These parts are interchangeable and have been made of brass metal. To assemble these parts standard threaded connections have been used to make a complete ejector. The convergent-divergent supersonic primary nozzle made-up of brass material, perfectly in line with CAS has been placed facing the center of the constant area section. There is total three primary nozzles, all having the same inlet area (A_{p1}), but different throat (A_{p2}) and exit (A_{p3}) areas. As the throat and the exit areas of the primary/motive nozzle are different, the nozzle divergence angle of all the three nozzles is also different. The constant area section (CAS) of the ejector has been made up of two different diameters and lengths. Dimensions of all the six ejectors used in the investigation are shown in table 1.

3.3 Measuring instruments and supporting devices

RTD temperature sensor having accuracy ±0.1°C have been used at various locations to measure the temperature. Bourdon-tube type pressure gauges accurate up to ±6kPa have been fitted to measure pressure at different locations in the facility. Voltage and current supplied to the compressor, and the heating elements in the generator and the evaporator are measured using analog voltmeters and ammeters with an uncertainty of ±0.5V and ±0.05A respectively. Total uncertainty of the experimental setup in the determination of COP of the system has been found to be less than 5%. The details of the measuring instruments used in the experimental setup are shown in table 2.

3.4 Experimental procedure and the operating parameters

The atmospheric temperature in hot weather conditions is generally higher than the saturation temperature of the refrigerant in the setup in off condition; hence the refrigerant in the set up exists in vapor phase at the time of starting. The liquid pump installed after the condenser is never designed to work with the vapor, so an auxiliary compressor has been used to increase pressure in the condenser to condense the vapor.

Table 1 Geometries of ejector used in the experiments

Sr. No	Ejector	Primary nozzle geometry			Secondary nozzle geometry			Area Ratio = (A_4/A_{p2})
		Inlet Area (A_{p1}) (mm^2)	Throat Area (A_{p2}) (mm^2)	Exit Area (A_{p3}) (mm^2)	CAS Area (A_4) (mm^2)	CAS's Length (L) (mm)	(NXP) (mm)	
1	I	31.7	3.14	10.3	31.7	80	2.54	10.08
2	II	31.7	3.14	10.3	31.7	120	2.54	10.08
3	III	31.7	4.91	15.9	31.7	80	2.54	6.45
4	IV	31.7	4.91	15.9	31.7	120	2.54	6.45
5	V	31.7	7.07	20.6	67.2	80	2.54	9.51
6	VI	31.7	7.07	20.6	67.2	120	2.54	9.51

Table 2 Instruments used in the experimental setup

Sr. No.	Parameter	Instrument type	Unit	Uncertainty	Operating Range
1.	Generator/ condenser pressure	Bourdon-tube type pressure gauge	PSI	± 2 PSI	0 – 500 PSI (Gauge pressure)
2.	Evaporator pressure	Bourdon-tube type pressure gauge	PSI	± 2 PSI	0 to 250 PSI (Absolute pressure)
3.	Temperature	Digital display unit with RTD sensor	°C	± 0.1°C	-50°C to 200°C
4.	Voltage	Analog	Volts	± 0.5 V	0 – 300 V, 50 Hz
5.	Current	Analog	Ampere	± 0.05 A	0 – 20 A, 50 Hz

As the pump started receiving liquid refrigerant, the auxiliary compressor was switched off from the system. Figure 4.3 shows a schematic diagram of the experimental setup. The setup takes approximately an hour to maintain the steady-state conditions with refrigerant R134a. The complete experimental procedure is explained using a flow chart as shown in the figure 4.4. The condensed refrigerant has been pumped to the generator where it got converted into high-pressure vapor refrigerant. The glycol solution in the generator has been maintained at 90°C and the temperature of the vapor refrigerant coming out of the generator has been maintained at 84°C. Superheating of 4°C has been ensured in all the experiments to consider heat loss in the connecting pipe to the ejector. The temperature of refrigerant in the evaporator has been varied from 10 to 17.5°C in a gap of 2.5°C. The temperature of the condenser was maintained by adjusting the mass flow rate of the tap water circulated through it and the sub-cooling of 1°C has been secured in the condenser for all the experiments.

4. Conclusion

The paper presents a 1-D mathematical model for designing and finding the geometries of various heat exchangers to be used in an ERS. The design of the heat exchanger is based on the geometry of the ejector calculated in the previous research paper of the sequence.

Acknowledgement

This research is funded under the "**Visvesvaraya Research Promotion Scheme**" funded by Dr. A. P. J. Abdul Kalam Technical University, Lucknow, India.

References

[1] Chunnanond K and Aphornrnratana S, Ejectors: Applications in refrigeration technology, Renewable and sustainable energy reviews,vol. 8, pp. 129-155, 2004.
[2] Eames IW, Aphornratana S and Haider H, A theoretical and experimental study of a small-scale steam jet refrigerator, International journal of refrigeration, vol. 18, no. 6, pp. 378-386, 1995.
[3] Aphornratana S and Eames IW, A small capacity steam-ejector refrigerator: experimental investigation of a system using ejector with movable primary nozzle, International journal of refrigeration, vol. 20, no. 5, pp. 352-358, 1997.
[4] Chunnanond K and Aphornratana S, An experimental investigation of a steam ejector refrigerator: the analysis of the pressure profile along the ejector, Applied thermal engineering, vol.24, pp. 311-322, 2004.
[5] Rungtrakoon N, Aphornratana S and Sriveerakul T, Experimental studies of a steam jet refrigeration cycle: Effect of the primary nozzle geometry to system performance, Experimental thermal and fluid science, vol. 35, pp. 676-683, 2011.
[6] Kumar V and Sachdeva G, 1-D model for finding geometry of a single phase ejector, Energy, vol. 165, vol. 75-92, 2018.

Performance analysis of ERS using R134a - An experimental investigation (Part 3) – Experimental results

Vikas Kumar[1*], Sandeep Tiwari[1], Gulshan Sachdeva[2] and Vaibhav Jain[3]

[1]Mechanical Engineering Department, Krishna Engineering College, Ghaziabad (India)
Corresponding author Email: vikas.burman21@gmail.com.
[2]Mechanical Engineering Department, National Institute of Technology, Kurukshetra (India)
E-mail: gulshan4you@gmail.com
[3]Mechanical Engineering Department, MAIT, New Delhi (India)
Email: vaibhavursathi@gmail.com

Abstract. The paper presents the experimental results of the ERS using R134a as a refrigerant. The performance analysis of the single-phase ejector is already analysed in the previous research paper of the series. Additionally, the design and dimension of the ERS is also calculated by the second research paper of the series. In this paper, the experimental results of the fabricated experimental setup are presented for various operating conditions.

Keywords: Ejector, experimental, COP.

1. Introduction

This paper is the continuation of the previous two research, this research paper contributes the experimental procedure and results of ERS. It also exhibits the fabrication of the experimental setup. A few researchers had also analyzed the ERS experimentally using different refrigerants. Figure 1 shows the photograph of the experimental setup of the ERS which is designed and analyzed in the previous two research papers.

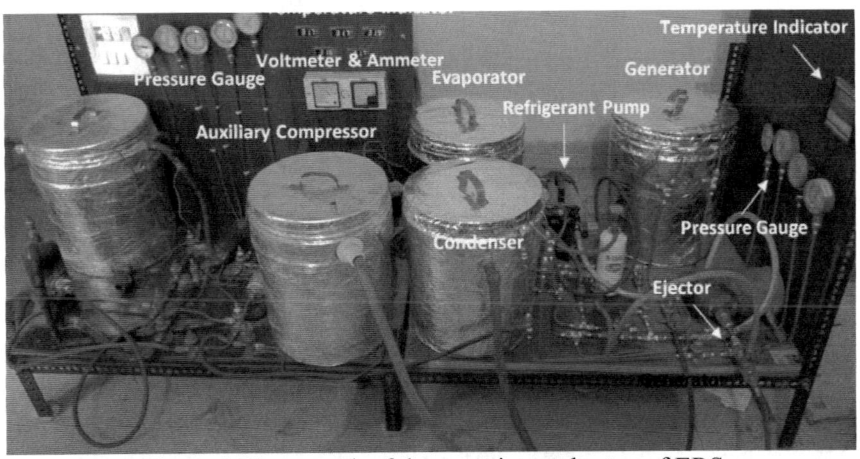

Figure 1 Photograph of the experimental setup of ERS

2. Results and discussions of the experimental observations

Performance comparison of the ERS has been carried out for six geometries of the ejector for various condenser and evaporator temperatures. Energy consumed by the liquid pump has been considered negligible in comparison to the energy required in the generator and is thus neglected. Therefore, COP of the ERS is evaluated as under:

$$COP = \frac{refrigeration\ effect}{Rate\ of\ heat\ supplied\ to\ the\ generator} \quad (1)$$

The refrigeration effect has been measured in terms of the voltage and amperage drawn by the heater. Similarly, rate of heat supplied to the glycol solution in the generator has been calculated in terms of electrical energy drawn by the heater of the generator. The performance of the ERS at the evaporator temperatures of 10, 12.5, 15 and 17.5°C has been evaluated for various condenser conditions. The co-efficient of performance calculated by using the ejector 'A' of area-ratio 10.0806 and CAS length 80mm is shown in the figure 2. The ejector 'A' yield COP of 0.50 at 17.5°C evaporator temperature and 26.8°C condenser temperature. This COP remained almost constant up to 30.4°C condenser temperature. The COP was found to be higher at 17.5°C evaporator temperature in comparison to 10°C. Double choking phenomenon, which is the basic characteristic of the single-phase ejector i.e., COP remains independent to the condenser pressure up to the critical pressure, can be seen in this figure. When the condenser pressure was increased above the critical condenser pressure, the ejector started working in single choking mode and the performance of the system was decreased significantly. The geometry of ejector 'A' provided critical condenser temperature of 32.6°C at 17.5°C evaporator temperature which decreased to 29.5°C at 10° evaporator temperature. It is concluded that the operating range and the COP of ERS increases with the increase in evaporator temperature.

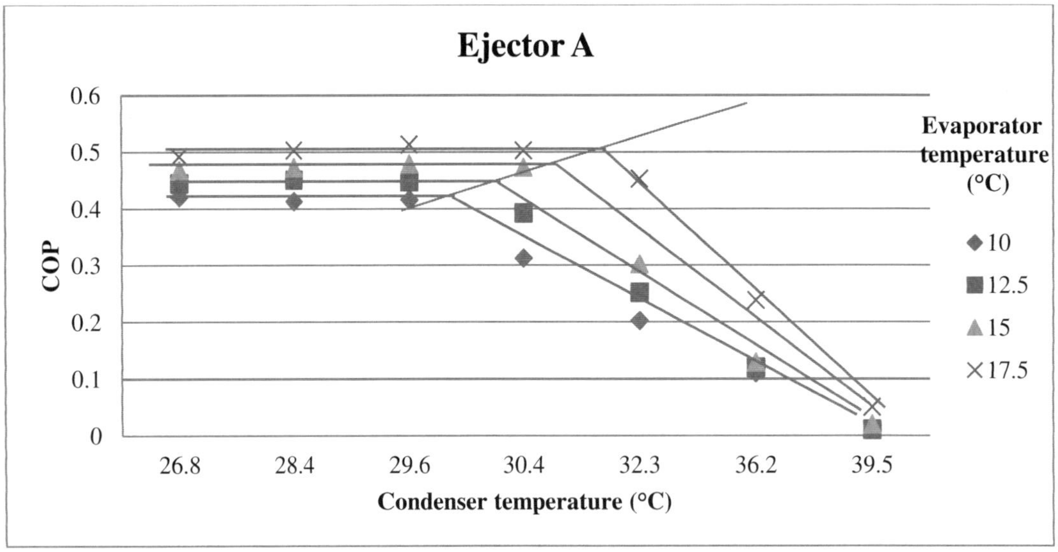

Figure 2 Variation of coefficient of performance with condenser temperature (°C) for ejector 'A' at various evaporator temperatures.

Ejector geometry 'B' is made of same throat and CAS diameter as of ejector 'A', so ejector 'B' is of the same area ratio 10.0806, but the length of CAS is 120mm, which is higher than that of ejector 'A'. Figure 3 shows similar trends, but with the lower coefficient of performance. Longer length of CAS marginally decreased the performance of the system because it increased the frictional losses inside the ejector. Figures 4 to 7 show similar trends in the performance with the ejector geometries 'C' to 'F'. An inclined straight line can be drawn by joining the critical condenser temperatures obtained at different evaporator temperatures. The slope of this line affects the working range of the ejector. If the slope is higher, decrease in the evaporator temperature does not decrease the condenser working pressure significantly. In other words, the ejector with higher slope should be used for having lower evaporator temperatures, but if the slope of the critical condenser line is less, it should be used with higher evaporator temperature to have more working range. This critical condenser curve suggests the use of ejector's geometry for the optimal performance. It is found that this curve can be used to find the critical condenser temperature for other evaporator temperatures, which are not considered in the study. All these experiments have been performed at fixed generator temperature of 84°C. A superheating of 4°C is used in all the experiments to compensate heat loss from the connecting pipe. The generator pressure has been maintained at 2520.96kPa that is the saturation pressure at 80°C temperature.

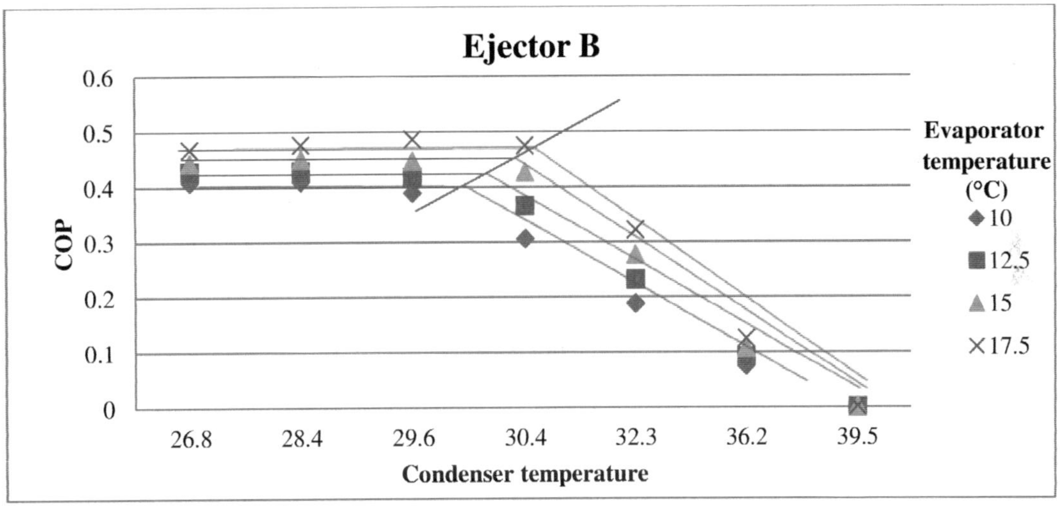

Figure 3 Variation of coefficient of performance with condenser temperature (°C) for ejector 'B' at various evaporator temperatures.

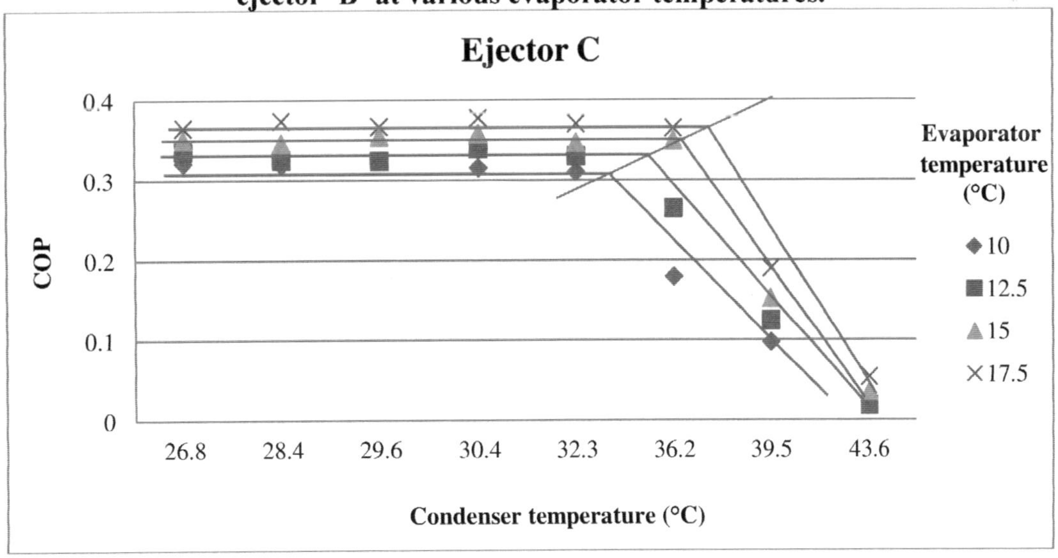

Figure 4 Variation of coefficient of performance with condenser temperature (°C) for ejector 'C' at various evaporator temperatures.

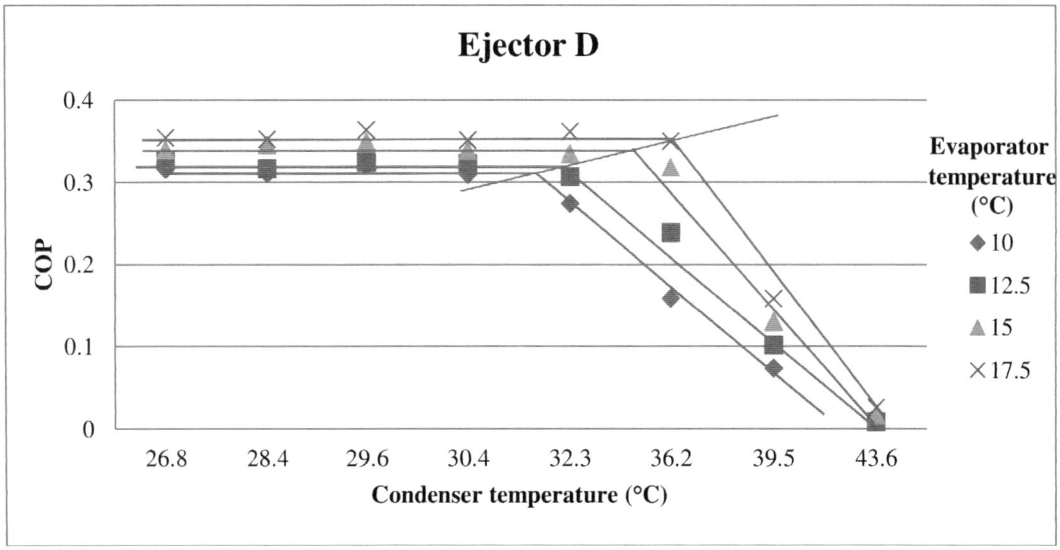

Figure 5 Variation of coefficient of performance with condenser temperature (°C) for ejector 'D' at various evaporator temperatures.

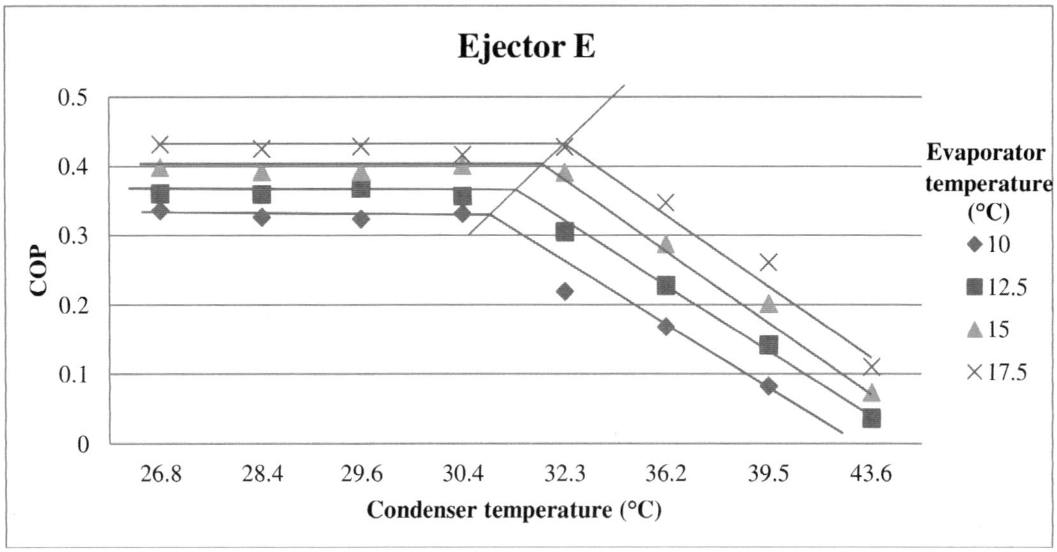

Figure 6 Variation of coefficient of performance with condenser temperature (°C) for ejector 'E' at various evaporator temperatures.

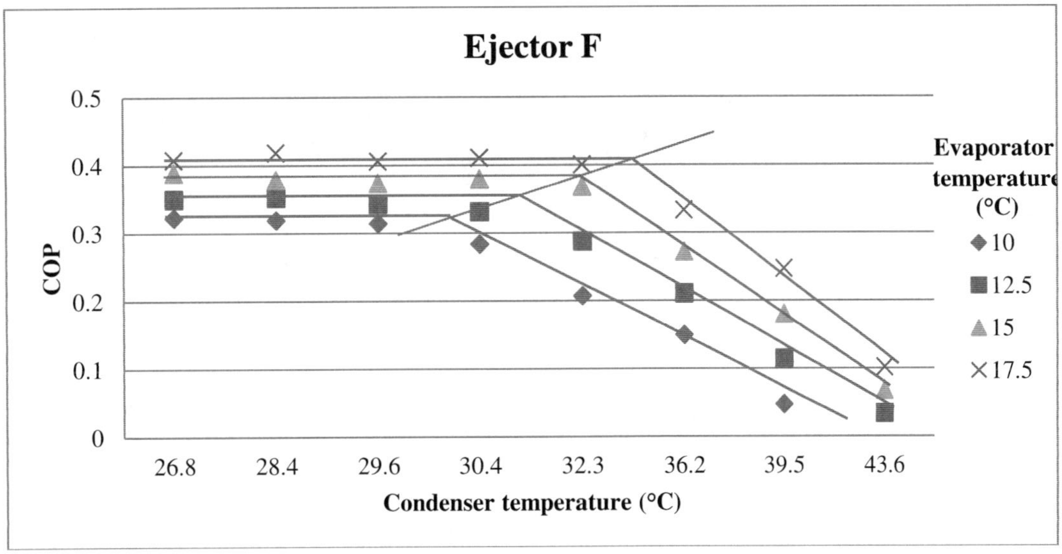

Figure 7 Variation of Coefficient of performance with condenser temperature (°C) for ejector 'F' at various evaporator temperatures.

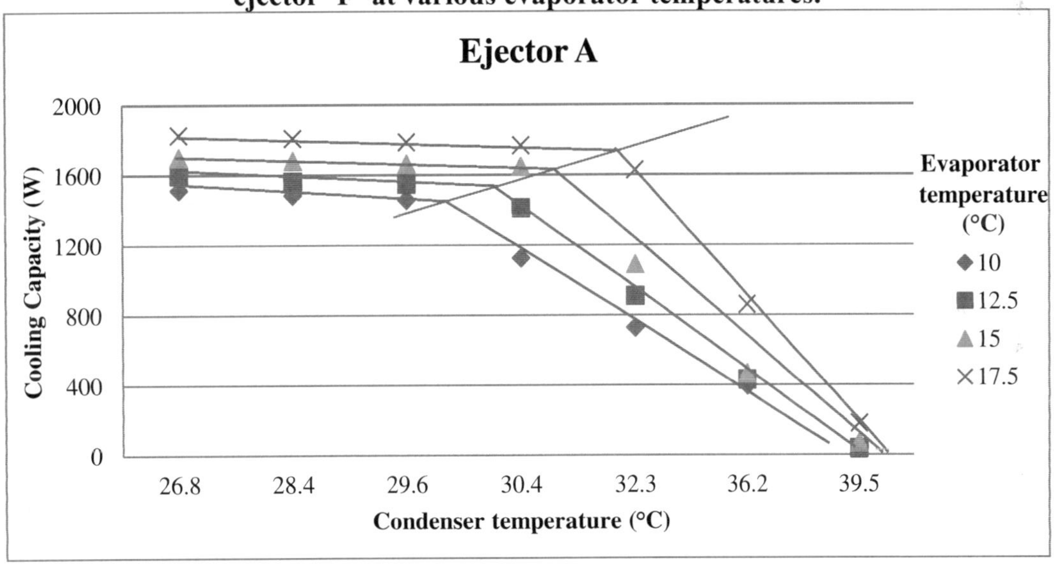

Figure 8 Variation of cooling capacity (W) with condenser pressure (°C) for ejector 'A' for various evaporator temperatures.

Figure 8 shows the cooling capacity produced in the system using ejector 'A' at various condenser and evaporator temperatures. The cooling capacity remained almost constant with the increase in temperature of the condenser up to the critical condenser temperature. In double choking mode, the increase in condenser temperature doesn't change the mass flow rate of the primary and secondary fluids, and therefore, the cooling capacity remains constant. Then after, the system started operating in single choking mode in which the back pressure to the ejector directly affected the mass flow rate of the entrained fluid and thus sharply decreased the cooling capacity. The maximum cooling capacity of 1828.68W and the critical condenser temperature of 32.3°C were achieved at 17.5°C evaporator

IOP Conf. Series: Materials Science and Engineering 1104 (2021) 012034 doi:10.1088/1757-899X/1104/1/012034

temperature. Decrease in the evaporator temperature from 17.5°C to 10°C reduced the cooling capacity and the critical condenser temperature of the system from 1828.68W to 1514.88W and 32.3°C to 29.8°C respectively. Figure 9 shows similar trends with the ejector 'B', but with the lower cooling capacity and critical condenser temperature. Longer length of CAS marginally decreased the cooling capacity of the system, because it increased the frictional losses inside the ejector. The frictional loss decreased the pressure of the mixed fluid at its exit and thus the critical condenser temperature.

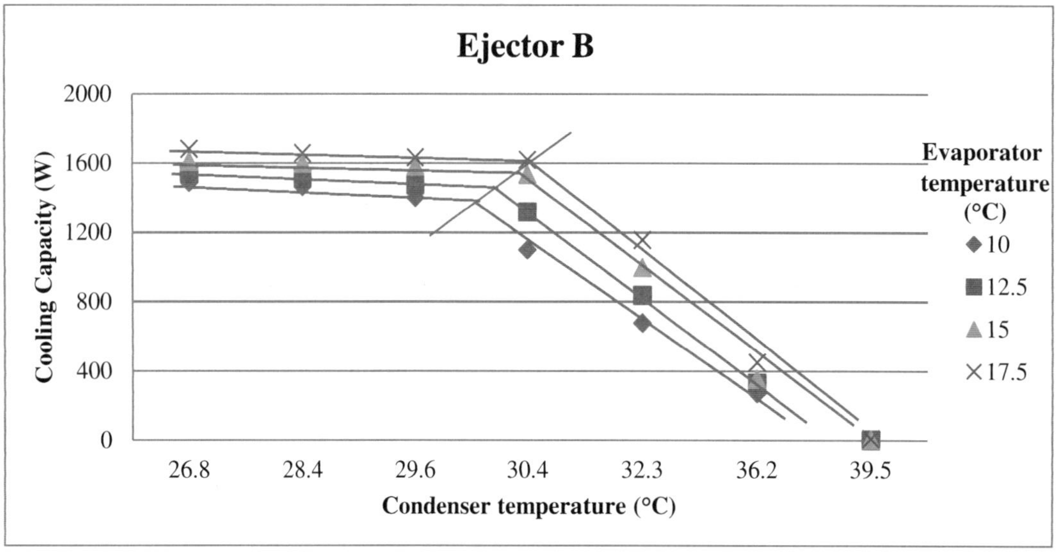

Figure 9 Variation of cooling capacity (W) with condenser pressure (°C) for ejector 'B' for various evaporator temperatures.

The thermal energy required in the generator depends upon the size of the primary nozzle used in the ejector. The ejectors 'A' and "B' has the primary nozzle of same size and requires 3600W heat energy in the generator. Similarly, for the ejector 'C' & 'D' and 'E' & 'F' it is 4000 and 4200W respectively.

Pressure lift ratio in an ejector refrigeration system is the ratio of critical condenser pressure to the evaporator pressure. It can be seen from figure 10 that the pressure lift ratio is higher at lower evaporator temperature. Lowering the evaporator temperature decreases the critical condenser temperature, but the rate of decrease of critical condenser temperature is less than that of evaporator temperature, therefore, the pressure lift ratio is found to increase with the decrease in evaporator temperature. The maximum and minimum value of pressure lift ratio for ejector 'A' is 1.83 and 1.55 at evaporator temperatures 10°C and 17.5°C respectively.

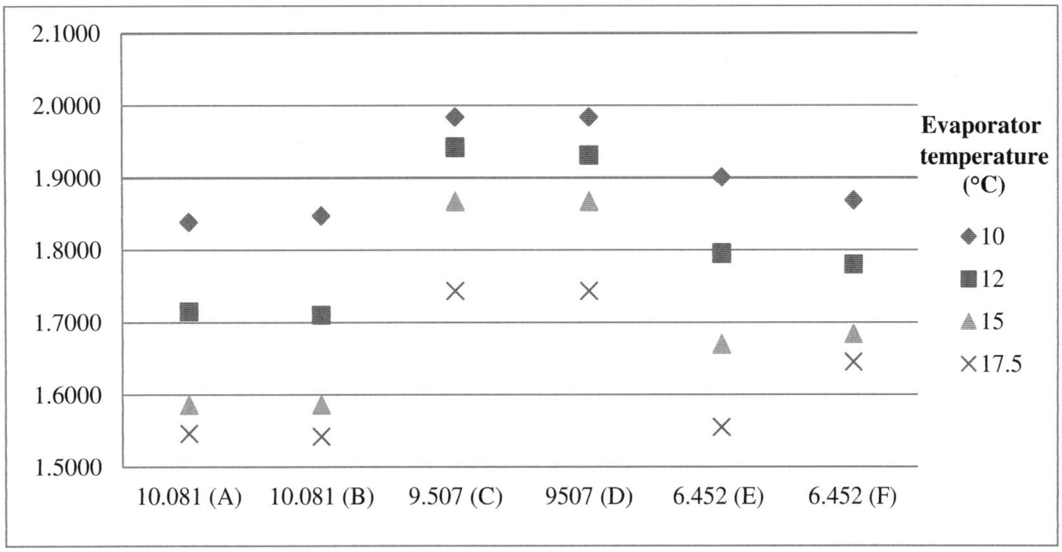

Figure 10 Variation of pressure ratio with respect to area ratio of all ejectors for different evaporator temperatures

3. Conclusion

The paper presents the experimental findings of the ERS experimental setup working on R134a refrigerant. The design and the geometry of the ejectors and the heat exchangers are already found in the previous research paper of the series. They are also validated by comparing the results of the Huang et al. [1]. The experimental results show that larger throat of the primary nozzle increases the critical condenser temperature of the ejector. Thus, the working range of the ERS also increases. But on the contrary, the COP of the system decreases and vice-versa.

Acknowledgement

This research is funded under the "**Visvesvaraya Research Promotion Scheme**" funded by Dr. A. P. J. Abdul Kalam Technical University, Lucknow, India.

References

[1] Huang BJ, Chang JM, Wang CP and Petrenko VA, A 1-D analysis of ejector performance, International Journal of Refrigeration, vol. 22, pp. 354-364, 1999.

[2] Chou SK, Yang PR, Yap C, Maximum mass flow ratio due to secondary flow choking in an ejector refrigeration system, International journal of refrigeration, vol. 24, pp. 486-499, 2001.

[3] Kumar V and Sachdeva G, 1-D model for finding geometry of a single phase ejector, Energy, vol. 165, vol. 75-92, 2018.

IOP Conf. Series: Materials Science and Engineering 1104 (2021) 012035 doi:10.1088/1757-899X/1104/1/012035

Exergy analysis of heat assisted ejector refrigeration system using R1234yf

Bharat Sharma[1], Gulshan Sachdeva[1] and Vikas Kumar[2]

[1]Department of Mechanical Engineering, National Institute of Technology Kurukshetra, Haryana, India.
Corresponding author E-mail: bharat_61900103@nitkkr.ac.in
[2]Department of Mechanical Engineering, Krishna Engineering College, Ghaziabad, India.

Abstract. An ejector refrigeration system (ERS) is introduced which operates on low-grade energy like solar energy, industrial waste heat etc. In this paper, the performance parameters and the conventional exergy analysis of ejector refrigeration system is measure with reference to ambient conditions. Exergy analysis was applied independently to each component of the system. The refrigerant R1234yf which has low Ozone depletion potential (ODP) and Global depletion potential (GWP) has been used to analysed the exergy analysis with the help of Engineering Equation solver software. The results shows that maximum exergy destruction in generator followed by ejector and other components.

Keywords: Ejector, Exergy analysis, refrigeration.

1. Introduction

As the economy of the world grows at fast rate, the living standard of the peoples also change which results in increase the demand of cool-fresh air used for refrigeration. Refrigerator-freezers are used for food and milk storage applications which uses the large amount of electricity. In this Ejector refrigeration system (ERS) single phase (vapor) ejector is mostly used [1] and works by using the low - grade waste energy for refrigeration. Earlier ejector is used as jet refrigeration system [2] but now it is most focus topic for research. Abdulateef et al. [3] explained the different technology used in solar-driven ejector refrigeration system and provided the summary of the various option for solar-driven cooling systems with corresponding refrigerants. Yan et. al. [4] works on hot pumped water which was driven by solar ejector compression heat pump system and concluded that ejector generated maximum exergy destruction by amount of 25.7% of the system. Performance parameter of the ERS with different refrigerants like R600a, R245fa and R141b are evaluated by Chen. Et. al. [5].

The researchers have focused their attention on ERS to enhance the performance of the system and by using environment friendly refrigerants. Many researchers have analysed the behaviour of the ejector using ideal and real gas equations for wide range of operating conditions and using R113, R141b, R134a, R142b, R600a, steam, air, propane etc. [6, 7, 8, 9, 10,11,12,13,14,15]. Huang et al. [7] proposed and developed a one-dimensional gas dynamic equation based mathematical model for evaluating and analysing the performance of a single-phase ejector working under double choking mode for wide range of operating conditions using R141b as a refrigerant. They found the area-ratio is the most critical dimension of an ejector for obtaining desired condenser pressure. Kumar and Sachdeva [16] proposed a mathematical model for calculating complete geometry of a single-phase ejector working under double choking conditions. The desired operating conditions and performance are the input parameters to the

IOP Conf. Series: Materials Science and Engineering 1104 (2021) 012035 doi:10.1088/1757-899X/1104/1/012035

proposed model and the complete geometry of the ejector is the output. Arbel et al [17] considered mainly three factors mixing, kinetic and shock wave to showed the exergy irreversibility. Exergy loss in each component of solar-driven ERS by Alexis [18]

2. System description and assumptions

The schematic diagram of the ERS is shown by fig. 1 which comprises of evaporator, condenser, generator, ejector, refrigerant pump and an expansion valve. The generator, condenser, evaporator use as a heat exchanger and water use as external fluid for heat transfer. This system is as similar as the vapor compression refrigeration system, the only difference is the ejector and pump which replaces the compressor that consumes lot of power as compared to the pumps.

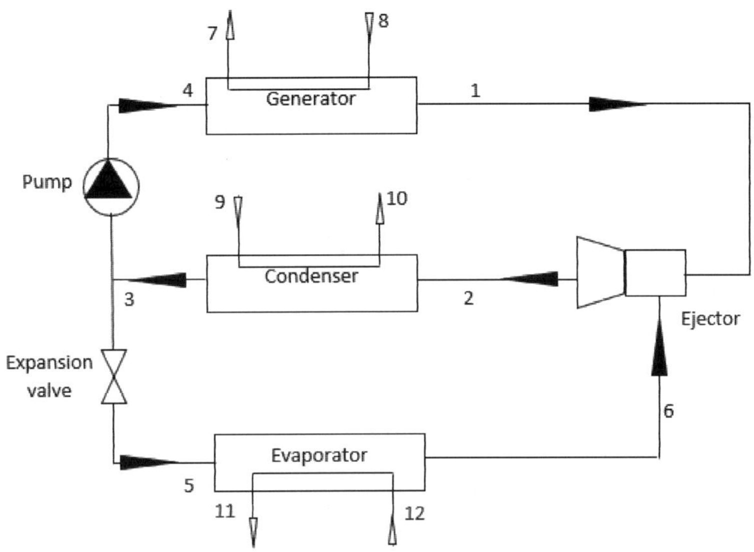

Figure 1. Schematic diagram of heat assisted ERS

An ejector in the ERS is the heart of system that works in place of compressor [19]. An ejector is comprising of four parts i.e., nozzle, suction section, mixing chamber and diffuser. The high-pressure motive flow refrigerant moves to the generator in liquid phase, where it takes heat from the waste heat and change into vapor. A high-pressure refrigerant expands through the primary nozzle at supersonic velocity and exits at high velocity and low pressure. Thus, creates suction effect for the low-pressure secondary refrigerant coming from the evaporator. Both refrigerant starts mixing according to constant pressure mixing in the constant area section and a normal shock develops at the end of mixing section because of the sudden back pressure developed by the condenser. Mixed refrigerants exit from diffuser section at subsonic velocity where it recovers the pressure, and discharge into condenser. The details related to the information of the system is available in Yapıcı et al [20]. Mixed refrigerant after leaving from condenser in condensed form where it separates into two different flows, one fluid flow goes through the pump and moves into generator and another part goes to the evaporator through expansion valve. Based on some assumptions, the p-h diagram is shown in Fig. 2.

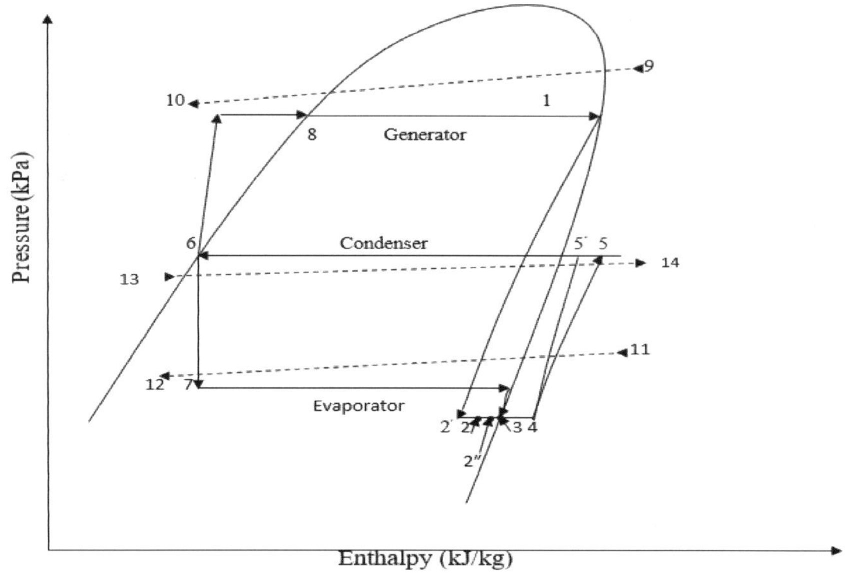

Figure 2. P-h diagram of an Ejector Refrigeration System

For analyzing the ERS system following assumptions are made:

- The steady state conditions are assumed in the ERS system and the heat and pressure losses in the heat exchangers is neglected.
- Steady state adiabatic one-dimensional fluid flow is assumed inside the ejector.
- Isentropic efficiencies in three section of the ejector (nozzle, diffuser and mixing) are 90%.
- Inlet and outlet temperature of external fluid are T_7= 100°C, T_8= 105°C, T_9= 27°C, T_{10}= 35°C, T_{11}= 10°C and T_{12}= 15°C.
- Neglect the frictional pressure loss effects.
- At inlet of the ejector, the flow is assumed to be stagnated.
- Consider the ambient temperature (T_o =25°C) and pressure (P_o =101.32kPa) as reference state.
- Cooling capacity is fixed at Q_{EV}=10kW.

3. Methodology

With the assumptions above, the energy analysis of ERS system is evaluated using the refrigerant R1234yf as working fluid proposed by (21).

3.1. COP and entrainment ratio of the ERS.

The capacity of the ERS is estimated by the entrainment ratio (μ). Entrainment ratio is the ratio of the entrained secondary mass flow rate ($\dot{m}_{eva}$) to the motive/primary mass flow rate ($\dot{m}_{gen}$). The entrainment ratio (μ) can be expressed as follows:

$$\mu = \frac{\dot{m}_{eva}}{\dot{m}_{gen}}$$

The ratio of cooling effect obtained to the heat energy supplied to the generator is known as Coefficient of performance (COP) and it can be written as follows:

$$COP = \frac{Q_{eva}}{Q_{gen}}$$

3.2. Conventional exergy analysis

During a process, when a system comes in equilibrium with its surroundings, the maximum possible useful work that brings this condition is know as exergy of the system [22]. The exergy consists of physical exergy ($\dot{E}^{PH}$), chemical exergy ($\dot{E}^{CH}$), kinetic exergy ($\dot{E}^{KE}$) and potential exergy ($\dot{E}^{PE}$) by neglecting other effects.

$$\dot{E} = \dot{E}^{PH} + \dot{E}^{CH} + \dot{E}^{KE} + \dot{E}^{PE}$$

Due to the absence of any chemical reaction the system, the chemical exergy is neglected. The kinetic exergy is assumed to be negligible and potential exergy is equivalent at the enter and exit of the ejector, the physical exergy ($\dot{E}^{PH}$) is written as:

$$\dot{E}_n = \dot{E}_n^{PH} = \dot{m}e_n^{PH} = \dot{m}\{(h_i - h_o) - T_o(s_i - s_o)\}$$

There is a further split of physical exergy based into thermal exergy (e^T) and mechanical exergy (e^M) due to temperature and pressure respectively.

$$e_n = e_n^{PH} = e_n^T + e_n^M$$

$$e_n = \{(h_n - h_{nx}) - T_o(s_n - s_{nx})\}_{P=c} + \{(h_{nx} - h_{no}) - T_o(s_{nx} - s_{no})\}_{T=c}$$

Where x is defined in above equation at the given pressure P and ambient temperature T_o.
For evaluating destruction exergy, the concept fuel-product is to calculate in terms of exergy and the expression in form of fuel-product for each component is given by [24].

$$\dot{E}_{F,K} = \dot{E}_{P,K} + \dot{E}_{D,K}$$

Where K refers to the k-th components in the system. $\dot{E}_{F,K}$ is term as fuel exergy, represents as maximum exergy to produce product exergy. $\dot{E}_{P,K}$ is term as product exergy which is the desired results attained by the components. Exergy destruction ($\dot{E}_{D,K}$) is the difference of the fuel and product exergy.
Exegetic efficiency (ε) is the ratio of the product and fuel exergy. It is expressed for each component as follows:

$$\varepsilon = \dot{E}_{P,K}/\dot{E}_{F,K}$$

Exergy destruction ratio (y) is ratio of the exergy destruction for each component to the total exergy destruction of the system. It is expressed as follows:

$$y = = \dot{E}_{D,K}/\dot{E}_{D,Total}$$

Table 1. Exergy calculation in each component of the ejector refrigeration system

Component	Fuel Exergy ($\dot{E}_{F,K}$)	Product Exergy($\dot{E}_{P,K}$)	Exergy Destruction ($\dot{E}_{D,K}$)
Generator (Gen)	$\dot{E}_{F,Gen} = \dot{m}_8(e_8 - e_7)$	$\dot{E}_{P,Gen} = \dot{m}_1(e_1 - e_4)$	$\dot{E}_{D,Gen} = \dot{E}_{F,Gen} - \dot{E}_{P,Gen}$
Evaporator (Eva)	$\dot{E}_{F,Eva} = \dot{m}_6(e_6 - e_5)$	$\dot{E}_{P,Eva} = \dot{m}_{12}(e_{12} - e_{11})$	$\dot{E}_{D,Eva} = \dot{E}_{F,Eva} - \dot{E}_{P,Eva}$
Condenser (Con)	$\dot{E}_{F,Con} = \dot{m}_2(e_2 - e_3)$	$\dot{E}_{P,Con} = \dot{m}_9(e_{10} - e_9)$	$\dot{E}_{D,Con} = \dot{E}_{E,Con} - \dot{E}_{P,Con}$
Ejector (Ejc)	$\dot{E}_{F,Ejc} = \dot{m}_2(e_1 - e_2)$	$\dot{E}_{P,Ejc} = \dot{m}_6(e_2 - e_6)$	$\dot{E}_{D,Ejc} = \dot{E}_{F,Ejc} - \dot{E}_{P,Ejc}$
Expansion Valve (EV)	$\dot{E}_{F,EV} = \dot{m}_5(e_3^M + e_3^T - e_5^M)$	$\dot{E}_{P,EV} = \dot{m}_5 e_3^T$	$\dot{E}_{D,EV} = \dot{E}_{F,EV} - \dot{E}_{P,EV}$
Pump (PUM)	$\dot{E}_{F,PUM} = W_{PUM}$	$\dot{E}_{P,PUM} = \dot{m}_4(e_4 - e_3)$	$\dot{E}_{D,PUM} = \dot{E}_{F,PUM} - \dot{E}_{P,PUM}$
System	$\dot{E}_{F,}Tot = \dot{E}_{F,}Gen + W_{PUM}$	$\dot{E}_{P,Tot} = \dot{E}_{P,EV}$	$\dot{E}_{D,Tot} = \sum \dot{E}_{D,K}$

4. Results and Discussion

All the results are calculated by the thermodynamic equations and properties of the refrigerant and external fluid are taken from the Engineering Equation Solver (EES). Thermodynamic properties at each position are present in the given table below.

Table 2. Thermodynamic properties at each position are present in the given table below.

Position	Refrigerant	$\dot{m}$ (kg/s)	P (kPa)	T (°C)	h (kJ/kg)	S (kJkg-k)	e^T(kJ/kg)	e^M(kJ/kg)	e (kJ/kg)
1	R1234yf	1.231	3080	90	393.6	1.577	21.16	40.05	61.21
2	R1234yf	2	907	35.5	383.9	1.604	4.657	38.91	43.56
3	R1234yf	2	907	35.5	247.2	1.163	38.91	0.607	38.3
4	R1234yf	1.231	3080	90	250	1.165	0.34	40.05	40.39
5	R1234yf	0.769	410.7	8	247.2	1.17	7.324	28.89	36.22
6	R1234yf	0.769	410.7	8	368.4	1.599	0.4804	28.89	29.37
7	Water	8.367	101.32	100	419.1	1.307	34.21	0.098	34.11
8	Water	8.367	101.32	105	440.2	1.363	38.59	0.098	38.49
9	Water	13.07	101.32	30	113.2	0.3949	0.032	0	0.032
10	Water	13.07	101.32	35	134.1	0.464	0.353	0	0.353
11	Water	4.454	101.32	10	42.09	0.151	1.602	0	1.602
12	Water	4.454	101.32	15	63.01	0.224	0.6962	0	0.6962

The below table shows the calculated value of the fuel-product exergy and destruction exergy which shows that more exergy destruction in generator followed by ejector and other components.

Table 3. Result obtained from the conventional exergy analysis

Component	Fuel Exergy ($\dot{E}_{F,K}$)	Product Exergy($\dot{E}_{P,K}$)	Exergy Destruction ($\dot{E}_{D,K}$)	Exegetic efficiency (ε) %	Exergy destruction ratio (y) %
Generator (Gen)	36.66	25.62	11.03	0.699	27.56
Evaporator (Eva)	5.265	4.032	1.233	0.765	13.14
Condenser (Con)	10.53	4.192	6.338	0.398	26.3
Ejector (Ejc)	21.71	10.92	10.79	0.5	54.2
Expansion Valve (EV)	7.237	5.635	1.602	0.7786	18.05
Pump (PUM)	3.38	2.56	0.82	0.757	8.44
System	40.04	4.03	36.01	0.1	89.93

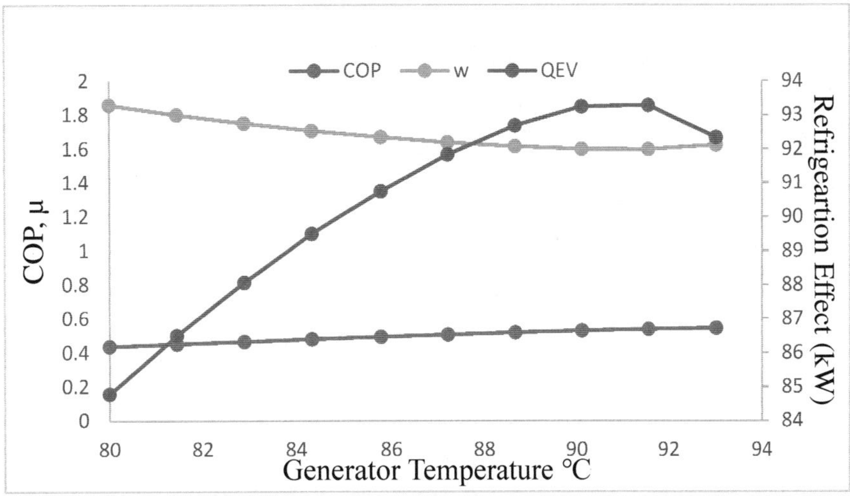

Figure 3. Effects of performance parameter with the variation of generator temperature.

Figure 3 shows the variation of generator temperature keeping other temperature constant, the refrigeration effect will increase from 80-93°C but decrease further. The COP will decrease and entrainment ratio will increase with increase of generator temperature. Figure 4 shows the difference between the generator and total exergy destruction with the variation of generator temperature. Both exergy destruction will decrease with variation of generator temperature.

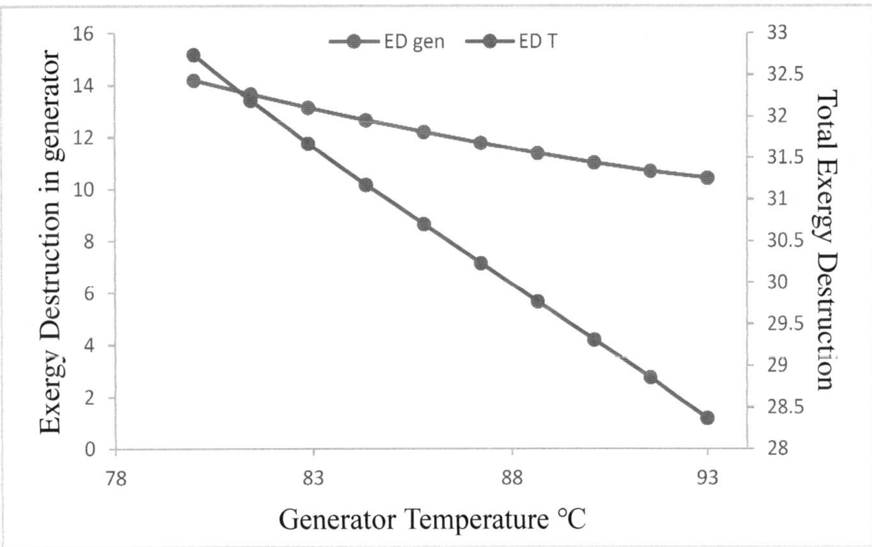

Figure 4. Effect of generator temperature on exergy destruction

IOP Conf. Series: Materials Science and Engineering 1104 (2021) 012035 doi:10.1088/1757-899X/1104/1/012035

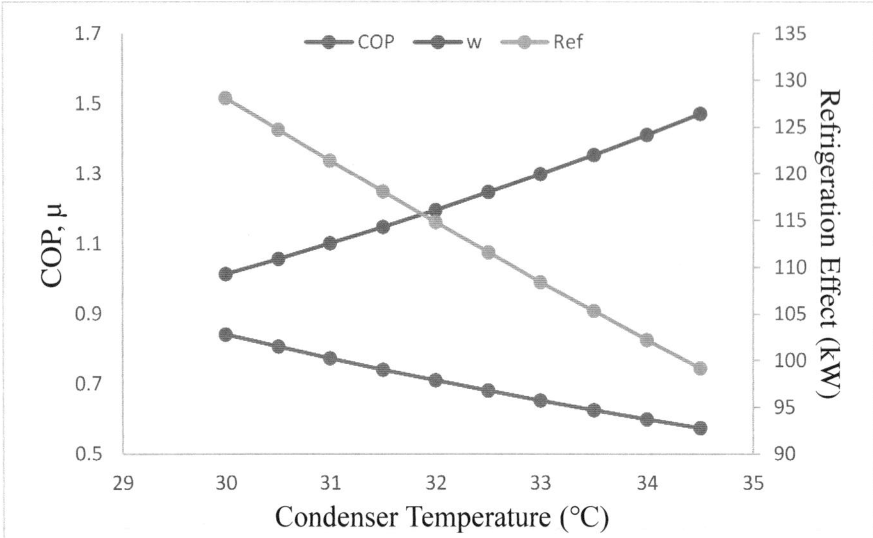

Figure 5. Effects of performance parameter with the variation of condenser temperature.

The performance parameter of the system with the variation of condenser temperature shown in figure 5 and 6, the entrainment ratio is increases with increase in condenser temperature between 30-35 °C but the COP and refrigeration effect will decrease because when temperature increase it will increase secondary mass flow rate. It also increases the exergy destruction of both system and condenser in below figure.

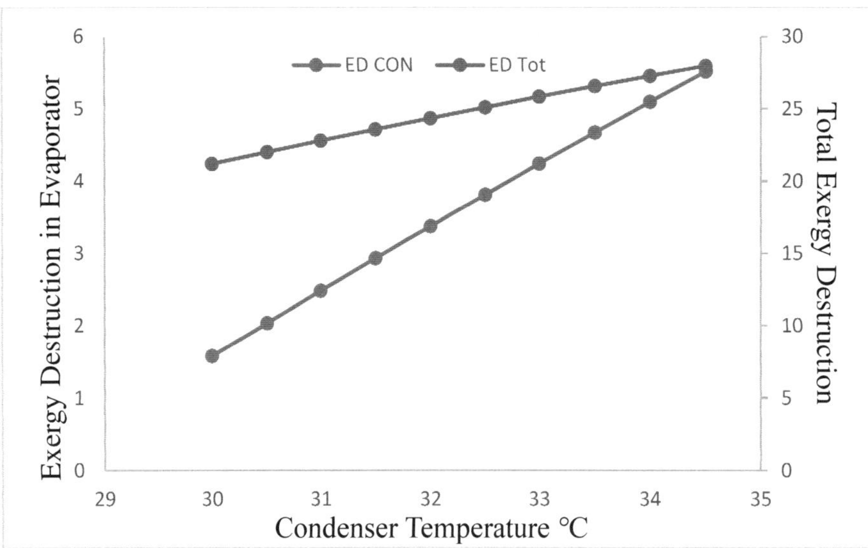

Figure 6. Effect of condenser temperature on exergy destruction

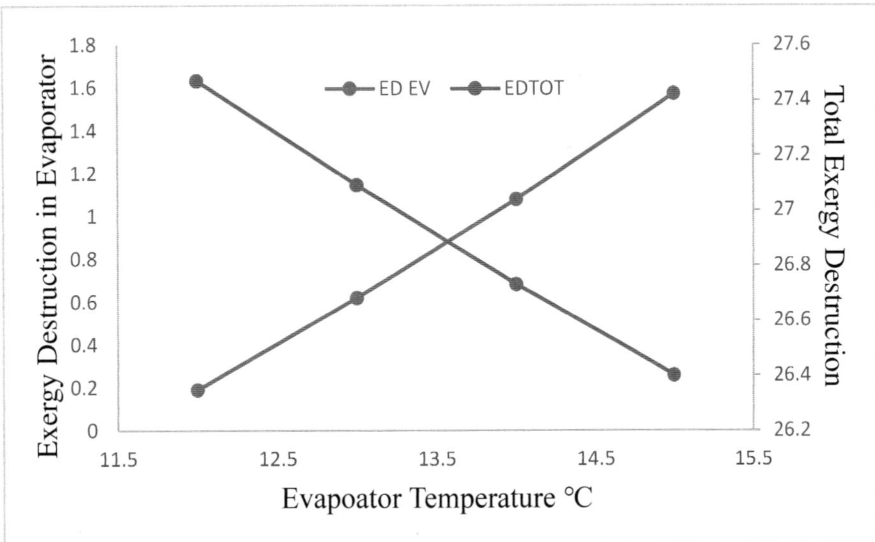

Figure 5. Effect of evaporator temperature on exergy destruction.

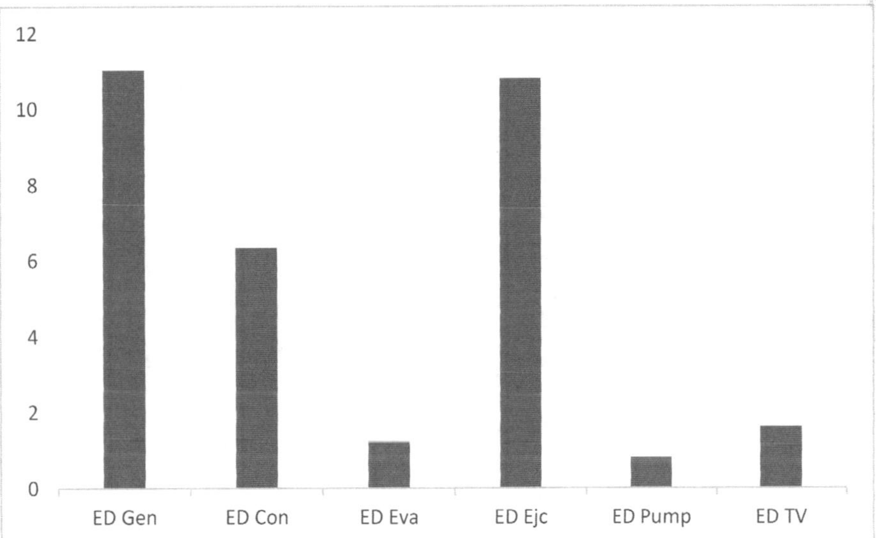

Figure 6. Results of exergy destruction in each component

The above figure 6. illustrates the exergy destruction in various component of the ERS system. It shows that maximum exergy destruction in generator then ejector and minimum in pump at generator temperature 93°C, condenser temperature 35°C and evaporator temperature 5°C.

Nomenclature

g	Acceleration due to gravity
h	Enthalpy
$\dot{m}$	Mass flow rate
Q	Heating Effect
R	Refrigeration Effect
V	Velocity
y	Exergy destruction ratio

Subscripts

Con	Condenser
d	Diffuser
e	Entrainment
Eva	Evaporator
Gen	Generator
n	Nozzle

Greek symbols

η	Efficiency
μ	Entrainment ratio

5. Conclusion

Conventional exergy analysis is used to analyze the ejector refrigeration system for a known operating condition. It is found from the analysis of conventional exergy that the highest exergy destruction occurs in generator followed by ejector and other components. The exergy destruction in heat exchanger like generator, evaporator and condenser can be reduced by alternate method but in ejector exergy destruction be more which cannot be decrease or eliminate by any process.

References

[1] ASHRAE Steam-Jet Refrigeration Equipment. In ASHRAE Equipment Handbook; ASHRAE: Atlanta, GA, USA, 1983; pp. 13.1–13.6.

[2] Decker, L.O. Consider the cold facts about steam-jet vacuum cooling. Chem. Eng. Prog. **1993**, 89, 1.

[3] Abdulateef, J.M.; Sopian, K.; Alghoul, M.A.; Sulaiman, M.Y. Review on solar-driven ejector refrigeration technologies. Renew. Sustain. Energy Rev. **2009**, 13, 1338–1349.

[4] Yan, G., *et al*., Energy and Exergy Efficiency Analysis of Solar Driven Ejector–Compressor Heat Pump Cycle, *Solar Energy*, 125 (2016), pp. 243–255

[5] Chen, J., *et al*., Parametric Analysis of Ejector Working Characteristics in the Refrigeration System, *Applied Thermal Engineering*, 69 (2014), pp. 130-142.

[6] Sun D, Comparative study of the performance of an ejector refrigeration cycle operating with various refrigerants, Energy conversion and management, vol. 40, pp. 873-884, 1999.

[7] Huang BJ, Chang JM, Wang CP and Petrenko VA, A 1-D analysis of ejector performance, International Journal of Refrigeration, vol. 22, pp. 354-364, 1999.

[8] Chen W, Liu M, Chong DT, Yan J, Little AB and Bartosiewicz Y, A –D model to predict ejector performance at critical and sub-critical operational regimes, International Journal of Refrigeration, vol. 36, pp. 1750-1761, 2013.

[9] He S, Li V and Wang RZ, Progress of mathematical modeling on ejectors, Renewable and sustainable energy reviews, vol. 13, pp. 1760-1780, 2009.

[10] Sharfi N and Boroomand M, An investigation of thermo-compressor design by analysis and experiment: part 1. Validation of numerical method, Energy conversation management, vol. 26, pp. 217–27, 2013.

[11] Tyagi KP and Murty KN, Ejector-compression systems for cooling utilizing low grade waste heat, Heat recovery systems, vol. 5, No. 6, pp. 545-550, 1985.

[12] Aly NH, Karameldin A and Shamloul MM, Modelling and simulation of steam jet ejectors, Desalination, vol. 123, pp. 1-8, 1999.

[13] Alexis GK and Katsanis JS, Performance characteristics of a methanol ejector refrigeration unit, Energy conversion and management, vol. 45, pp. 2729-2744, 2004.

[14] Huang BJ and Chang JM, Empirical correlation for ejector design, International journal of refrigeration, vol. 22, pp. 379-388, 1999.

[15] Ruangtrakoona N and Aphornratana S, Design of steam ejector in a refrigeration application

based on thermodynamic performance analysis, Sustainable Energy Technologies and Assessments, vol. 31, pp. 369–382, 2019.
[16] Kumar V and Sachdeva G, 1-D model for finding geometry of a single phase ejector, Energy, vol. 165, vol. 75-92, 2018.
[17] Arbel, A., *et al.*, Ejector Irreversibility Characteristics. *Journal of Fluids Engineering*, 125 (2003), pp. 121–129
[18] Alexis G.K., Exergy Analysis of Ejector-Refrigeration Cycle Using Water As Working Fluid, *International Journal of Energy Research*, 29 (2005), pp. 95–105.
[19] Chunnanond K and Aphornratana S 2004 Ejectors: applications in refrigeration technology Renew Sustain Energy Rev. 8 pp 129–55.
[20] Rafet Yapici, Experimental investigation of performance of vapor ejector refrigeration system using refrigerant R123, Energy Conversion and Management, 45(5) (2008) pp. 953-961.
[21] Sharma B et al 2019 *J. Phys.:* Conf. Sries **1290** 012128
[22] Chen J, Havtun H and Palm B, 2015 Conventional and advanced exergy analysis of an ejector refrigeration system, *Appl. Energy.* **144** pp 139-151.

IOP Conf. Series: Materials Science and Engineering 1104 (2021) 012036 doi:10.1088/1757-899X/1104/1/012036

Generative Design Method (GDM) – A State of Art

Rajneesh Jaisawal[1], Vandana Agrawal[2]

[1]Master's student, CAD&M, Dept. of Mechanical Engineering, Motilal Nehru National Institute of Technology (MNNIT) Allahabad, India
[2]Assistant Professor, Dept. of Mechanical Engineering, Motilal Nehru National Institute of Technology (MNNIT) Allahabad, India

Email: jaiswalrajneesh60@gmail.com , vandana@mnnit.ac.in

Abstract. This paper presents the review of generative design method (GDM). The main aim of generative design system is to create new design that produces number of feasible distinct design with the help of computing power and manufacturing capability. A generative design process has mainly three important components, which are a structured design framework, a method of generating dissimilarity, and a way of determining preferable outcomes. The aim of this paper is to review the different generative design method (GDM), its role in the product design and development especially in conceptual design phase, which is a time consuming, and complex design process, and to assist designer in creating innovative design of existing design. Based on the literature survey, the various classification involved in generative design method, and its applications in different fields have been discussed.

1. Introduction

In modern days, the industrial/engineering products design is important task for designer according to the increasing demand of costumer not only based on functional performance but also the appearance of the products in daily life. Therefore, in the initial product design process (i.e. also known as conceptual design stage) to explore alternative design by hand sketching is time-consuming process. Designer may not be able to predict all possible variation of shapes of the product and in this stage use of computer-aided design (i.e. CAD) is very limited; it is mainly utilized in later stage to analyze, validate and fabricate the design [1]. Here in the modern era, there are so many advancement in the field of optimization techniques, design simulation, parametric design technique, and extensive application of computer programming. Therefore, it allows design engineer a new way of exploring design as compared to old days [2]. Therefore, a generative design method came in to existence to provide distinct design/shape of the product based on the consumer preferences.

In any product development process, conceptual design stage is an important phase in an engineering design of product. This phase is complex one and large amount of time is required when we talk about the appearance of the product based on consumer needs for generating number of designs. Therefore, based on design requirements (i.e. design parameters, its range and geometric constraints) of the product, it is important to have system that should provide number of distinct optimal design/shapes in design space and the method is known as generative design method.

2. Applications

Generative design method was initially used in the area of Architectural design but it is now used in various applications, like industrial design, jewelry design, consumer product design etc. Its few applications and advantages are given below.

- It has application in conceptual or initial design phase as it provides number of design shapes. For example it can be used in ornament industry mainly in jewellery design to generate number of designs with less effort.
- It can be integrated with Design for manufacturing (DFM) software to ensure manufacturability.
- It enables the design engineer to explore multiple ideas and make better-informed decision during early design stage.
- It reduces the process time, as designer has to explore the different design by like traditional method like 2D drawing and hand sketching etc.

Due to its wide applications, various generative design methods have been proposed. In the present paper, a state of art of it is presented for the same. The paper is organized as follows: section 3 provides the literature review related to generative design method, and its classification. Section 4 gives the applications of different generative design techniques, Section 5 provides comparison of different technique and finally section 6 gives the conclusion.

3. Literature Review

A large volume of literature is available explaining the generative design method (GDM), its classification and applications in the various fields. In this paper, a review of literature has been carried out related to engineering applications. Here sub-section 3.1 gives the definition of generative design method and its major components.

3.1. Generative Design Method (GDM)

Sometimes, product design phase is critical and time-consuming process, which create burden on the designer to explore innovative design a case for example there may be when appearance is valuable to customer. In such cases, generative design method can be utilized for generating 2D profile as well as 3D CAD models.

Frazer [3] pioneered the concept of "generative design" in year 1970's. The main aim of generative design system is to create new design process that produces number of feasible distinct design with the help of computing power and manufacturing capability as given by Shea [4]. The number of design alternatives for the products may be in the form of 2D sketch, image, drawings or 3D parametric CAD models. Generative design process contains three major components (1) A structured design framework, (2) A method of generating dissimilarity, (3) A way of determining preferable outcomes.

3.2. Classification of GDM

The generative design method can be classified in five different categories based on the literature survey, which are constrained parametric search strategy [6,29], evolutionary algorithm [10-12,26,30], iterated local search (ILS) [7-9], shape grammar [13,15,16], and topology optimization tool [1,27,28]. Each of the method is described briefly in the following subsection as given below.

3.2.1. GDM based on constrained parametric search. Sivam Krish [6] proposed a generative design method to explore various design in conceptual design phase. This method is totally designer driven method and it does not limit imagination of designer, which is the major advantage of this technique. Overall proposed method can be understood from flowchart of steps involve in generative design process as shown in figure 1. Here the CAD based generative design system is presented where genetic model and geometric constraints replace the algorithm and source code respectively. Due to this reason, this proposed generative design system is more designer centric.

Major limitation of proposed GDM is that this is an exhaustive search method, as it requires multiple genotype to explore large design space because single genotype cannot cover entire design space.

One more method has been proposed by Bentley [29] based on this classification which is known as genetic algorithm designer (GADES) method which utilizes the above concept and genetic algorithm technique.

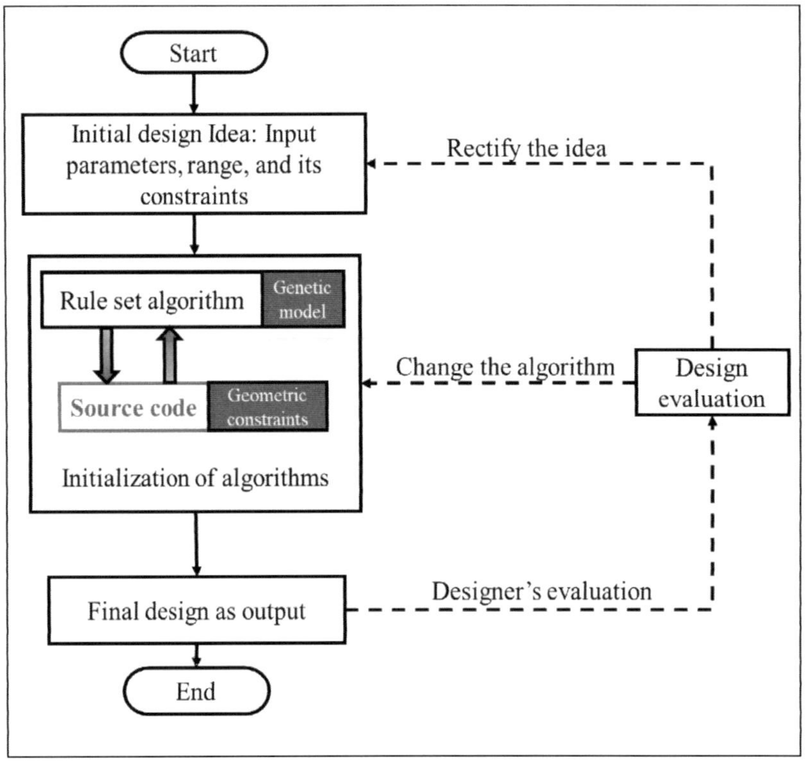

Figure 1. Flowchart of Generative design process.

3.2.2. GDM based on iterated local search (ILS). Khan et al. [7] proposed a sampling technique, which is an extended version of Latin hypercube sampling (ELHS) to explore the design alternative of the CAD models in constrained and high dimensional design space. In the ELHS technique, number of trials is same as number of interval and in any trail; only one design is taken from each sub-space based on LH-rule [7]. Among all designs (feasible and infeasible both), a feasible design is considered based on enumeration in each trial and one who provide the minimum cost value is selected. To generate distinct designs, a similarity constraint (i.e. similarity threshold) is utilized based on Euclidian distance. A candidate solution is selected based on one who has the minimum cost value and satisfy the LH-rule [7]. There is a limitation of this sampling technique that the designs obtained do not always provide the space filling property as it is dependent on input design.

Gunpinar et al. [8] proposed a technique, which is known as shape sampling approach to explore the various shape in the constraint shape space within pre-defined product specification. This technique is based on generative design method that utilizes particle tracing (PT) algorithm [23,24]. Here sampling quality is computed based on two factors which are space filling and computation time so it requires that a sample should have good space filling property and less computation time.

Dogan et al. [9] proposed a generative sampling system to generate number of 2D profiles based on given constraints within pre-defined design space. It proposes for 2D problem (i.e. image as input) which utilizes a sampling algorithm by programming for design space. Number of different samples are generated by minimizing Audze-Eglais potential energy [14] for having uniform distribution of samples. Distinct samples are generated while maintaining curve degree, continuity and concavity property. Samples are produced by modifying the profile curve and the sampling algorithm finds the ones within the bounds specified by user-provided similarity thresholds satisfying the conditions and constraints.

3.2.3. GDM based on evolutionary optimization algorithm. Khan et al. [10] proposed a sampling-TLBO technique (S-TLBO), which is generative design technique to generate all possible design option for CAD models in constrained and unconstrained design space within the pre-defined design constraints, so that designer or customer can think about all possible designs. This S-TLBO technique can be utilized for high dimensional problem with good space filling property and provides proper sampling within highly irregular environment of infeasible designs. S-TLBO technique is an extension of basic TLBO (Teaching learning based optimization i.e. evolutionary optimization technique) [22] from single optimal solution to multi-optimal solution. TLBO algorithm utilizes the philosophy of teaching and learning in the class, which is a popular among meta-heuristics optimization algorithms such as genetic algorithm (GA) [18], particle swarm optimization (PSO) [19], artificial bee colony (ABC) [20] and covariance matrix adaptation evolution strategy (CMA-ES) [21] due to effective performance and simplicity in use.

Khan et al. [11] proposed a generative design technique, which is known as space-filling-GDT (Sf-GDT) to generate optimal design alternative for CAD models in design space within pre-defined design constraints. In this work, Jaya algorithm [17] which is utilized for effective search strategy and advantage of using this algorithm that does not require any tuning of specific parameter as in the case of genetic algorithm. To produce distinct designs, a weighted grid method is integrated with proposed GDT. This proposed method is also able to explore hybrid design space, which consists both continuous and discrete parameters. In this research paper, an example of ceiling lamp is taken while considering both criteria space filling and non-collapsing feature to produce distinct optimal designs as shown in figure 2.

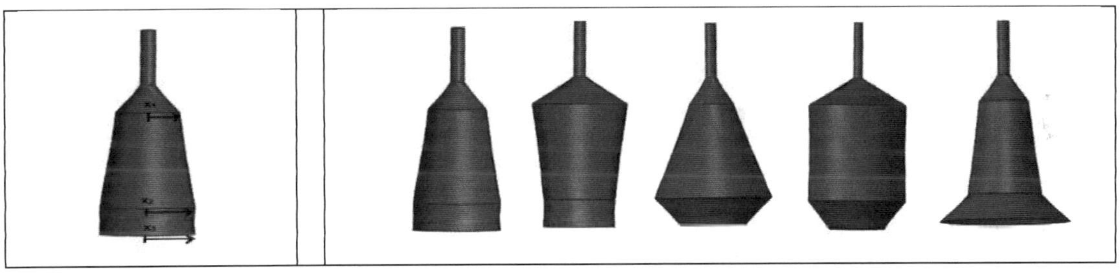

(a) (b)

Figure 2. Shapes generation of ceiling lamp of three-design parameter by Sf-GDT. (a) Input parametric CAD model. (b) Generated 3D CAD models.

Moriguchi et al. [12] proposed a generative design technique which is also known as SSA design sampler (SSA-DS) to generate the number of creative designs in the pre-determined design space. This is automatic search technique based on spatial simulated annealing (SSA) [25] to produce N number of designs in constrained/unconstrained design space. Here uniform distribution of generated designs (i.e. space filling criteria) is achieved by minimizing Audze-Eglais potential energy [14]. Several other research work has been carried out using this approach as [26, 30].

3.2.4. GDM based on shape grammar. Kielarova et al. [13] proposed a generative design system, which utilizes the shape grammar based GDT and a set of shape rules to explore various design and these set of shape rules are based on shape transformation rule. The proposed method was applied on jewellery ring design. However, this technique has major drawback that it requires an expert for defining set of transformation steps for each object separately.

Another approach using the concept of shape grammar and evolutionary algorithm by Kieralova et al. [15]. This method was applied on jewellery design of ring. Pugliese et al. [16] proposed a method, which is based on shape grammar and utilizes to extract the model of Harley-Davidson brand motorcycle.

3.2.5. GDM based on topology optimization tool. Kazi et al. [1] proposed a generative design tool (i.e. known as Dreamsketch tool) which is based on topology optimization to explore the various shapes using 2D digital sketching skills. Further topology optimization is done to get feasible optimized shape. The algorithm was applied on connecting rod using Autodesk Inventor. This topology optimization tool is based on minimization of stiffness while fixing the volume to certain value or fraction and minimization of weight under stress or displacement as constraints. Several other research work has been carried out using this approach [27, 28].

4. Applications of Different techniques

This section explains in brief the area where the various generative design method can be applied for analysis of different shapes of any product within certain defined constraint in initial design phase of product development process.

Here constrained parametric search based GDM is a designer driven technique to explore the solution space (or design space) to obtain number of different design by utilizing genotype. This proposed method is based on exhaustive search strategy due to maximum involvement of design engineer/designer and application is limited in daily usable products due to need of more time and difficult for normal person other than designer.

An evolutionary algorithm based GDM is automatic search strategy to explore entire design space within pre-defined customer specifications. This proposed generative design method can be utilized by a customer or designer easily and experts are not required. This technique uses evolutionary optimization algorithm such as genetic algorithm (GA) [18], Spatial simulated annealing (SSA) technique [25], teaching learning based optimization (TLBO) [22], Jaya algorithm [17], particle tracing (PT) algorithm [23,24] etc.

In shape grammar based GDM, shape grammar is used to create generative design technique for design exploration based on shape grammar rules. However, implementing shape grammar is still a problem for generalized problems in the present days. So fixed shape structure is considered via analysis of simple shape grammar for consequences analysis. Here it is proved that capability of shape grammar is limited by choosing finite fixed structure. Shape grammar is generally used for 2D design space and rarely used for 3D design exploration problems.

In topology optimization tool [5] based GDM, which has input as problem definition and output as a single or number of optimal solutions. This GDM utilizes mathematical methods, which optimally distribute the material within a pre-known design space. Its main objective function is to maximize or minimize the stiffness and minimize the weight while considering the volume fixed and stress as constraints. In the modern day, number of software are currently used by various industries on the large scale like Altair's OptiStrut, efiForm, and Abaqus 6.14 etc.

5. Comparison of Different techniques

As compared to evolutionary algorithm like spatial simulated annealing (SSA) [12], Particle tracing [8] technique based GDM (i.e. iterated local search technique) provides better space filling property and shorter computation time. One major limitation of PT algorithm is that it does not always guarantee to obtain feasible shape in shape space and distribution of infeasible region is highly irregular. This above problem has been removed in evolutionary algorithm based GDM e.g. Jaya algorithm, TLBO based GDM etc.

As compared to shape grammar, constrained parametric search based GDM does not need any special external rules and grammar encoding. However, this constrained parametric search cannot be utilized for the engineering problems, which are computable, and correlation between problem and solution space is possible as compared to evolutionary algorithm based GDM technique.

Topology optimization tool [5] based GDM provides different design based on material removal from the object while minimizing stiffness, keeping volume fixed, and other constraints criteria as compared to other method.

There are also other methods [31-34] Which provides number of different shapes as compared to generative design method but generative design method have advantage of utilizing non collapsing and space filling criteria to provide distinct and optimal designs.

6. Conclusion

This paper proposes a review on generative design techniques, their advantages, limitations, and applications. GDM is mainly used for generating number of designs and the design best suited according to user is selected. GDM is mainly classified in to five categories which are based on (1) Constrained parametric search strategy, (2) Evolutionary optimization algorithm like GA, ABC, SSA, CMA-ES, Jaya algorithm, TLBO technique etc., (3) Iterated local search technique, (4) Shape grammar rule, (5) Topology optimization tool.

Generative design method provides number of distinct shape without any proper model set and also generate design alternatives considering non-collapsing and space filling property as compared to other method of shape exploration.

Acknowledgments

Authors would like to offer special thanks to Mr. Siddhartha, PhD Scholar at MNNIT Allahabad, for his valuable advises on writing research paper, to Prof. Mukul Shukla, Professor at MNNIT Allahabad, India, for his help on understanding advanced optimization techniques.

References

[1] Kazi R H, Grossman T, Cheong H, Hashemi A and Fitzmaurice G 2017 Dreamsketch: Early stage 3d design explorations with sketching and generative design *Proc. of the 30th Annual ACM Symp. on User Interface Software and Technol. (UIST), ACM,* (New York: Association for Computing Machinery) pp 401-14.

[2] Umetani N, Igarashi T and Mitra N J 2012 Guided exploration of physically valid shapes for furniture design *ACM Trans. Graph.* **31** pp 1-86.

[3] Frazer J H 2002 Creative design and the generative evolutionary paradigm Bentley *Creative evolutionary systems* ed P J and Corne D W (San Francisco:Morgan Kaufmann) pp 253-74.

[4] Shea K, Aish R and Gourtovaia M 2005 Towards integrated performance-driven generative design tools *Autom. Constr.* **14** pp 253–64.

[5] Nana A, Cuilliere J C and Francois V 2016 Towards adaptive topology optimization *Adv. Eng. Softw.* **100** pp 290–307.

[6] Krish S 2011 A practical generative design method *Comput. Aided Des.* **43** pp 88-100.

[7] Khan S and Gunpinar E 2017 An extended Latin hypercube sampling approach for CAD model generation *Anadolu University Journal of Science and Technology A- Applied Sciences and Engineering* **18** pp 301-14.

[8] Gunpinar E and Gunpinar S 2018 A shape sampling technique via particle tracing for CAD models *Graph. Models* **96** pp 11–29.

[9] Dogan K M, Suzuki H, Gunpinar E and Kim M 2019 A generative sampling system for profile designs with shape constraints and user evaluation *Comput. Aided Des.* **111** pp 93-112.

[10] Khan S and Gunpinar E 2018 Sampling CAD models via an extended teaching– learning-based optimization technique *Comput. Aided Des.* **100** pp 52–67.

[11] Khan S and Awan MJ 2018 A generative design technique for exploring shape variations *Adv.*

Eng. Inform. **38** pp 712–24.
[12] Khan S, Gunpinar E and Moriguchi M 2017 Customer-centered design sampling for cad products using spatial simulated annealing *Proc. of CAD'17* (Okayama) pp 100-03.
[13] Kielarova S W, Pradujphongphet P and Bohez E L J 2013 An Approach of Generative Design System: Jewelry Design Application *2013 IEEE Int. Conf. on Industrial Engineering and Engineering Management* (Bangkok:IEEE) pp 1329–33.
[14] Audze P and Eglais V 1977 New approach for planning out of experiments *Problems of Dynamics and Strengths* **35** pp 104-7.
[15] Kieralova S W, Pradujphongphet P and Bohez E L J 2015 New interactive-generative design system: hybrid of shape grammar and evolutionary design – an application of jewellery design *Int. Conf. in Swarm Intelligence ICSI 2015: Advances in Swarm and Computational Intelligence* pp 302-13.
[16] Pugliese M J and Cagan J 2002 Capturing a rebel: modelling the Harley-Davidson brand through a motorcycle shape grammar *Res. Eng. Des.* **13** pp 139-56.
[17] Rao R V and Waghmare G 2017 A new optimization algorithm for solving complex constrained design optimization problems *Eng. Optim.* **49**(1) pp 60–83.
[18] Haldurai L, Madhubala T and Rajalakshmi R 2016 A Study on Genetic Algorithm and its Applications *Int. J. Comput. Sci. Eng.* **4(**10) pp 139-43.
[19] Poli R, Kennedy J and Blackwell T 2007 Particle swarm optimization:An overview *Swarm Intelligence* **1(**1) pp 33-57.
[20] Karaboga D and Basturk B 2008 On the performance of artificial bee colony (ABC) algorithm *Appl. Soft Comput.* **8**(1) pp 687-97.
[21] Igel C, Hansen N and Roth S 2007 Covariance matrix adaptation for multi-objective optimization *Evol. Comput.* **15**(1) pp 1-28.
[22] Yu K, Wang X and Wang Z 2016 Constrained optimization based on improved teaching–learning-based optimization algorithm *Inf. Sci.* **352** pp 61-78.
[23] Gunpinar E, Suzuki H, Ohtake Y and Moriguchi M 2013 Generation of bi-monotone patches from quadrilateral mesh for reverse engineering *Comput. Aided Des.* **45**(2) pp 440–50.
[24] Gunpinar E, Moriguchi M, Suzuki H and Ohtake Y 2014 Feature-aware partitions from motorcycle graph *Comput. Aided Des.* **47** pp 85–95.
[25] Alrefaei M H and Diabat A H 2009 A simulated annealing technique for multi-objective simulation optimization *Appl. Math. Comput.* **215**(8) pp 3029–35.
[26] Chen B, Pan Y, Wang J, Fu Z, Zeng Z, Zhou Y and Zhang Y 2013 Even sampling designs generation by efficient spatial simulated annealing *Math. Comput. Model.* **58**(3-4) pp 670–6.
[27] Hoosmand A and Campbell M I 2016 Truss layout design and optimization using a generative synthesis approach *Comput. struct.* **163** pp 1-28.
[28] Sangeun O, Yongsu J, Seongsin K, Ikjin L and Namwoo K 2018 Deep Generative Design: Integration of Topology Optimization and Generative Models *Int. Design Eng. Technical Conf. & Computers and Information in Eng. Conf.(IDETC/CIE)* (Quebec City:ASME).
[29] Bentley P J 1999 From coffee tables to hospitals: generic evolutionary design *Evolutionary design by computers*, ed Bentley P J (San Francisco:Morgan Kaufmann) pp 405-23.
[30] Caladas L 2006 GENE_ARCH: an evolution-based generative design system for sustainable architecture *Lect. Notes Comput. Sci.* **4200** p 109.
[31] Kalogerakis E, Chaudhuri S, Koller D and Koltun V 2012 A probabilistic model for component-based shape synthesis *ACM Trans.Graph.* **55** pp 1–11.
[32] Kleiman Y, Fish N, Lanir J and Cohen-Or D 2013 Dynamic maps for exploring and browsing shapes *Proc. of the 11th Eurographics /ACMSIGGRAPH Symp. on Geometry Processing* (Goslar:Eurographics Association) pp 187–96.
[33] Averkiou M, Kim V G, Zheng Y and Mitra N J 2014 Shapesynth: Parameterizing model collections for coupled shape exploration and synthesis *Comput. Graph. Forum* **33**(2) pp 125–34.

IOP Conf. Series: Materials Science and Engineering 1104 (2021) 012036 doi:10.1088/1757-899X/1104/1/012036

[34] Schulz A, Xu J, Zhu B, Zheng C, Grinspun E and Wojciech M 2017 Interactive design space exploration and optimization for cad models," *ACM Trans. Graph* **36**(4) 157.

The study of Recoil Reducing Compact Air Pressure Device (CAPD)

Abhimanyu Saxena, M Dixit, M Rashid, S C Nigam, S K Singh
Dronacharya Group of Institutions, B-27, Knowledge Park – III, Greater Noida, U.P India.
Corresponding Author: abhimanyu.sxn18@gmail.com, shaileshsonu10@gmail.com

Abstract: Defense organization is one of the most reputed organization where best of the officers and soldiers serve for the safety on one's country, without complaining about anything and giving their all in to protect us from various threats that could lead to disintegration of the country.
Thus, the main motive to develop this device is to help our defense services to get the best, while they are giving their best for the country. The idea is to develop a device that would counter the recoil of the gun used by the officers and soldiers, to make them more efficient. Keeping in mind that the renewable resources of energy are depleting at a fast pace and pollution is increasing day by day leading to global warming, we have taken air as our resource to produce the desired effect. The device would be developed using aluminium, compact single cylinder air compressor and arduino programming to measure the pressure required for each shot fired from the gun. The principles of Law of conservation of momentum are applied to reduce the recoil of the gun to 80 %, thus helping in better aiming and precision. By installing this device, the demerits that makes AK47 a primitive gun would be subdued and without buying the expensive advanced guns, the arsenal would be upgraded at minimal cost possible.

Keywords: Conservation of momentum, Single stage reciprocating air compressor, Recoil of gun, Recoil reducing device, Recoil reducing compact air device, Indian Defense Services, AK47.

1.0 INTRODUCTION

The idea was developed to overcome the problem of budget of Defense Services (only 2.1% of GDP), which unable them to upgrade the arsenal time to time and hence they have to adjust with the weapons of primitive technology. Where as other countries enhance the technology timely thus making their defense sector stronger, which will give them an upper hand during a war as well as to combat the terrorists.

The most used gun in the Indian Defense Services which constitutes of Indian Army, NSG, CRPF etc is AK47.

AK47 is a really good gun, but it has become primitive after various advanced guns being inducted by the US, USSR, China, Japan, & Finland. The major drawback of this gun is high recoil velocity, which makes it difficult to handle and often leads to injuries.

Thus, in order to help our Defense Services we have come up with an idea to develop a Compact Air Pressure Device which would help in minimizing the recoil of the gun by developing a device which would use air to produce the required pressure

The device would be a simple single cylinder air pressure device of a very compact size and extremely light in weight which would produce a counter recoil force to keep the gun balanced. In this way the recoil will get reduced, thus providing better aiming, precision and control of the gun.

There are around 2 Lakhs guns already present in our Defense Sector and other 7 Lakhs are going to be inducted in a year. Thus, to meet the requirements of around 9 Lakhs gun, the production of this device would also help in creating work for a number of unemployed people as well as will help in increasing in the GDP of the country.

2.0 LITERATURE REVIEW

S.No	Author Name	Major findings	Year	Reference
1.	Vijaykumar F.Pipalia Dipesh D. Shukla Niraj C. Mehta	Investigation on reciprocating air compressor - In understanding the mechanism enough to enable development of reliable heat transfer correlations and at the application level to enable development of easy, quick and reliable procedures for predicting its impact on compressor performance.	December, 2015	[1]
2.	Manoj Bansidas Agravat	Importance of Conservation of momentum E= mc2 equation to momentum a component of the total energy equation (related to sum of special relativity energy and momentum) 1) A decrease of time and velocity shows momentum will increase short term. 2) An increase of time and velocity shows momentum decrease short term. If conservation of energy fails to be true, this inference is important.	November, 2014	[2]
3.	Harper, W.H;Ellis, P.H;Hanlon ,W.H;Merkey, R.P	The effects of Recoil on shooter performance- 1) Isolate and quantify recoil characteristics that effect aiming error. 2) Investigate and measure how those factors affect the soldier's perception and tolerance of recoil. 3) Evaluate the effectiveness of recoil-mitigating device.	May, 1996	[3]
4.	Simon Collyer , Clive Warren	Aim, Fire, Aim –Project planning style in dynamic environment - The make-static approach may be applied where safety and risk minimization is an	August, 2010	[4]

	Bronwyn Hemsley Chris Stevens	imperative or change adaption offers little advantage		
5.	David Streiner, Geoffrey Norman	Precision - The term precision refers to how precisely an object of study is measured. The article ends with a call for the use of the more traditional terms as better reflecting the process of scale development and the uses to which they are put.	2010	[5]
6.	C-G Wang, W-H Xu, C.Xu	Influence of different automatic action of gun on its recoil effect in man-gun system - The research results show that the rules of the changes of gun recoil force and the shooter' recoil velocity and displacement with recoiling time, and the number of occurrences and time of the maximum are not same for two different kinds of automatic gun.	May, 2014	[6]
7.	Sun-Young Park; Jae-Sung Bae; Jai-Hyuk Hwang; Kuk-Jeong Kang	Forward momentum of a soft recoil system- A soft-recoil or FOOB (Fire-Out-Of-Battery) system can reduce the recoil force considerably. Its firing sequency is different from that of a conventional or FIB (Fire-In-Battery) system.	January, 2010	[7]
8.	Michael Courtney; Amy Courtney	Testing bullets at reduce velocity - Testing bullet performance at lower velocity presents a challenge because smokeless powder charges cannot be safely lowered to arbitrary levels in cartridge arms, and extending the range to very large distances (so that air resistance slows the bullet to the desired level) is impractical because the required accuracy is probably not available.	January, 2009	[8]
9.	Jan Hjelmgren	Dynamic measurement of pressure - develop fundamental reference transducers for periodic pressure sources. The approach employed in most cases, and felt to be unsatisfactory by most people, is to choose a measurement system with a bandwidth many times that of the pressure signal to be measured.	2002	[9]

10.	Arun Vijayakumar D; Om Prakash; R. K Sharma	Estimation of Air density and its importance in barometric pressure measurement - The air density in a controlled environment at NPL, India is found to be varying within 1.135 kg/m3 and 1.155 kg/m3, the lowest during summer and the highest during winter months.	February, 2012	[10]
11.	Florian Ion Tiberiu Petrescu; Taher Abu-Lebdeh; Antonio Apicella	Oscillating Sliding mechanism - A mechanism with an oscillating slider is a mechanism quite often used in the technique. It has some advantages over other similar mechanisms being able to work with accelerations and reduced forces under the same conditions with another similar mechanism.	February, 2018	[11]
12.	P.K Stefanopoulus; G.F Hadjigeorgiou; Konstantina Filippkis; Dimirios Gyftokostas	Gunshot Wounds – A review of ballistics related to penetrating trauma - Ballistic factors such as the impact velocity and energy should not be considered in isolation, as their specific effects are determined by the interaction between the projectile and tissues.	December, 2014	[12]
13.	Harri Palocheimo; Michael Adebayo Omidiora	Feasible study on compressed air energy storage system for portable electrical and electronic device - The objective for this feasibility study was to assess the viability of introducing small-scale CAES (Compressed Air Energy Storage) system for mobile device charging and lighting in rural off-grid areas of the world and especially in the developing countries, where existing energy infrastructure is very limited.	July, 2009	[13]
14.	Kelvin Russell	Recent advances in small arms recoil reduction - This work presents the final patent (in a three-part series of patents) in the area of rifle operating group-based recoil reduction.	September, 2014	[14]

15.	Hiroaki Kajikawa; Tokihiko Kobata	Pressure gauge Calibration Applying O-A-O pressurization to reference gauge - The reference gauge was pressurized by 0-A-0 pressurization, whereas the test gauge could be pressurized through various procedures according to the user's requirements and the measurement conditions of the test gauge at industrial sites.	April, 2014	[15]

3.0 MATERIAL

Aluminium – As we can see from the calculations done in **section 7.3**, the discharge pressure that is required to control the recoil of gun is $P_d = 0.0158$ psi. So, the material that we can use could be potentially aluminium. The major reason to choose aluminium is that -

It is a very light metal with specific weight of $2.7 g/cm^3$, about that of steel that makes it easy to handle and install. It naturally generates a protective thin oxide coating which keeps the metal from making further contact with the environment. It is ductile and has low melting point i.e. 660.3^oC. It shows tensile strength as temperature drops. As it is non-magnetic in nature, it can be used in making weapons and shields and mainly due to its sound and shock absorption, non-sparking, and recyclability properties we will use aluminium.[16]

4.0 PRINCIPLE

CONSERVATION OF MOMENTUM :

To calculate the recoil, we apply the law of conservation of momentum and this is derived from Newton's third law. Law of conservation of momentum states that every applied force has an equal and opposite reaction. [2]

$$\mathbf{F_i = -F_r}$$

Where, F_i = imparted force and F_r = Reaction force

In a typical recoil situation, the release of the body of smaller mass (body 1) has an impact on a larger body (body 2). If both the bodies start from rest, the law of conservation of momentum states that,

$$\mathbf{M_1V_1 = -M_2V_2}$$

Now, our main aim is to reduce the recoil velocity by adding a compressor that can discharge pressure in the opposite direction. For all this to happen, the law of conservation of momentum should be followed.

$$\mathbf{M_1V_1 = M_2V_2 + M_3V_3}$$

Where, M_3 is the mass of the compressor and V_3 is discharge velocity.

5.0 METHODOLOGY

Basically, when we are talking about the methodology of the project, we come across two different principles which are:

i. conversation of momentum
ii. single stage reciprocating compressor.

Initially, we have calculated the recoil velocity of the gun, with the help of the data extracted from different internet sources and then we will calculate the discharge pressure of our device in psi.

The basic idea of the project is to install a reciprocating compressor on the gun which will eject out the required amount of pressure thus producing force to counter the recoil velocity of the gun, hence stabilising it using the principle of conservation of momentum.

The suction and discharge effect are created due to the pressure difference inside the compressor. Initially, the atmospheric pressure of the device due to which the suction effect is created and after consuming the fluid inside the compressor the internal pressure increases and the atmospheric pressure decreases due to which we get high discharge velocity.

The installation will be done beside the scope using a sliding mechanism, which is by now the most appropriate location to install the device, the location might be changed after more rigorous experiments.

6.0 EXPERIMENTAL ANALAYSIS

6.1 NOTATIONS

- M_1 = Mass of Bullet
- M_2 = Total Mass of Gun
- M_G = Mass of Gun
- M_M = Mass of Magazine
- $M_2 = M_g + M_m + (30 \times M_1)$
- M_3 = Mass of CAPD
- All the masses are in Grams (g).
- V_1 = Velocity of Bullet
- V_2 = Velocity of Gun (Recoil)
- V_{2N} = Velocity of Gun (Recoil) after installing CAPD.
- V_3 = Velocity of air discharge by CAPD.
- All the velocities are in Meter / Second (m/s)

6.2 DATA

IOP Conf. Series: Materials Science and Engineering 1104 (2021) 012037 doi:10.1088/1757-899X/1104/1/012037

Data has been taken in accordance with AKM Rifle.

- $M_1 = 16.3$ g
- $M_2 = 4389$ g
- $M_G = 3470$ g
- $M_M = 430$ g
- $M_3 = 700$ g (Approximately)
- $V_1 = 715$ m/
- $V_2 = 2.65$ m/s

6.3 CALCULATIONS

- Momentum Conservation before Installing CAPD.

$M_1 \times V_1 = M_2 \times V_2$

- Momentum Conservation after Installing CAPD.

$M_1 \times V_1 = (M_2 \times V_{2N}) + (M_3 \times V_3)$

- Reducing V_2 to 20% we get V_{2N}.

$V_{2N} = 20/100 \times V_2 \Rightarrow 20/100 \times 2.65$

$\mathbf{V_{2N} = 0.53\ m/s}$

$M_1 \times V_1 = M_2 \times V_{2N} + M_3 \times V_3$

$16.3 \times 715 = (4389 \times 0.53) + (V_3 \times 700)$

$(11654.5 - 2326.17) / 700 = V_3$

$\mathbf{V_3 = 13.32\ m/s}$

- $P_d = 1/2\ \rho\ V_3^2$
- P_d = Dynamic pressure produced by CAPD in Pascal (Pa).
- $\rho = 1.225$ kg/m^3 (Density of Air).
- $V_3 = 13.32$ m/s
- $P_d = 109$ Pa
- Pa = 0.00014503773 psi
- $p_d = 0.0158$ psi

8.4 RESULTS

- $\mathbf{V_{2N} = 0.53\ m/s}$
- $\mathbf{V_3 = 13.32\ m/s}$
- $\mathbf{p_d = 0.0158\ psi}$

7.0 RESULTS & DISCUSSION

After making the desired calculations we reached to a conclusion that the pressure which needs to be released from the air pressure device is approximately 0.0158 psi per shot fired.

The calculations are done by extracting the mass of gun, bullet and magazine and velocity of the bullet of AK47 from the internet so, the pressure required is accurate under our consideration which is quite less and can be released from a small compressor as well. So, the size of the compressor will be small and the weight of the air pressure device will also be less and this will not lead to any extra load on the army men.

After, calculating the forces we have come to a conclusion that to maintain the equilibrium of the gun based on the principle of conservation of momentum, we will place the air pressure device just below the barrel of the gun and to attach the air pressure device we will be using a sliding mechanism by developing and fixing a slider of desired shape on the gun with the help of various machining processes such as welding, grinding, shaping and turning and CNC machining.

This device will help the Indian armed forces to use the gun with more precision and can potentially be a good idea under the Make in India initiative as our defence equipment are mostly purchased from The United States and Russian Federation. Under a small defence budget this Idea will be a game changer if it works after all the testing of the prototype is done and the project

These are the 3 important values that we obtained after doing the calculation:

1. New Recoil Velocity = 0.53 m/s
2. Velocity of air expelled by the air pressure device = 13.32 m/s
3. Pressure to be developed to counter the recoil = 0.0158 psi

10.0 CONCLUSION

Hence our objective is to create a Compact Air Pressure Device of around **700 grams** using **aluminium** which would be able to produce a pressure of **109 pascal** or **0.0158 psi.**

This device would help in reducing the recoil of gun from **2.56 m/s** to **0.53 m/s** when a bullet is fired, thus the efficiency gained is **80%.**

This would in turn help in increasing the aiming and precision of the rifle, thus helping our Defense Services to get the best output in defending the border and protecting the country from various threats.

11.0 REFERENCE

1. Pipalia, Vijaykumar F., Dipesh D. Shukla, and Niraj C. Mehta. "Investigation on Reciprocating Air Compressors-A Review." *International journal of Recent scientific research* 6.12 (2015): 7735-7739.
2. Agravat, Manoj. "Importance of conservation of momentum." *Science* 2.
3. Harper, W. H., et al. *The Effects of Recoil on Shooter Performance*. No. ARL-TR-382. ARMY RESEARCH LAB ABERDEEN PROVING GROUND MD, 1996.
4. Collyer, Simon, Clive Warren, Bronwyn Hemsley, and Chris Stevens. "Aim, fire, aim—Project planning styles in dynamic environments." *Project Management Journal* 41, no. 4 (2010): 108-121.
5. Streiner, D. L., & Norman, G. R. (2006). "Precision" and "accuracy": two terms that are neither. *Journal of clinical epidemiology*, *59*(4), 327-330.

6. WANG, C. G., XU, W. H., & XU, C. (2014). Research on the influence of different automatic action of gun on its recoil effect in man-gun system. *Acta Armamentarii*, 05.
7. Park, Sun-Young, Jae-Sung Bae, Jai-Hyuk Hwang, Kuk-Jeong Kang, and Sang-Tae Ahn. "A Study on the Forward Momentum of a Soft Recoil System." *Journal of the Korea Institute of Military Science and Technology* 13, no. 6 (2010): 976-981.
8. Courtney, Michael, and Amy Courtney. "A method for testing bullets at reduced velocity." *arXiv preprint arXiv:0812.4934* (2008).
9. Hjelmgren, Jan. *Dynamic measurement of pressure. - A literature survey*. 2002.
10. Vijayakumar, D. A., Prakash, O., & Sharma, R. K. (2012). Establishment of a force balanced piston gauge for very low gauge and absolute pressure measurements at NPL, India. In *Journal of Physics: Conference Series* (Vol. 390, No. 1, p. 012013). IOP Publishing.
11. Petrescu, F. I., Abu-Lebdeh, T., & Apicella, A. (2018). Study of an Oscillating Sliding Mechanism. *American Journal of Engineering and Applied Sciences*, *11*(2), 870-880.
12. Stefanopoulos, P. K., Hadjigeorgiou, G. F., Filippakis, K., & Gyftokostas, D. (2014). Gunshot wounds: a review of ballistics related to penetrating trauma. *Journal of Acute Disease*, *3*(3), 178-185.
13. Paloheimo, H., & Omidiora, M. (2009, June). A feasibility study on Compressed Air Energy Storage system for portable electrical and electronic devices. In *2009 International Conference on Clean Electrical Power* (pp. 355-362). IEEE.
14. Russell, K. (2014). Recent Advances in Small Arms Recoil Reduction. *Recent Patents on Mechanical Engineering*, *7*(3), 241-246.
15. Kajikawa, H., & Kobata, T. (2016). Pressure gauge calibration applying 0-A-0 pressurization to reference gauge. *Acta Imeko*, *5*(1), 59-63.
16. Dispinar, D., & Campbell, J. (2007). Effect of casting conditions on aluminium metal quality. *Journal of materials processing technology*, *182*(1-3), 405-410.
17. Rottman, Gordon L. *The AK-47: Kalashnikov-series assault rifles*. Bloomsbury Publishing, 2011.

Continual improvement through ISO 9001 – An evidence based audit conducted in 200 organizations

Karri Naveen[1], Rohan Senanayake[2], Chithirai Pon Selvan[3]

[1]Research Scholar, Mechanical Engineering, Lincoln University College, Malaysia
Email:phdstudent.naveen@lincoln.edu.my
[2]Principal Supervisor, HOD, Mechanical Engineering, Lincoln University College, Malaysia
Email: drrohan@lincoln.edu.my
[3]Co-Supervisor, Head of School – Science and Engineering, Curtin University, UAE,
Email: pon.selvan@curtindubai.ac.ae

Abstract. Continual improvement is the most significant element of ISO 9001. Inadequate research studies are found to make the assessment of Continual Improvement of certified organizations through an onsite audit. Researcher has spent 386 mandays onsite conducting audit of 200 organization to determine the conformance to the criteria laid in ISO 9001:2015. This research study is based on primary data obtained through process observation, reviewing documented information and interviewing concern professionals including the top management. The key factors determined for continual improvement were analyzed to confirm effectiveness, adequacy and suitability of the quality management system. The outcome of the audit was measured against the conformity assessment variable influencing the continual improvement of the quality management system. Research data was analyzed by using the most simple statistical software "Jamovi". The compliance with the criteria for outcome of the management review, non conformity and corrective action were mapped against the five point Likert scale.

1. Introduction

ISO 9001 Certification needs no special introduction as last year (2019) alone 8,83,521 Certificates were issued across the world. The growing importance and worldwide acceptance of the Quality Management System has also raised serious concerns on the degree of implementation of the criteria as per ISO 9001. Assessment of ISO 9001 does not levy any provision to grade the level of compliance against the requirements of the standard. Therefore its very challenging to attribute the continual improvement through quality management system certification. And hence this research study is based on the primary data gathered through onsite audit with evidence based approach to determine the degree of compliance with the continual improvement element. There are various tools and techniques for continual improvement which are practiced across the world. Some continual improvement tools and techniques were emerged from the actual efforts of world-class industry like: A3, 5S, Catchball, DMAIC, Gemba Walks, Kanban, 5 Why, etc.

IOP Conf. Series: Materials Science and Engineering 1104 (2021) 012038 doi:10.1088/1757-899X/1104/1/012038

2. Research Gap

Most of the research studies in the area of Quality management system are based on the feedback obtained from the sampling organization. The major constrain in this type of methodology is lack of objective evidence. Research pertaining to quality management system would not yield intended output if the data is gathered by mere collection of feedback without objective evidence.

3. Research Methodology

A formal onsite audit is performed in 200 organizations as per ISO 19011. The questionnaire is prepared to determine the conformance to the requirements of the ISO 9001:2015 standard. 386 Mandays were spent onsite to verify the conformance to ISO 9001 criteria. The outcome of this evidence based approach is analysed by using five point Likert scale to quantify the research data.

4. Reliability Analysis

A reliability analysis was carried out 227 questions that were assessed by researcher which were measured through a 5 point Likert type scale. Cronbach's alpha calculated for the audit questionnaire to reach acceptable reliability ($\alpha = 0.82$) indicating high consistency.

Scale Reliability Statistics

	Cronbach's α
scale	0.820

5. Results

The outcome of the research study through onsite audit was quantified by using statistical tool called Jamovi.

Table-1 presents the results Management Review conformity assessment variable. The Mean is 3.62 and the Median is 4.0 with standard deviation being 0.57. The results indicate the effectiveness of the management review in accordance with the agenda determined in the standard and the outcome of the reviews. The results were inclined towards 3 point on Likert scale which indicate the Management Reviews were conducted as a formality to fulfill the audit requirement and did not yield any fruitful benefit in effectiveness of the quality management system.

Table: 1, Management Review

Descriptives

	Management Review
N	200
Missing	0
Mean	3.62
Median	4.00
Standard deviation	0.527
Minimum	3
Maximum	5

Table-2 represent the frequency of the non conformity on a Likert scale; 4 point resulting 56.5% and the least being 5 point contributing mere 1.0%. Handling non conformity and initiating corrective action was not implemented to its full potential as majority of the organizations attributed this requirement as "Failure" and felt offensive to accept or declare the deviations from the requirement of the standard.

Table-2, Non Conformity and corrective action

Frequencies of Non conformity and corrective action

Levels	Counts	% of Total	Cumulative %
3	85	42.5 %	42.5 %
4	113	56.5 %	99.0 %
5	2	1.0 %	100.0 %

Graph-1 indicate the continual improvement by implementing the quality management system. More than half of the sample does not conform to this requirement and hence continual improvement could not be attributed to the quality management system.

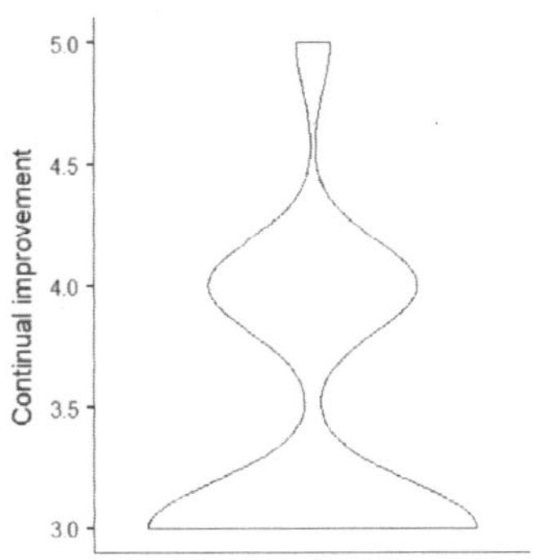

Graph-1, Continual Improvement

6. Conclusions

1) Effectiveness, adequacy and suitability of the ISO 9001:2015 could be determined by considering outcome of management review, analysis and evaluation.
2) The organization could also address opportunities and needs of the interested parties for continual improvement.
3) Ineffective implementation of ISO 9001 is the major contributing factor for holding the continual improvement of certified organization.

7. Scope for future work

1) Three year performance trend can be studied and analyzed for ISO 9001 certified organizations to establish correlation between continual improvement and the quality management system.
2) Non conformity management and competency requirement for initiating corrective actions could be studied for ISO 9001 certified organizations.
3) Outcome of the management review could be studied to determine correlation with continual improvement.

References

[1] Bravi, L., Murmura, F., & Santos, G. (2019). The ISO 9001:2015 Quality Management System Standard: Companies' Drivers, Benefits and Barriers to Its Implementation. *Quality Innovation Prosperity/Kvalita Inovácia Prosperita*, 23(2), 64-82. doi: 10.12776/ QIP.V23I2.1277

[2] Carnerud, D. (2018). 25 years of quality management research – outlines and trends. *International Journal of Quality & Reliability Management,* 35(1), 208-231. doi 10.1108/IJQRM-01-2017-0013.

[3] Chiarini, A. (2017). Risk-based thinking according to ISO 9001:2015 standard and the risk sources European manufacturing SMEs intend to manage. *The TQM Journal*, 29(2), 310- 323. https://doi.org/10.1108/TQM-04-2016-0038.

[4] Domingues, P., Sampaio, P., & Arezes, P.M. (2016). Integrated management systems assessment: a maturity model proposal. *Journal of Cleaner Production*, 124, 164-174.

[5] Gluck, A., Aston, B., Briggs, S., Cianfrani, C., Desai, D., Palmes, P., Robitaille, D., & West, J. (2015). Keep Calm and Prepare for ISO 9001:2015. *Quality Progress.*

[6] Cronbach, L. J. (1951). Coefficient alpha and the internal structure of tests. *Psychometrika*, 16, 297-334.

[7] Croft, N. H. (2012). *ISO 9001:2015 and beyond – Preparing for the next 25 years of quality management standards.* http://www.iso.org/iso/news.htm?refid=Ref1633.

[8] International Organization for Standardization, (2015). ISO 9001:2015 - *Quality management systems - Requirements.*

[9] International Organization for Standardization, ISO Survey 2019. Available at: http://www.iso.org.

[10] Fonseca, L. (2015b). Relationship between ISO 9001 certification maturity and EFQM Business Excellence Model results. *Quality, Innovation and Prosperity*, 19(1), 85-102.

[11] ISO/TR 8550:1994, Guide for selection of an acceptance sampling system, scheme or plan for inspection of discrete items in lots.

[12] ISO 7966:1993, Acceptance control charts

IOP Conf. Series: Materials Science and Engineering 1104 (2021) 012039 doi:10.1088/1757-899X/1104/1/012039

PRECIPITATION BEHAVIOUR OF LOW CARBON MICROALLOYED STEEL

[1]Harikesh Singh, [2]Ashish Kumar Singh, [3]Zakir Ali Hyderi, [4]Prof. Dr.M.K. Banerjee

[1,2,3] Department of Mechanical Engineering, [4]Department of Metallurgical Engineering, [1,2,3] IIMT College of Engineering, Greater Noida, [4]MNIT, Jaipur

Email : harikeshmnit@gmail.com

Abstract. In this paper precipitation behaviour of low carbon micro alloyed steel is seen by various heat treatment cycles at different temperature ranges. Precipitation occurs during the ageing cycles and as we can conclude from harness graph and microstructural study that precipitation of alloys like NbC, SiC, CrC, NiC, VC, MoC, BC precipitates on higher temperatures, therefore hardness tends to increase at some range of high temperature ageing. After their precipitation occurs there got a saturation in the system and again hardness tends to fall. Although the hardness result of all the samples do not show very significant difference, but as a whole the test shows rise, saturation and fall in hardness values which is due to various precipitation temperatures of various alloying elements involved.

1. INTRODUCTION

At the end of the process of Quenching, second phase precipitation of microstructure occurs. At ambient temperature Precipitation with time is known natural ageing/maturing, whereas at elevated temperature precipitation is known as artificial ageing/maturing. After solution treatment hardness is comparatively less. The utmost strength and hardness develops when mixture is matured at a suitable temperature which varies between 120^0C and 200^0 C. In few conditions the ageing temperature can be 300^0C. Maturing duration may range from 4 hours to 24 hours.
Unconstrained decay of supersaturately well buid arrangement happens during maturing treatment. Aeging temperature and level of supersaturation assume an incredible function on the previous properties of the amalgam. More the level of immersion and the maturing temperature , more serious will be the maturing. Increased maturing temperature is embraced when steady stage is required along with dimension security. The cycle is additionally Termed as settling maturing. Other than mechanical properties, physico- substance properties are additionally affected by maturing. This occurs because of themetastable microstucture of the alloy that is framed during maturing of supersaturately strong arrangement got by arrangement treatment. Ascend in temperature varies the nuclear situations with relating variation in the powers related with interatomic bonds. Simultaneously circulation of second stage particles likewise changes. With variety in the assembling strategy of the segments, it shows variation in mechanical properties on higher temperatures after warmth heat treatment. It has been seen that the squeezed segments are a lot of better than the cast parts in this regard. Then again, the conduct of cast versus squeezed segments of silumin type compounds is the polar opposite. The distinction in conduct is because of the idea of encouraging stages at high temperatures.
1.1 A few stages are related with the cycle of precipitation solidifying like
(1) Formation of particles of the accelerating stage comprises of revision of iotas inside the gem cross section.
(2) Formation of progress structures as changed Guinier-Preston zones and middle stages. This may offer ascent to most extreme reinforcing in the compound.
(3) Development of stable stage from progress stages of particles haveing regular limits with the grain of the grid.

(4) development of the bigger grains to the detriment of the adjacent more modest grains. Because of this pressure help happens in the cross section as a rule at maturing elevated temperatures, which effects considerable increase in ductility and decrease in strength of alloy[1].

2. LITERATURE REVIEW

2.1. STEEL

The word steel is utilized for some, extraordinary amalgams of iron. The combinations differ in manner they are formed and in the extents of the material alloyed to the iron, both. All prepares, in any case, contain modest quantities of carbon and manganese. As such, it very well may be termed that steel is a glasslike amalgam of I-C and a few different components, which solidifies over its basic temperature. Like expressed above, there do exist a few kinds of prepares however commonly these order in two classifications are plain carbon and alloyed steels[2].

2.1.1. PLAIN CARBON STEELS

Plain carbon steel finds various applications and around 90% of absolute steel creation falls into this class. The purposes behind quite a wide pertinence are its minimal effort combined with sensibly great usefulness, castability and machinability. The properties of carbon prepares can be changed fundamentally by basic mechanical, warm and surface medicines. These prepares, in different structures, are broadly utilized for general designing purposes. The different applications incorporate wires and wire items, rounded items, sheets and strips, railroad rails, projected and fashioned items, cutting devices, and so on All steels contain silicon and manganese. Hence, it becomes necessary to define limits for silicon and manganese in order to differentiate between carbon steels and alloy steels. The American Iron and Steel Institute (AISI defines carbon steel as that steel for which no minimum content is specified or required for aluminium, chromium, cobalt, niobium, molybdenum, nickel, titanium, tungsten, vanadium, or zirconium, or any other element added lo obtain desired alloying effect; when the specified minimum for copper does not exceed 0.40 per cent; or when the maximum content specified for any of the following elements does not exceed the percentage noted against them;

Manganese 1.65 percent

Silicon0.60 percent

Copper 0.60 percent

In simple words, a steel containing less than per cent manganese and 0,60 per cent silicon is referred to as carbon steel[3].

2.1.2. CLASSIFICATION CARBON STEELS

Plain carbon steels can be divided into various classes based on carbon content, application and steel manufacturing method. Arrangement on the bases of carbon content is famous and habitually utilized. For the most part, with an expansion in the carbon content from 0.01 to 1.5% in the combination, its quality and hardness increments yet such an increment past 1.5% causes apparent decrease in the flexibility and capliability of the steel. As indicated by carbon content, carbon prepares might be delegated:

a. Low C steels (carbon less than 0.30%)
b. Medium C steels (carbon between 0.30 to 0.60%)
c. High C steels (carbon greater than 0.60%)
d. Hypoeutectoid steels (carbon less than 0.80%),
e. Eutectoid steels (carbon = 0.80%)
f. Hypereutectoid steels (carbon greater than 0.80%)

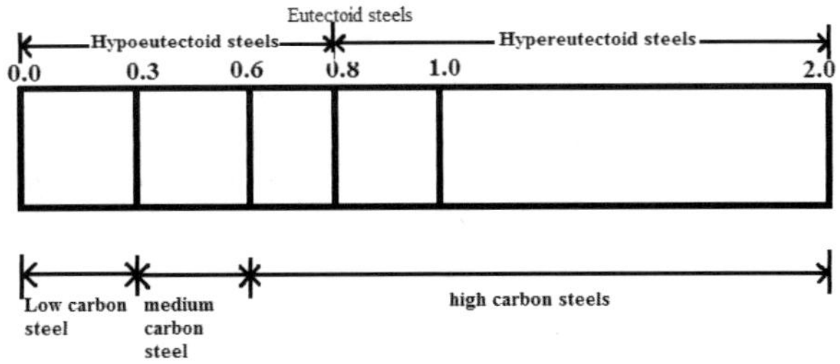

Fig 1. Carbon content in carbon steels

Low C steel or mellow steel, constitutes carbon upto 0.30% reacts to moderate heat treatment as progress in flexibility is considered however has nul impact in regard of its properties.
Medium C steels having carbon constitue going from 0.30to 0.60% improves machin ability by heat treatment. It should likewise be noticed that the steel is particularly versatile for producing and where hardness is alluring.
High C prepared, is steel consisting carbon in the scope of 0.60 to 2.0% and particularly classed as high carbon steel. In completely heat treat condition, it is extremely hard and it will withstand higher wear, shear and will in this way be exposed to minimum distortion.
In view of uses, steel might be named auxiliary steel, constructional steel, apparatus steel, and so on These prepares can additionally be partitioned based on explicit applications, for example rail steel, spring steel, heater steel, sheet steel, cold heading steel, and so forth
Steel fabricating technique greatly affects the properties and nature of steel. Subsequently it has been a training to specify the steel fabricating technique up to mid- eighties; electrically liquefied prepares were utilized to be indicated by setting letter "E" before the AISI/SAE assignment of prepares. Characterization dependent on steel fabricating technique isn't famous these days because of progressions in steel making innovation, Earlier, in view of steel making strategies, prepares were known as corrosive bessemer prepared steel, open hearth prepared steel, electric prepared steel and L.D. prepared steel etc.

2.2. ALLOY STEELS

The restrictions of plain carbon prepares can be limited by the expansion of at least one components to plain carbon prepares. Carbon prepares accordingly alloyed with components are alluded to as amalgam prepares. Hence composite steel is the one that contains determined amount(s) of alloying element(s) and additionally more than 1.65 percent manganese, 0.60 percent copper as well as percent silicon.
The properties of composite prepares rely upon both carbon and alloying components. Nature and amount(s) of alloying element(s) direct the designing conduct of amalgam prepares. Alloying components not just limit/wipe out the restrictions of plain carbon prepares yet in addition confer/improve explicit attributes, for example, obstruction against consumption and oxidation, and attractive and electrical properties.
cobalt, copper, titanium, tungsten, lead, manganese, molybdenum, nickel, silicon, Aluminium, boron, chromium and vanadium are few common mixing elements added to Steels [3].

2.2.1. CLASSIFICATION OF ALLOY STEELS

Alloy steels can be classified in to various classes depending on various parameters. Some popular bases for alloy steels classification are:

a. Amount of alloying element(s).
b. Principal alloying element(s).
c. Application of steels.
d. Microstructure of steels.

Based on amount of alloying element/(s) available in steels, alloy steel also referred as low alloy steel, medium alloy steels and high alloy steels. In general, a low alloy steels contains up to 5% total alloy content while total alloy content of a high alloy exceeds 10%. For a Steels can be grouped as pearlitic steels, ferritic steels, martensitic steels, bainitic steels, and austenitic steels on the bases of microstructure of steels[3].

Table1: Table summarizing the various effects of different alloying elements in the steel.

Alloying/mixing element	Cr	Co	Mn	Mo	Ni	Ti	W	V
Hardenability	+2		+2	+2	+1	+2	+2	+3
High temperature strength	+1	+1		+2		+1	+2	+2
Ductility and toughness			+1					
Abrasion resistance	+1			+1		+1	+2	+1
Fine austenite grain size				+1			+1	+3
Corrosion resistance				+1	+1			

2.2.2. HOW DO MIXING/ALLOYING HELPS IN ACHIEVING OF MARKABLE IMPROVEMENT IN PROPERTIES OFSTEEL?

Truth be told, the properties of compounds are very reliant on the connection between synthetic creation, preparing and their microstructure. For example, at whatever point a component is added to an unadulterated metal, the last modifies the size of the cross section structure of the metal and relying upon the composite framed, it can likewise change its grid type. In some cases metals do respond together to shape intermetallic mixes with exceptionally complex cross section structure. Such mixes soften at a fixed temperature and have a lower conductivity and pliability however more prominent quality and hardness than a combination of face focused body, focused or hexagonal grid structure. Hence, combinations increment quality and hardness of metal by changing its structure. Besides, as expressed above, alloying empowers the development of fine grain size since it favors the ability of the metal to be hardened by quenching in oil or air rather than in water. Indeed, oil is a cooling agent offering slow cooling rate and thus the grain form more regularly with time and hence they are finer[2,3,5].

2.2.3. COMPOSITION VARIANCE OF FEW ELEMENT AND THEIR EFFECT ON THE FINAL ALLOY

Likewise alloying elements are used in steel so as to improve its properties and this is in the scope of rendering the steel appropriate for different uses. Alloying is not only a matter of adding elements to steel, the quantity of alloying elements added is of prior importance so as to achieve the required hardness, toughness or machinability of steel.

Table 2: shows the different composition variance used in alloying element with their effect on the final steel[2,3,5].

Alloying element	Range/varience	Effect on properties
Carbon(C)	0to25 % 0.25to0.70% 0.70to1.50% More than1.50%	Increases heat treatment and ductility. Increases machinability. improvement in strength and hardness. Decrease in ductility and malleability.
Silicon	Initial to0.30% 0.30to1%	Improves good casting fluidity. Improves heat resistance.

Copper	0.20to0.50% High amount	Increases steel resistance to atmospheric corrosion. Unwanted to steel as it affects surface finish.
Chromium	0to5% upto14%	Quenching speed reduction, increased toughness, and wear resistance Resistance to corrosion improves, critical temperature.
Manganese	1.65to2.10% 10%to14%	Increases electrical resistance and magnetic property of steel and reduces its coefficient of expansion. Increases hardness and toughness.
Tungsten	3to6%	Increases cutting hardness and abrasion resistance.
Vanadium	0to0.05% Less than 0.25% 1 %	Hardenability is Increased. Induces resistance to softening at high temperatures. Retain hardness at high temperature.
Nickel	0to5% 8to12% 15to25% 25to35% 36%	provides refined grain structure and causes hardening abilities. Resistance to low temperature impact. High magnetic properties are imparted to steel. Corrosion resistance is increased. Invar is obtained, which has a low coefficient of temperature.
Molybdenum	0.15to0.25%	Uses in combination with chromium, it improves ultimate strength of steel without affecting ductility or workability.
Lead	0.15to0.35%	Improves machinability.
Boron	0.0005to0.03%	Improves hardenability of steel.

2.2.4. LOW CARBON MICROALLOYED STEELS

It holds an important category of steels estimated to be around 12% of total world steel production. These steels contain 0.07 to 0.12% carbon, up to 2% manganese and small additions of niobium, vanadium and titanium (usually max. 0.10% and total microalloying elements generally less than 0.15 %.) in various alloying. These steels stands in terms of performance and cost, between carbon steel and low alloy steel. These are used in major steel market sector in many parts of the world and their development has played an important role in the expansion of certain key industries such as oil and gas extraction, construction, and transportation[6]. The first microalloying element to be widely used was vanadium added to C Mn steels in the USA as reported by Bullens in 1916. Indeed early in the 20th century Hemp Ford made major utilisation of vanadium steels for the construction of its Mode1 T, the first mass produced motor car. Small titanium additions can also improve the strength of steel and this was first exploited in Germany in 1921. Micro titanium and micro vanadium additions began to be used in China & India in the 1950s and 1960s for large diameter oil and gas pipelines. The requirement was high strength as compared to mild carbon steel, combined with improved toughness and good weldability. The material is preferably produced by a thermomechanical (hot-cold) rolling process, which maximizes grain refinement as a basis for improved mechanical properties.
Machinability is improved than QandT steels because of their more uniform hardness and their ferrite-pearlite microstructure. Because microalloyed steels are not quenched and tempered they are not susceptible to quench cracking, nor do they need to be straightened or stress relieved [7]. Molybdenum & vanadium has an important role in the initial development. The addition of 0.1-0.2% molybdenum produces a fine grain structure of acicular ferrite and substantially enhances the precipitation hardening effects achieved with the other alloying elements. Consequently, an estimated 2 million tons of Mo containing micro alloy steels for pipelines have been produced worldwide during the 1970s. Low carbon microalloyed steels, are designed to provide better mechanical properties and/or greater resistance to atmospheric corrosion than conventional carbon steels. They are not considered to be alloy steels in the normal sense because they are designed to meet specific mechanical properties rather than a chemical composition[8].

Applications:

- In Rail road construction
- In Off-shore/Platforms formation
- In Bridge element construction
- In Suspension part production
- In preparing building Structural components
- In Heavy Equipment produce
- In Tubular Components production
- In Vehicles/Transportation

Table 3. Chemical composition of low carbon micro alloy steel sample (PB53306)

Alloying element	Maximum	Sample composition
C	0.090	0.061
Mn	1.70	1.59
S	.008	0.002
P	0.018	0.007
Si	0.45	0.193
Cr	0.350	0.215
Ni	0.50	0.013

Cu	.200	0.010
Ti	0.025	0.017
V	0.070	0.046
Nb	.100	0.048
Mo	0.25	0.00
Al	0.050	0.047
N	0.0120	0.0038
Ca	0.005	0.0031
B	0.0005	0.0003
Nb+Ti	.150	0.111
Al/N 2:1		12.37
Sn	0.015	0.001
As	0.020	0.002
Sb	0.010	0.0003
Pb	.010	0.0001
Bl	0.010	0.0011
CE(HW)	-	-
CE(PCM)	0.210	0.16

3. GRAIN SIZE REDUCTION IN STEEL

It is known that mechanical properties of metals depend on their grain size. The Hall-Petch (σ0= σ1+ KyD-1/2) formula is an example of the reliance in which the yield stress of polycrystalline materials is invert to the grain size of material. Where σ0 is the flow stress, σ1 the fractional stress resisting the motion of a dislocation, Ky the extent to which dislocations are collided at grain barriers, and D is the grain average dia. This suggests, the finer the grain size is the greater the value will be of flow stress. Grain size reduction is the only commercially available conventional technique which increases ductility and strength simultaneously [8].

Figure 3. shows variation in the yield strength of LCS(low C steel) as a function of its grain size. It can be seen that reduction of ferrite grain size varies from 10μm - 0.5μm, increases the yield strength by 400MPa without any addition of alloying elements. The increase in strength can be seen due to the significant numbers increase of grain boundaries/unit volume.

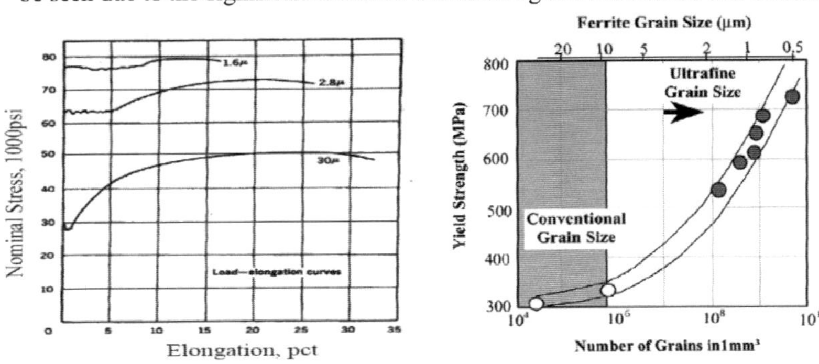

Fig 3. yield & tensile strength with different grain size of steels

Grain size shows effects on transition temperature. Grain coarsening effects the transition temperature and makes steel more suitable to fail by brittle fracture. Grain size effect on the transition temperature is more affective in the of low carbon steels. It is not easy to derive a mathematical relation in grain size and creep strength, because variation in grain size for creep study or high temp strength, is always followed by other changes. Coarser grain steel has better creep/high temperature strength above the equi cohesive temperature. Below equicohesive temperature, finer grain steels shows better high temperature strength. It has been observed that cast steels have incresed high temperature strength over forged steels. Coarse grain size of cast steel is believed to be responsible for the phenomenon. Similar reason is given for better high temperature strength of silicon deoxidized steels as compared to aluminum deoxidized steels. Fatigue strength does not exhibit any basic relationship with grain size. However, coarse grained steels have lowe fatigue strength as compared to fine grained steels. This is true provided the temperature of the test piece is not so high that creep is a dominating marvel. Coarse grained steel has preferred hardenability over fine grained steel. The explanation behind this is that coarse grained steel has less grain limits. Grain limits are where the pace of dissemination is high. Thusly, arrangement of pearlite, which is a dispersion controlled cycle, begins at grain limits. With more modest grain limit region in coarse grained prepares, special development of martensite from austenite happens on cooling. It will additionally be applicable to take note of that pearlite arrangement continues by nucleation and development instrument. Austenitic grain limits go about as nucleation locales for pearlite arrangement.

Fine grain steels have poor machin ability than coarse grain steels. Coarse grained steel has reduced toughness and responsible for providing small discontinuous chips during machining. In low C Mn-Si steels the DBTT, the temperature at which 50 % of the fracture area is covered by brittle area, can be lowered with grain size reduction. It can be said that the grain size reduction will increase fatigue and wear properties of low carbon steels. Grain refinement increases s toughness as well as improves strengthof the materials and provides scope for resarch. If grain refinement is done to the nano meter level, the nanocrystalline metals have improved mechanical properties in comparison to their coarse-grained metals, results in various applications[1,9].

4. EXPERIMENTAL PROCEDURE

Various processes has been carried out in the presented project in order to know about the precipitation behaviour in low carbon micro alloyed steel whose chemical composition is given in table 3. Some steps during process are likewise:

a. Samples were cut with the help of circular saw and hand saw to avoid over heating during cutting action.
b. Afterwards samples were heated to a austenitising temperature of 11500C in a muffle furnace for 1hour holding time.
c. Then samples were quenched in a ice brine solution.
d. Hardness of the as quenched sample were measured by means of Vicker's hardness test.
e. Afterwards samples were aged at various temperatures like 100,200, 300,400,500,600,625,650,675,700oC for holding time of 1hour each.
f. Then hardness of the samples were measured by means of Vicker's hardness test and a graph were plotted between Vicker's hardness number (VH) and temperature according to the results to know the effect of precipitation on hardness of sample.
g. Thereafter microstructure study were carried out on the samples having most significant difference in hardness numbers by means of optical microscopy.

5. RESULTS AND DISCUSSION

As we can take two phases of result according to our experiment:

5.1. ACCORDING TO THE HARDNESS NUMBER

Table 4. Shows various differently treated samples at varying temperature and their hardness by Vicker's hardness testing machine.

S.No.	Sample type	Ageing Temperature (^{0}C)	VHN
1	Before processing	NIL	205
2	Aged Samples for 1 hour each	100	225
		200	218
		300	219
		400	225
		500	235
		600	252
		625	220
		650	238
		675	210
		700	229
3	As quench samples (1150°C/water quenched)	NIL	215

We take the values having significant difference during ageing cycle like VH 218,235,252 and 238. We took the values and plotted a graph between hardness number and temperature of ageing cycles which is further shown in fig 4. Also, we can see a difference in hardness after quenching.

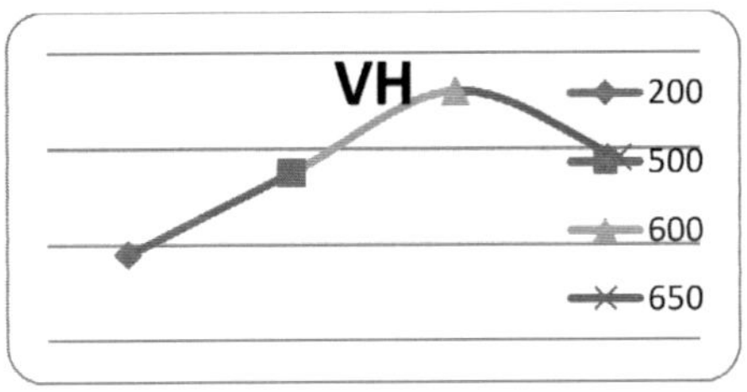

Fig 4. Graph between Vicker's hardness number and ageing temperature

As the graph shows there, we got an increase in hardness to an ageing temperature limit and falls back again. This is due to high temperature precipitation of some alloys like NbC.

5.2. ACCORDING TO MICROSTRUCTURAL STUDY

The microstructures of selected samples are:

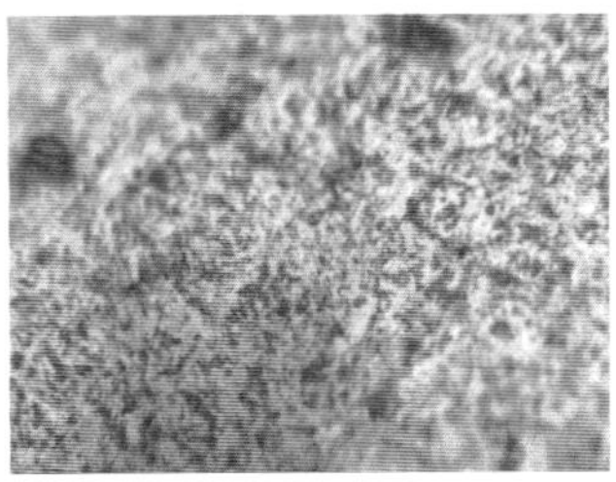

Fig 5. microstructure of aged sample at 200^{0}C For 1hour at 1000x magnification

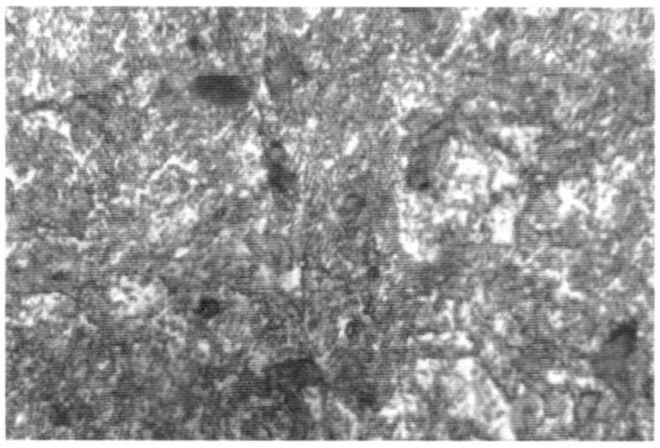

Fig 6. microstructure of aged sample at 500^{0}C For 1hour 1000x magnification.

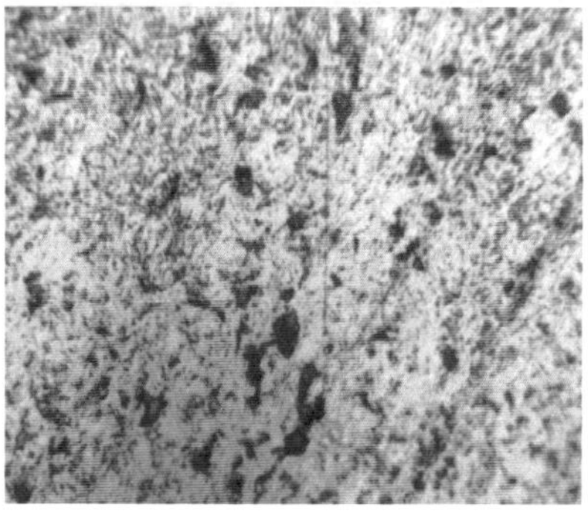

Fig 7. microstructure of aged sample at 600^{0}C For 1hour 1000x magnification.

Fig 8. microstructure of aged sample at 650^{0}C For 1hour 1000x magnification.

As we can see from the images above that there are precipitated particles in the microstructure and also the precipitation of high temperature precipitating alloy like NbC in the microstructure. Hardness increases up to the ageing temperature of 6000C.There are 10% Pearlite, a little amount of Upper Bainite, Ferrite and some corrosion product due to straining at the grain boundaries in the microstructure. Then after the temperature of 6000C, precipitation of high temperature precipitants occurs and there got a saturation and hardness tends to decrease due to coarsening of precipitates.

6. CONCLUSION

The study shows that precipitation occurs during the ageing cycles and as we can conclude from harness graph and microstructural study that precipitation of alloys like NbC, SiC, CrC, NiC, VC, MoC, BC precipitates on higher temperatures, therefore hardness tends to increase at some range of high temperature ageing. After their precipitation occurs there got a saturation in the system and again hardness tends to fall. Although the hardness result of all the samples do not show very significant difference, but as a whole the test shows rise, saturation and fall in hardness values which is due to various precipitation temperatures of various alloying elements involved.

REFERENCES:

[1] T.V.Rajan, C.P.Sharma and Ashok Sharma; Heat treatment: principles and techniques, PHI Learning; pg 285-286.
[2] Séblin. B / Jahazeeah. Y / Sujeebun. S / Manohar /Wong Ky. B Steel Material Science-MECH 2104.
[3] *C.P. Sharma,* Engineering Materials: Properties And Applications Of Metals And Alloys By Prentice Hall India Pvt., Limited.
[4] Justin Furness, Steels - The Structure of Engineering Steels, Materials Information Service, May 8, 2001.
[5] Kenneth G. Budinski, Michael K. Budinski, *Engineering Materials*: Properties and Selection, Prentice Hall, 2010.
[6] Degarmo, E. Paul; Black, J T.; Kohser, Ronald A. (2003), Materials and Processes in Manufacturing (9th ed.), Wiley, ISBN 0-471-65653-4.
[7] Michael Korchynsky, A New Role for Microalloyed Steels – Adding Economic Value, Strategic Minerals Corporation, Pittsburgh. Pennsylvania, USA
[8] Rune Lagneborg, Tadeusz Siwecki, Stanislaw Zajac and Bevis Hutchinson, The Role Of Vanadium In Microalloyed Steels, Swedish Institute for Metals Research S-11428 Stockholm, Sweden.
[9] Hamid Azizi-Alizamini, Austenite Foramtion And Grain Refinement In C-Mn Steels, The University Of British Columbia, Dec. 2010. (Ph.D thesis).

Review of Recent Advancements Made In Flexible Energy Storage Devices

H Kumar[1], K Kumar [2]and H K Bhardwaj [3]
[1] Department of Mechanical Engineering, National Institute of Technology, Delhi, India. Email: hrishkec@gmail.com
[2] Department of Mechanical Engineering, Meerut Institute of Engineering and Technology, Meerut, India. Email: krishna.kumar@miet.ac.in
[3] Department of Mechanical Engineering, Meerut Institute of Engineering and Technology, Meerut, India. Email: harsh.bhardwaj@miet.ac.in

ABSTRACT: There are unlimited potentials have been seen in the fields of condition health monitoring, wearable consumer electronics and human motion tracking for the design of soft exosuits. Energy storage devices are the key element in all above mentioned fields, which can be of high capacitance, light weighted and flexible too. Fiber based supercapacitors are based fit to these. In last few decades' different kind of fiber based supercapacitors have been developed, among them carbon nano-tube (CNT) based fiber electrodes showed that their capacitance, energy density and cycle life is higher as compared to conventional wire based electrodes. In this paper our aim is to review the advances made in the field of different type of fiber basedsupercapacitors.

Keywords: Flexible energy storage devices, supercapacitors, carbon nano-tubes.

1. INTRODUCTION

In supercapacitors energy is stored within the spacein-between an electrode and the electrolyte. It has been seen that the conversion of a simple cloth fabric into energy storing supercapacitor is an advancement. A one-step hydrothermal synthesis process is used on natural cotton thread by Zhou et al [1] to develop a multi- walled carbon nanotubes cotton thread. Because of their good electrochemical and mechanical properties, the multi-walled carbon nanotubes supercapacitors, which were cotton thread based, can be directly knitted into fabrics without changing their parent structures. Fiber electrodes (GHs/MWCNTs-CT) and polyvinyl alcohol-H3PO4 gel electrolyte based supercapacitors exhibit high capacitive performanceandlongcyclelife(95.51%capacitanceperformanceafter80000chare-discharge cycles).

The supercapacitors, which we used in apparel electronics, faces irreversible deformationduetolongtermlocalstressconcentrationwhichleadstostructuralfatigue.Huang et al [2] first time developed wire shaped shape memory functions embedded supercapacitors called shape memory supercapacitor (SMSC). SMSC's remember their predesigned shape, when they reached their trigger temperature they restore serious plastic deformations and regain their original shapes within seconds. This advance feature demonstrated unlimited potential in wearable electronic devices other than energy storage like temperaturealarm.

Deng et al [3]wrappingalignedcarbonnanotube(CNT)ontoashapememorypolyurethane(SMP)substrat e to develop a shape memory fiber shaped supercapacitor (SFSC).When SFSC deformed under bending and stretching and exceeds the thermal transition temperature, they restore their shape without affecting their electrochemical performances.

Graphene fiber supercapacitor called RGO-GO-RGOfiber was designed and fabricated by Hu et al [4].During their development process they used laser irradiation of graphene oxide fiber (GO) to obtain a specificreductiongrapheneoxidefiber(RGO). In a single layer of graphene fiber RGO is integrated as electrode and GO as separator. This all in one graphene fiber has improved mechanicalflexibility,lightweightandhighperformancesothatitcanbeweavedintextilefor wearableelectronics.

Graphene has high thermal conductivity and molybdenum disulfide (inorganic compound of molybdenum and sulfur) has high electrochemical conductivity. Wang et al [5] intercalated graphene sheets and molybdenum disulfide sheets which provide a large ion accessible surface area results in increase of excessive specific capacitance of 368 F/cm^3.

Yang et al [6]stacked chemically converted graphene (CCG) sheets face to face. During their study they find out that water can serve as an effective spacer between the CCG sheets and prevent CCG restacking. If we talk about the macroscopic level CCG sheets, these are available in Self-stacked solvated graphene (SSG) film form, whose energy density, power density and operation rate is enhanced as compared to CCG.

First time Gopalsamy et al [7] prepared a metal oxide graphene fiber (Bi2O3 NT-graphenefiber)byawetspinningprocess.Thishybridfibersupercapacitorhasastablecycling life of about 1000 cycles and this is compact in size thus reduce the space requirement as compared to other type of compositesupercapacitor.

Woven cotton and polyester fabrics are widely used in textile industries now a days. Screen printing technique was used byJost et al [8] to impregnate the porous carbon materials into woven cotton and polyester fabric. They also examined the porous carbon materials behaviour by using various other techniques such as cyclic voltammetry, galvanostatic cycling and electrochemical impedance spectroscopy. Cotton and polyester based carbon electrodes are light in weight and have small electric resistance.

Wire shaped supercapacitor (WSS) gain much attention among researchers. Among various fiber electrodes carbon nanotube (CNT) fiber is the one which is light in weight, chemically inert and have high thermal and electrical conductivity. Xu et al [9] used spandex fiber as substrate and polyvinyl alcohol (H2SO4) gel as electrolyte and separator to produce a highly elastic and stretchable carbon nanotube fiber.

As per the necessity to fabricate that kind of supercapacitors which can sustain high tensile strains along with high capacitance performance Wang et al [10] developed a ultra-stretchable wire shaped supercapacitor to sustain tensile strain up to 850% For this purpose two flexible carbon nanotube graphene manganese oxide fibers are helically wound around a super elastic core fiber.

Le et al [11] mainly worked in field of coaxial supercapacitor. They designed a Carbon micro-fibers (CMFs) coaxial supercapacitor with the help ofmulti-walled carbon nanotubes which are bundled as a core electrode and carbon nanofiber paper which is used as an outer electrode. This coaxial supercapacitor uses both type of carbon fibers micro andnano.

Graphenefiberbasedsupercapacitorshavelimitedelectrochemicalperformancedueto its serious stacking and hydrophobicity in aqueous medium. Chen et al [12] dispersed hydrophilic poly vinyl alcohol into a non-liquid crystalline graphene oxide and after that they used wet spinning and chemical reduction process for the development of a hybrid fiber.Their research work resulted a higher toughness, capacitance andhydrophilicitypoly vinyl alcohol (PVA) or graphene hybrid fiber.

According to Liu et al [13] the electrode structure also affect the performance of fiber based supercapacitor. Mn2O3 cubic arrays on carbon fiber structure have been fabricated into straight,bentandcoiledfibersupercapacitorsamongwhichcoiledfibersupercapacitorshowed excellent cyclic life time and highercapacitance.

A traditional hydrothermal method was used by Gu et al [14] to grow CuCo2O4 nano wire arrays on Ni wires. As compared to solid state fiber supercapacitors the energy and power density of these supercapacitors is higher.

The energy density of Fiber based supercapacitors is very low in comparison to conventional batteries.To overcome this drawback in energy density and to achieve superior electrochemical performances Li et al [15]started their work for the development of a fiber shaped asymmetric supercapacitor (also called FASC). They used CNT-ZnO-NWs-MnO2 fibers as positive electrode and carbon nano tube fibers as negative electrodefor the fabrication of FASC fibers. It has been judged that over 1000 bending cycles FASCs retained 96.7% of initialcapacitance.

First time Senthilkumar et al [16] used copper hexacyanoferrate coated carbon fibers (CuHCF@CFs) as the positive electrode and banana peel derived porous carbon coated carbon fibers (PC@CFs) as thenegative electrode for the development of hybrid fiber supercapacitor (HFSC). Improved specific capacitance of 19.2 F g^{-1} and energy density of 10.6 W h kg^{-1} made these fibersto be suitable for wearable electronic devices.

Due to their tiny size and flexibility yarn supercapacitors reflects a great potential in textile for wearable electronics. Yarn supercapacitors have low energy density and since naked fibers are used for the fabrication of yarn supercapacitors (YSC) there may be short circuit when they come close to each other. A coaxial wet spinning assembly approach was proposedbyKouetal[17]topreparepolyelectrolytewrappedgraphene/carbonnanotube/their mixture core-sheath fibers. The prepared YSCs has high energy density and no short circuit problem.

A lot of research work has been done in the fields of flexible electrodes (FEs) and flexible supercapacitors (FSCs). Dong et al [19] divided FEs/FSCs into three main categories: fiber-like FEs/FSCs, paper-like FEs/FSCs and 3-D porous SCs and their corresponding FSCs. Fiber-like FEs/FSCs were subdivided into categories: plastic fiber supported FEs/FSCs, metal fiber supported FEs/FSCs, CF bundle supported FEs/FSCs, natural fiber supported FEs/FSCs, CNT yarn supported FEs/FSCs, GN fiber supported FEs/FSCs and electrochemically active fibersupportedFEs/FSCs.PaperlikeFEs/FSCsweresubdividedintocategories:freestandingpaper-like FEs and flexible substrate supported paper-like FEs. 3-D porous SCs and their corresponding FSCs subdivided into categories: textile FEs/FSCs and sponge-likeFEs/FSCs.

Dai et al [19] worked for the development of a flexible solid state micro supercapacitor, to which they named pen-ink-carbon-fiber(Ink-CF). Their teamsimply coated the surface of carbon fibers (CFs) with ink. The developed Ink-CFs exhibit improved electrochemical performances such asat 5μA current, their specific capacitance is 4.31 mF cm^{-2} and power density of $3.8x10^{-7}$ Wh cm^{-2}.

Electrodeposition method is used by Noh et al [20] to directly grow the metal oxides of MnO2 and MoO3 on flexible carbon fibers (CFs). For the development of a highperformanceasymmetric supercapacitor (ASC), they assembled CF/MnO2 as positive electrode and CF/MoO3 as negative electrode. The fabricated ASC has high energy density of 2.70-1.78 μWh cm^{-2} and a power density of 0.53-8.30 $mWcm^{-2}$.

To achieve a high energy density of 2 mWh cm^{-3} and power density of 11 W cm^{-3} under a highoperating voltage of 1.6 V, Jin et al [21] developed a fiber shaped asymmetric supercapacitor (FASC). These fibers have an advantage that their capacitance does not affect much when they embedded in a glove.For the fabrication of such fibers, Polyaniline (PANI) coated carbon fiber thread (PANI@CFT) is used as positive electrode and functionalized carbon fiber thread (FCFT) is used as negative electrode.FASC fibers have an advantage that even after stretching 100%, they maintain the same capacitance.

Urethane elastic fiber core spun yarns (UY) was used bySun et al [22] in a two-step process of carbonnanotubes(CNTs)dippingand polypyrrole (PPy) electrodepositionfor the fabrication of a highly stretchableyarnsupercapacitor. The obtained yarn supercapacitor is able to sustain a high capacitance of 69 mF cm^{-2} under strain rate of80%.

Considering workingmechanism and electrodematerial(carbon/metaloxide/conductive polymer), Xue et al [23] reviewed the recent achievements in flexible and wearable supercapacitors,especially in the area of yarn/fiber shaped and planer supercapacitors.

They also discussed the design of 1D yarn/fiber shaped flexible electrodes and 2D flexible electrodes.

Liu et al [24] fabricated a flexible wire-shaped fiber asymmetric supercapacitor (WFASC). They simplyused MnO2-PPy-carbon fiber composite as positive electrode and V2O5-PANI-carbon fiber composite as negative electrode for the fabrication of WFASC. It has been examined that V2O5 and MnO2 has a work function difference with a large potential of 2.0 V. These supercapacitors exhibit a high aerial capacitance of 0.613 F cm^{-2} andshow a maximum energy density of 0.340 mWh cm^{-2}.Their power density is of the order of 30 mW cm^{-2}.

Pu et al [25] proposed a self-charging power textile with triboelectric nano-generator (TENG)fabrictoharvestmechanicalenergyofhumanmotionandyarnsupercapacitortostore human motion energy which were integrated in a simple cloth. To achieve the high specific capacitance of 72.1 mF cm^{-2} , energy density of 1.6 mWh cm^{-2} and power density of 2.42 mW cm^{-2} with a capacitance retention of 96% in 10000cycles,a common fabric yarn is need to be converted into highly conductive yarn, for which electrolysis deposition of Ni coating is followed by reduced graphene oxide (rGO) fibers coating.

Liu et al [26] worked in the area of enhancementof energy density and power density of 1D flexible supercapacitor, for which they developed a yarn supercapacitor. Theyintegrated graphene metallic textile composite electrode with low costgraphenesheetsimmobilizedonthesurfaceofNicoatedcottonyarnsascurrentcollector. In addition they used electrolysis process todeposit the Graphene on the surface of Ni cotton yarns.

To fulfill the need of a high performance electrode for redox supercapacitors, Lee et al [27] used Gradient biscrolling technique.For the fast transportation of ions they used biscrolling technique, wherehundreds of layers of conducting polymer carbonnanotubesheets(CNTs)arescrolledinto20μmdiameteryarn.Toinvestigatetheelectrochemicalperformanceofbiscrolledyarnelectrodes in solid and liquid electrolyte two identically prepared electrodes wereused.

Lu et al [28] used electrodeposition method to develop a high performance fiber based supercapacitor for wearable supercapacitors. To develop a hybrid carbon nanotube (CNT)/graphene/polyaniline (PANI) fiber, electrodeposition method is used, in which CNT/ graphene fiber which wasfurther coated with polyvinyl alcohol (PVA) phosphoric acid during wet spun process. The obtained hybrid fiber exhibit excellent elastic strain of 800%.

2. CONCLUSION:

After made review of the previous work undertaken in the fields of fiber based supercapacitors, it has been shown that these supercapacitors have many advances in terms of high capacitance, high energy density, long cycle life and easy to fabricate into woven to be used in wearable electronics. Carbon nano-tube based supercapacitors gain much popularity among researchers. To increase the utility and to achieve the remaining challenges a lot of works need to be done in future.

REFERENCES

1) Zhou Q,Jia C, Ye X, Tang Z and Wan Z (2016). A knittable fiber-shaped supercapacitor based on natural cotton thread for wearable electronics. *Journal of Power Sources***327**365-373.
2) Huang Y, Zhu M, Pei Z, Xue Q, Huang Y and Zhi C (2016). A shape memory supercapacitor and its application in smart energy storage textile. *JournalofMaterialsChemistryA***4**1290-1297.
3) Deng J, Zhang Y, Zhao Y, Chen P, Cheng X and Peng H. A shape memory supercapacitor fiber. *Angewandte Chemie*https://doi.org/10.1002/anie.201508293.
4) Hu Y, Cheng H, Zhao F, Chen N, Jiang L, Feng Z and Qu L (2014). All-in-one graphene

based supercapacitor. *Nanoscale***6**6448-6451.

5) Wang B, Wu Q, Sun H, Zhang J, Ren J, Luo Y, Wang M and Peng H (2017). An intercalated graphene/ (molybdenum disulfide) hybrid fiber for capacitive energy storage. *Journal of Materials Chemistry A***5**925-930.

6) Yang X, Zhu J, Qiu L and Li D (2011). Bioinspired effective prevention of restacking in multilayered graphene films: towards the next generation of high- performance supercapacitors. *Advanced Materials***23**2833-2838.

7) Gopalsamy K, Xu Z, Zheng B, Huang T, Kou L, Zhao X and Gao C (2014). Bismuth oxide nanotube graphene fiber based flexible supercapacitors. *Nanoscale***6** 8595-8600.

8) JostK,PerezCR,McDonoughJK,PresserV,Heon M,DionG and GogotsiY(2011). Carbon coated textile for flexible energy storage. *Energy and Environmental Science***4**5060-5067.

9) XuP,GuT,CaoZ,WeiB,YuJ,LiF,ByunJH,LuW,Li Q and ChouTW. Carbon nanotube fiber based stretchable wire-shaped supercapacitors. *Advanced Energy Materials*https://doi.org/10.1002/aenm.201300759.

10) Wang H, Wang C, Jian M, Wang Q, Xia K, Yin Z, Zhang M, Liang X and Zhang Y (2018). Superelastic wire-shaped supercapacitor sustaining 850% tensile strain based on carbon nanotube graphene fiber. *Nano Research***11** (5)2347-2356.

11) LeVT,KimH,GhoshA,KimJ,ChangJ,VuQ A,PhamDT,LeeJH,KimSW andLeeYH (2013). Coaxial fiber supercapacitor using all carbon material electrodes. *ACSNano***7** (7) 5940-5947.

12) Chen S, Ma W, Xiang H, Cheng Y, Yang S, Weng W and Zhu M (2016). Conductive, tough, hydrophilic poly (vinyl alcohol)/graphene hybrid fibers for wearable supercapacitors. *JournalofPowerSources***319**271-280.

13) LiuB,LiuB,WangX,ChenD,FanZandShenG(2014).Constructingoptimized wire electrodes for fiber supercapacitors. *NanoEnergy***10**99-107.

14) Gu S, Lou Z, Ma X and Shen G.CuCo2O4 nanowires grown on a Ni wire for high performance, flexible fiber supercapacitors. *ChemElectroChem* https://doi.org/10.1002/celc.201500020.

15) LiY,YanX,ZhengX,SiH,LiM,LiuY,SunY,JiangYandZhangY(2016). Fiber-shaped asymmetric supercapacitors with ultrahigh energy density for flexible/wearable energy storage. *JournalofMaterialsChemistryA***4**17704-17711.

16) Senthilkumar ST, Kim J, Wang Y, Huang H and Kim Y (2016). Flexible and wearable fiber shaped high voltage supercapacitors based on copper hexacyanoferrate andporouscarboncoatedcarbonfiberelectrodes.*JournalofMaterialsChemistryA***4** 4934-4940.

17) Kou L, Huang T, Zheng B, Han Y, Zhao X, Gopalsamy K, Sun H and Gao C(2014). Co-axial wet spun yarn supercapacitors for high-energy density and safe wearable electronics. *Nature Communications***5**3754.

18) Dong L, Xu C, Li Y, Huang ZH, Kang F, Yang QH and Zhao X (2016). Flexible electrodes and supercapacitors for wearable energy storage: a review by category. *Journal of Material Chemistry A***4**4659-4685.

19) DaiS,GuoH,WangM,LiuJ,WangG,HuCandXiY(2014).Flexiblemicro-supercapacitorbasedonpenink-carbonfiberthread.*JournalofMaterialsChemistryA***2** 19665-19669.

20) Noh J, Yoon CM, Kim YK and Jang J (2017). High performance asymmetric supercapacitor twisted from carbon fiber/MnO2 and carbon fiber/MoO3. *Carbon***116** 470-478.

21) Jin H, Zhou L, Mak C L, Huang H, Tang WM and Chan HLW (2015). High performance fiber-shaped supercapacitors using carbon fiber thread (CFT) @polyaniline and functionalized CFT electrodes for wearable/stretchable electronics. *NanoEnergy***11**662-670.

22) Sun J, Huang Y, Fu C, Wang Z, Huang Y, Zhu M, Zhi C and Hu H (2016). High performance stretchable yarn supercapacitor based on PPy@CNT@urethane elastic fiber core spun yarn. *NanoEnergy***27** 230-237.

23) Xue Q, Sun J, Huang Y, Zhu M, Pei Z, Li H, Wang Y, Li N, Zhang H and Zhi C. Recent progress on flexible and wearable supercapacitors. *AdvancedScienceNews*https://doi.org/10.1002/smll.201701827.

24) LiuW,LiuN,ShiY,ChenY,YangC,TaoJ,WangS,WangY,SuJ,LiLandGaoY(2015).Wireshap edflexibleasymmetricsupercapacitorbasedoncarbon fibercoatedwithmetaloxideandpolymer.*JournalofMaterialsChemistryA***3**13461- 13467.

25) Pu X, Li L, Li M, Jiang C, Du C, Zhao Z, Hu W and Wang ZL. Wearable self-charging power textile based on flexible yarn supercapacitors and fabric nanogenerators. *Advanced Materials*https://doi.org/10.1002/adma.201504403.

26) Liu L, Yu Y, Yan C, Lia K and Zheng Z (2015). Wearable energy-dense and power-dense supercapacitor yarns enabled by graphene-metallic textile composite electrodes. *Nature communications***6**7260.

27) LeeJA,ShinMK,KimSH,ChoHU,SpinksGM,WallaceGG,LimaMD, LeproX,KozlovME,BaughmanRHandKimSJ(2013).Ultrafastchargeanddischarge biscrolled yarn supercapacitors for textiles and micro devices. *Nature Communications***4** 1970.

28) Lu Z, Foroughi J, Wang C, Long H and Wallace GG. Superelastic hybrid CNT/graphene fibers for wearable energy storage. *Advanced Energy Materials* https://doi.org/10.1002/aenm.2017020.

AUTHOR INDEX